THE
BIOLOGY
COLORING
BOOK

Also available in the Coloring Concepts series:

Zoology Coloring Book by L. M. Elson

Human Brain Coloring Book by M. C. Diamond, A. B. Scheibel, and L. M. Elson

Marine Biology Coloring Book by T. M. Niesen

Human Evolution Coloring Book by A. L. Zihlman

Botany Coloring Book by P. Young

Computer Concepts Coloring Book by B. M. Glotzer

THE
BIOLOGY
COLORING
BOOK

by
Robert D. Griffin

Illustrations by
Cinthea Vadala

BARNES & NOBLE BOOKS
A DIVISION OF HARPER & ROW, PUBLISHERS
New York, Cambridge,
Philadelphia, San Francisco, Washington,
London, Mexico City, São Paulo, Sydney

This book was produced by Coloring Concepts, Inc.
P.O. Box 324, Oakville, CA 94562

The book editors were Joan W. Elson and Jeanne Flagg.
Layout and design were by Joan W. Elson, Robert D. Griffin, and Lawrence M. Elson.
The copy editor was B. F. Emmer.
Type was set by ComCom, a Division of Haddon Craftsmen.
Page makeup and production coordination were by C. Linda Dingler and Abigail Sturges.
The proofreader was Bernie Borok.

Plate 57: Adapted from *Biology of the Cell,* Second Edition, by Stephen L. Wolfe, © 1981 by Wadsworth, Inc. Reprinted by permission of the publisher.

ISBN: 0-06-460307-5

88 89 90 MPC 10 9 8 7 6 5 4 3

ROBERT D. GRIFFIN, M.A., has been teaching biology for 27 years. For the last 19 years he has been on the faculty of City College of San Francisco, where he has taught biology courses ranging from anatomy to zoology. Previously he had extensive experience teaching in junior and senior high schools and in special science programs for gifted elementary school students. His principal academic interest is in innovative methods of teaching biological concepts and reasoning to both liberal arts students and biology majors.

CINTHEA VADALA received her education at Music and Art High School and Hunter College. She is the illustrator of *The Zoology Coloring Book, The Human Brain Coloring Book,* and *The Computer Concepts Coloring Book.* She is now a free-lance scientific illustrator living in the San Francisco Bay area.

To my parents, my first and best teachers, who taught me early the most important of all lessons, that learning is a joy sufficient unto itself, and

To my sons, Kurt, Garrett, and Bruce, who have been such a joy to teach and have made this book worth the effort.

CONTENTS

PREFACE

This is a book about life. In particular, it is a book about life processes. It deals with the cell and its role as the basic unit of life and with how cells obtain the energy for life. It deals with how cells are put together to form the tissues, organs, and organ systems that make up an individual living organism and how that organism interacts with its environment, including the other living organisms in it. And it deals with how hereditary traits are passed from parent to offspring, how those traits are carried as a code in the structure of molecules, and how that code is translated into action.

In writing this book I have attempted to develop each of its topics in small steps starting at the very beginning, so that anyone can start coloring and learn some biology without needing a teacher or any other textbook. At the same time, I have kept in mind the basic information and the sequence in which it is introduced in most high school and college textbooks. Students taking their first biology course will find this book extremely helpful in understanding the concepts, visualizing anatomical structures, and fixing these concepts and structures in memory. If you find some topics and details irrelevant to your needs, simply skip those particular plates.

If this is your first Coloring Concepts book, you are in for a pleasant surprise. The coloring activity is not some sort of happy playtime but an integral part of what has proved to be a highly effective learning method. For example, with this book you will never find yourself arriving at the bottom of a page and realizing that your mind was elsewhere and you have no idea what you just read. Instead, you will hardly read a sentence or two before it is time to stop and color something, and that something will be directly related to what you just read. First you will color a heading or a title, which requires you to take a close look at the word and even its spelling. Then, with the same color, you will color the associated parts of the illustration. Not only does this physical activity make it much more difficult for your mind to wander to some other topic, but it also requires the activity of

the parts of your brain that are involved in movement and in perception of color and shape. (Even if you are one of the many who suffer from red-green color blindness, you will be able to find many color combinations that are clearly distinguishable to your eye.) As you probably know, the more areas of your brain you involve simultaneously in trying to learn something, the more easily you will understand and remember that material. In addition, when you have finished coloring a plate, you will have an aesthetically pleasing picture, color-coded for rapid, effective review at a later time, such as just before an examination.

Considerable effort has gone into making this book an accurate and effective teaching device. You are likely to gain the most from it if you use it in the way in which it was intended. Be sure to read the general coloring instructions that follow this preface and the individual coloring notes in boldfaced type in the text accompanying each plate.

Very few books are the work of a single person, and this one is no exception. I am deeply indebted to Mr. Al Lilleberg, who reviewed the first 22 plates; to my colleague at City College, Mrs. Valerie Meehan, who reviewed the chemistry section; and to Dr. Jack Tomlinson, who reviewed approximately 80 percent of the book. Their recommendations improved the clarity and precision of the book in many places. Any deficiencies that remain are, of course, entirely my responsibility.

A coloring book is nothing without illustrations to be colored, and in this case they convey much of the information, so they must be exactly right and must be closely coordinated with the text. I am very grateful for Cinthea Vadala's artistic talents and her patience in redrawing things I changed my mind about or failed to make clear the first time around. She has taken some incredibly rough and sketchy outlines and converted them into illustrations that are beautiful as well as scientifically accurate.

I owe a large debt of gratitude to Larry Elson, longtime colleague and friend, who talked me into writing this book in the first place. He has been a source of good advice and abundant encouragement all along and was of immense help in developing both the art layout and the text for the final section on human organ systems. I owe even more gratitude to Joan Warrington Elson, who has been my editor. Not only did she perform with skill (and charity) all the usual tasks of an editor, pointing out ambiguities, inconsistencies, and typographical errors, but she has been an especially valuable source of encouragement and down-to-earth help.

Last, but certainly not least, I am deeply grateful to my sons, Kurt, Garrett, and Bruce, for countless hours of transcribing, retyping, library research, and general errand running, as well as for getting along pretty much without a father for the past two and a half years, and to my wife, Marta, who has endured much for a like period.

HOW TO USE THIS BOOK:
COLORING INSTRUCTIONS

1. This is a book of illustrations (plates) and related text pages. You color each labeled structure the same color as its name (title). The title and structure are linked by identical letters (subscripts). In doing this, you will be able to relate identically colored title and structure at a glance. Structural relationships become apparent as visual orientation is developed. These insights, plus the opportunity to display a number of colors in a visually pleasing pattern, provide a rewarding learning experience.

2. You will need coloring instruments. Colored felt-tip pens or colored pencils are recommended. Laundry markers (with waterproof colors) and crayons are not recommended: the former because they stain through the paper, the latter because they are coarse and messy and produce unnatural colors.

3. The organization of illustrations and text is based on the author's overall perspective of the subject and may follow, in some instances, the order of presentation of a formal course of instruction on the subject. To achieve maximum benefit of instruction, you should color the plates in the order presented, at least within each group or section. Some plates may seem intimidating at first glance, even after reviewing the coloring notes and instructions. However, once you begin coloring the plate in order of presentation of titles and reading the text, the illustrations will begin to have meaning, and relationships of different parts will become clear.

4. As you come to each plate, look over the entire illustration and note the arrangement and order of titles. Scan the coloring instructions (printed in boldfaced type) for further guidance. Be sure to color in the order given by the instructions. Most of the time this means starting at the top of the plate with A and coloring in alphabetical order. Contemplate a number of color arrangements before starting. In some cases, you may want to color related forms with different shades of the same color; in other cases, contrast is desirable. One of the most important considerations is to link the structure and its title (printed in large outline or blank letters) with the same color. If the struc-

ture to be colored has parts taking several colors, you might color its title as a mosaic of the same colors. It is recommended that you color the title first and then its related structure. If the identifying subscript lies within the structure to be colored and is obscured by the color used, you may have trouble finding its related title unless you colored it first.

5. In some cases, a plate of illustrations may require more colors than you have in your possession. Forced to use a color twice on the same plate, you must take care to prevent confusion in identification and review by employing them on well-separated areas. On occasion, you may be asked to use colors on a plate that were used for the same structure on a previous, related plate. In this case, save the colors until you reach the appropriate title.

6. Symbols used throughout the book are explained below. Once you understand and master the mechanics, you will find room for considerable creativity in coloring each plate. Now turn to any plate and note:

a. Areas to be colored are separated from adjacent areas by heavy outlines. Lighter lines represent background, suggest texture, or define form and (in the absence of "do not color" symbols) should be colored over. Some boundaries between coloring zones may be represented by dotted lines. These represent a division of names or titles and indicate that a physical boundary may not exist or, at best, is not clearly visible.

b. As a general rule, large areas should be colored with light colors and dark colors should be used for small areas. Take care with very dark colors: they obscure detail, identifying subscripts, and texture lines or stippling. In some cases, a structure will be identified by two subscripts (e.g., A + D). In this case, you may wish to use crosshatching or stripes of the two colors.

c. Any outline-lettered word followed by a small capitalized letter (subscript) should be colored. In most cases, there will be a related structure or area to color. If not, the word functions as a heading and is colored black (●) or gray (★).

d. In the event that structures are duplicated on a plate, as in left and right parts or serial parts, only one may be labeled with a subscript. Without boundary restrictions or instructions to the contrary, these like structures should all be given the same color.

e. In looking over a number of plates, you will see some of the following symbols:

● = color black; generally reserved for headings

★ = color gray; generally reserved for headings

–¦– = do not color

A^1, A^2, etc. = identical letter with different exponents implies that parts so labeled are sufficiently related to receive same color or shades of the same color, unless otherwise specified in the bold-faced coloring instructions

7. The title of a structure to be colored on the related (facing) plate is set in *italics* where it first appears in the text. This enables you to spot quickly in the text the title of a structure to be colored.

THE
BIOLOGY
COLORING
BOOK

1
BEING ALIVE

Biology is the study of life, so in this first coloring plate, let's take a simpleminded look at what we mean by being alive before we get into the complexities. It is generally accepted that a rabbit and a plant are alive while a rock isn't, so a comparison may be useful.

Color the rabbit (A), the plant (B), and the rock (C) in the center of the plate. Color the heading Movement gray, and color the related structures in that section. (The -¦- symbol means do not color.) Use green for the plant and any colors that seem appropriate for the rabbit and the rock. (Yellow should be reserved for the sun.) These same colors will be used throughout the plate, but it will be best to color one section at a time and then read the explanation. Retain A, B, and C colors for each section, adding new colors as directed.

We need only to see a *rabbit* run across a field to be convinced that it is alive. If we saw a *rock* run across a field, we would be amazed. We might even exclaim, "Good grief, it's alive!" So one obvious feature of being alive is movement. We wouldn't expect a *plant* to run across a field, but careful observation does disclose slow movement, as when flowers open or close or leaves turn to follow the movement of the sun.

Color the heading Organization, titles D, E, and F, and their related structures. It is a tradition among biologists to use red for arteries and blue for veins. The heart can be a reddish brown.

Another feature of being alive is a very high level of organization. If we look closely at a rock, we can see a certain amount of organization: any particular kind of rock will be composed of the same minerals in the same proportions as other rocks of that kind. But the way they are arranged is very different from one rock to the next. Any rabbit, however, has a structure virtually identical to that of any other rabbit. Each one has its *heart, veins, arteries,* and other organs in almost exactly the same places. Plants, too, have their organs: leaves, stems, roots, and flowers, although there is more variation in their arrangement. Closer examination shows that each of these organs is made up of several different tissues. These tissues are made of tiny units called cells, and the cells, although they are microscopic, have their own highly organized internal structure.

Color the Homeostasis heading and section.

Another feature of living things is the tendency to maintain constant conditions inside the body. This tendency is called homeostasis, which means "remaining the same." In very hot weather the rabbit keeps its ears erect and spread out to radiate away the maximum amount of heat, while in cool weather the ears are kept closer to the body to conserve heat. Plants do not regulate their temperature much, but in very hot, dry weather the guard cells close the numerous holes (stomata) in the leaves to reduce the evaporation of water.

Color the Energy Utilization heading, title G, and the associated structures in the section. Use yellow for the sun.

If the *sun* shines on a rock, the rock will absorb a little bit of heat energy, but it doesn't do anything with it. A plant will absorb some heat energy too, but it will also absorb much of the light energy falling on it and will convert it into the chemical energy of carbohydrates, proteins, and fats, which can be used to keep the plant alive and make more leaves, stems, and roots. The rabbit will eat parts of the plant and use that energy to keep itself alive and to run, grow new fur, and do other jobs.

Color the heading Reproduction and the related section, including titles A¹, B¹, and B². Color the seed (B¹) and the seedlings (B²) green. Color the juvenile rabbits (A¹) the rabbit color.

Reproduction is so characteristic of living things that it hardly needs comment. We are not surprised when two rabbits produce baby rabbits or when plants produce *seeds* that grow into more plants. If two of our favorite rocks produced baby rocks, however, we would be very surprised.

Color the Growth and Development heading and section, including titles and structures A², A³, and B³.

Although some living things consist of only a single cell that reproduces by dividing in half, there is always a period of growth sooner or later. In animals and plants, which consist of numbers of cells, growth from a fertilized egg cell (called an *ovum* in animals and an *ovule* in plants), is accompanied by a process of development. During this process, cells not only increase in number but also change themselves into different kinds of cells and form the various tissues and organs that make up the new individual.

BEING ALIVE.

MOVEMENT.★

ORGANIZATION.★
HEART D
ARTERIES E
VEINS F

HOMEO
STASIS.★

GROWTH
AND
DEVELOPMENT.★
OVUM A²
EMBRYO A³

FLOWER

LEAF

ROOTS

GUARD CELLS

OVULE B³

RABBIT A
PLANT B
ROCK C

STOMA OPEN

REPRODUCTION.★
JUVENILE A¹

ENERGY
UTILIZATION.★

SEED B¹
SEEDLING B²

SUN G

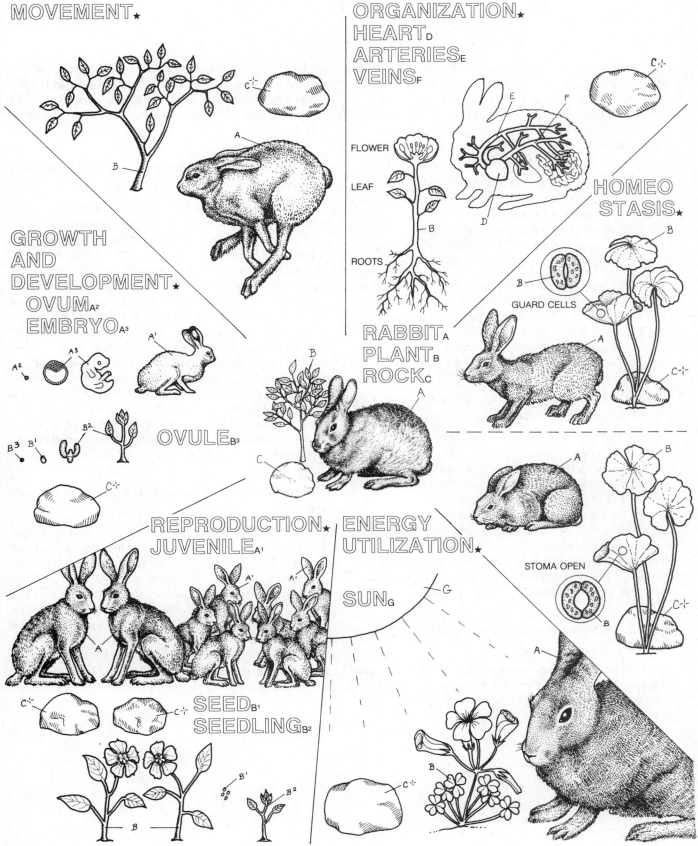

2
HOW SCIENCE BEGAN

As we begin our study of biology, it seems worthwhile to take a brief look at what a science is and how science began. Contrary to what some people believe, science is not a mysterious study reserved for geniuses; it is a systematic way of learning based on a trait common to all humans: curiosity.

Select a pale color and use it to color the title Curiosity and the scene below it. We are not interested here in the details of the scene, only in the fact that it illustrates curiosity, so you should color it all in a single color that is pale enough not to obscure the picture.

The first early humans to discover a naturally occurring fire, such as a tree ignited by lightning, were undoubtedly filled with *curiosity*. They might have been afraid, but if they had even the beginnings of human intellect, their curiosity would quickly have overcome their fear, and they would have started to examine the fire closely.

With a different pale color, color the title Experimentation and its scene.

Gradually our observers would have noticed that the tree itself was being consumed by the fire and that the fire was spreading along the tree to places that weren't burning originally. Before long they would have started to wonder if the fire could be spread to another tree or to a branch of another tree. Almost immediately they would find a branch and try to burn it. Little by little, *experimentation* would continue with the testing of all kinds of substances to see what would burn and what would not burn. As their observations began to be general rules— "wood burns, rocks don't"—science was born.

With another pale color, color the title Magic and its scene.

With time, a few people who came to know more than their fellows found that they could pass off much of their knowledge as *magic,* and they became witch doctors, seers, and oracles. For example, the person who first discovered substances that would flare up into brilliantly colored flames when they were thrown into a fire could easily convince others that he or she had magical powers over fire.

With another pale color, color the title Greek Science and its scene.

As humans mastered the use of fire, agriculture, and domestic animals, some leisure time became available, and some people began to pursue knowledge just for the sake of knowing things. This became especially true in ancient *Greece.* Some early Greeks—particularly Aristotle—were careful observers of nature, but Euclid's success in developing an entire system of geometry by deduction from a few "self-evident" statements called axioms led to the practice of doing all science that way. Experimentation came to be regarded as irrelevant. The imperfect world of reality could not be allowed to spoil the beauty of the perfect world of ideas. Questions about nature were to be answered not by studying nature itself but by asking a great authority, who arrived at the answers by the power of pure reason. This attitude lasted nearly 2000 years. Until the Renaissance, most important questions were answered with "Aristotle says . . ." or "Euclid says . . ."

Color the title Galileo and its scene with another pale color.

Although the thirteenth-century philosopher Roger Bacon wrote of the importance of experiment, it was not until the late sixteenth century that Galileo's extensive experiments and forceful argument established what we regard as "true science." *Galileo* is honored today as the "father of science," and scientists are still guided by Galileo's admonition, "All knowledge is vain without the confirmation of experiment." Galileo overstated the case somewhat, since there are legitimate fields of knowledge in which experiments are essentially impossible. But in science, the experiment is the final test of the validity of a scientific idea.

HOW SCIENCE BEGAN.

CURIOSITY_A

EXPERIMENTATION_B

MAGIC_C

GREEK SCIENCE_D

GALILEO_E

Science today still begins with curiosity leading to observation. Almost immediately upon observing something new, a scientist—or any other curious person—will find one or more questions coming to mind.

Color the headings Observation and Question gray. Color titles A and B and their scenes. The robots represent hypotheses.

Once a question is raised, an answer is looked for. From Galileo's time onward, scientists have made a habit of regarding every answer as tentative until it has been confirmed by experiment. Such a tentative answer is called a *working hypothesis* (plural: hypotheses), to emphasize that it is still unreliable and is being worked on.

As science progressed, it became clear that even the working hypothesis method had some pitfalls. First, anyone who has an idea that seems to be a good one has a tendency to develop a certain affection for that "brainchild." This can lead to failing to recognize its shortcomings, even when one is trying very hard to be honest. The solution to that shortcoming is the method of *multiple working hypotheses*. In this method there is a deliberate attempt to develop a "family" of hypotheses to cover all the possible answers to a question. The affection must be divided among several hypotheses, and a person is inclined to test and evaluate the hypotheses more honestly.

Color title A¹ and B¹ and the associated scenes.

In these scenes, the hurdles the little hypotheses are jumping over represent tests by experiment. Although it doesn't always happen, it is awfully easy when you are fond of a hypothesis to set up a test that really doesn't challenge it very severely, as illustrated by the rather easy hurdle in the *nonchallenging test* scene. On the other hand, if there are many hypotheses, there is a definite tendency to want to reduce their number, so more severe tests are designed, and they are usually designed specifically to eliminate hypotheses rather than to support them. When one or more hypotheses survive deliberate attempts to eliminate them, we can begin to have some genuine confidence in them.

Read the poem that follows and color the picture of each blind man's hypothesis and its corresponding title (C¹ through C⁶) as you come to it. The elephant is not to be colored.

This poem was written by John Godfrey Saxe (1816–1887).

> It was six men of Indostan
> To learning much inclined,
> Who went to see the Elephant
> (Though all of them were blind),
> That each by observation
> Might satisfy his mind.
> The First approached the Elephant,
> And happening to fall
> Against his broad and sturdy side,
> At once began to bawl:
> "God bless me! but the Elephant
> Is very like a wall!"
> The Second, feeling of the tusk,
> Cried, "Ho! what have we here
> So very round and smooth and sharp?
> To me 'tis mighty clear
> This wonder of an Elephant
> Is very like a spear!"
> The Third approached the animal,
> And happening to take
> The squirming trunk within his hands,
> Thus boldly up and spake:
> "I see," quoth he, "the Elephant
> Is very like a snake."
> The Fourth reached out an eager hand,
> And felt about the knee.
> "What most this wondrous beast is like
> Is mighty plain," quoth he;
> "'Tis clear enough the Elephant
> Is very like a tree."
> The Fifth, who chanced to touch the ear,
> Said: "E'en the blindest man
> Can tell what this resembles most;
> Deny the fact who can,
> This marvel of an Elephant
> Is very like a fan!"
> The Sixth no sooner had begun
> About the beast to grope,
> Then, seizing on the swinging tail
> That fell within his scope,
> "I see," quoth he, "the Elephant
> Is very like a rope."
> And so these men of Indostan
> Disputed loud and long.
> Each in his own opinion
> Exceeding stiff and strong,
> Though each was partly in the right,
> And all were in the wrong!

Although the poet emphasized the disagreement, the poem illustrates another value of the method of multiple working hypotheses. Nature is not always as simple as we wish it to be, and sometimes several hypotheses are all true simultaneously.

SCIENTIFIC METHOD TODAY.

OBSERVATION.★

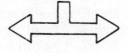

WORKING HYPOTHESIS_A

QUESTION.★

MULTIPLE WORKING HYPOTHESES_B

NONCHALLENGING TEST_A1

CHALLENGING MULTIPLE TESTS_B1

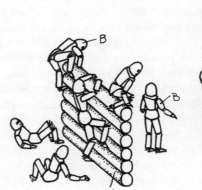

REALITY_C+
HYPOTHESIS 1_C1
HYPOTHESIS 2_C2
HYPOTHESIS 3_C3
HYPOTHESIS 4_C4
HYPOTHESIS 5_C5
HYPOTHESIS 6_C6

4
MATTER AND ELEMENTS

Although chemistry is a subject that some people try to avoid, we can't get very far in our study of biology before we need to consider a few of the basic principles of chemistry. After all, living things are made of matter, and matter follows the laws of chemistry. Even the activities we regard as unique to living things are the result of chemical reactions.

Color the headings Earth and States of Matter. Color titles A, B, and C and the solid, the liquid, and the gas in the illustration of the earth and the cross section of the landscape. Do not use red, white, or yellow, which need to be reserved for the lower part of the plate. Note that clouds are used to represent gas, though in reality clouds are actually composed of invisible water vapor (gas) and visible water droplets (liquid).

The stuff that makes up the earth and everything else in the universe is called matter. Matter exists in three principal states: *solid, liquid,* and *gas.* The land of the earth is largely solid; the oceans, lakes, and rivers are largely liquid; the atmosphere is largely gas but contains some microscopic droplets of liquid and particles of solid suspended in it. If the pressure or temperature within a substance is changed enough, the substance can change from one state to another. When we get water cold enough, for example, it turns into a solid, which we call ice. If we add heat energy to it, it becomes a liquid. If more heat is added to it, it becomes water vapor, a gas. Substances we think of as solid simply require temperatures much higher than we normally experience in order to change their states. (A fourth state of matter, called plasma, exists only at the extremely high temperatures found in stars.)

Color the "magnified" view of the particles of a solid, liquid, and gas in the circles below the landscape.

Matter is composed of incredibly tiny particles called atoms and molecules. Even the highest-powered microscopes will not allow us to see these particles as they are shown in the "magnified" view of this plate, but scientists have a great deal of indirect evidence that they exist and behave in the ways that we will discuss here. These particles are believed to be in constant motion, except at absolute zero ($-273°C$ or $-459.4°F$), a temperature that has never quite been reached but at which all motion of parti-

cles should theoretically stop. In a solid, the particles are strongly attached to one another, and their only motion is vibration. A solid, therefore, has a fixed volume (if the temperature doesn't change) and a fixed shape. As we add heat energy, the particles vibrate faster and harder and finally vibrate so hard that they break loose from one another and enter the liquid state, where they are free to flow over one another. A liquid, then, has a fixed volume but takes the shape of whatever container it is in. As more heat energy is added, the particles move faster and faster until they break entirely away from one another and form a gas. In the gas state the particles are very far apart, and the gas has neither a fixed shape nor a fixed volume. It will expand to fill any space that is available to it.

Color the headings Elements and Compound. Color the titles, the atoms, and the symbols in the remainder of the plate. The colors that are conventionally used in textbooks and molecular models are white or yellow for hydrogen and red for oxygen. Use both colors for the water molecule (two Hs and one O) and its empirical formula.

All matter, living or nonliving, is made up of one or more of only 106 fundamental substances called elements. Only 92 of these elements occur naturally; the other 14 have been produced artificially, and they break down in very tiny fractions of a second.

An element is a substance that cannot be broken down into anything simpler by ordinary chemical means. (Elements can be broken down by the immense forces inside stars or accelerators, but such forces are not considered "ordinary.") The smallest particle of an element that can be obtained by those ordinary means is called an atom. Although it is possible for a single atom to exist separately from other atoms, *hydrogen, oxygen,* and a few other elements have atoms that tend to pair up. When two or more atoms attach to one another, they make up a molecule. If the atoms are of the same kind, we have a molecule of an element, such as the hydrogen molecule or the oxygen molecule. If the atoms are of different kinds, they form a molecule of a compound. As you will see in the next few plates, a compound can have properties that are very different from those of the elements that make it up. Thus two atoms of hydrogen (a gas) and one atom of oxygen (another gas) will form a molecule of water (which becomes a liquid as soon as it cools). A compound always has a fixed ratio of the elements making it up, which is expressed in the *empirical formula* for that compound.

MATTER AND ELEMENTS.

EARTH★

STATES OF MATTER★
SOLID A
LIQUID B
GAS C

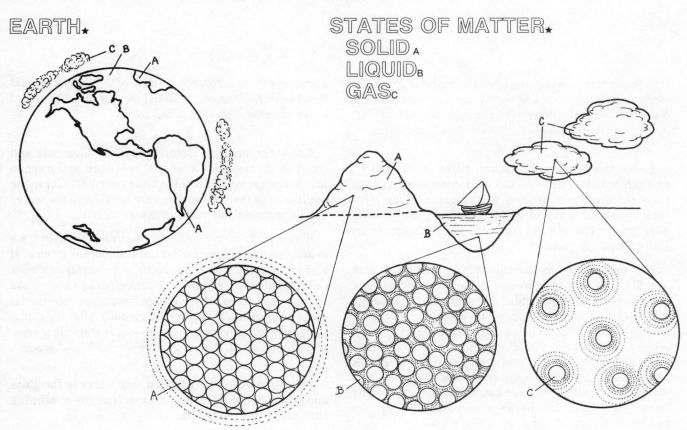

ELEMENTS★
HYDROGEN ATOM/
SYMBOL H
OXYGEN ATOM/
SYMBOL O

COMPOUND★
EMPIRICAL
FORMULA H_2O

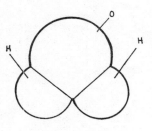

5
COMPOUNDS AND MIXTURES

It is important to recognize from the beginning that the elements making up a compound are not merely mixed together. This plate illustrates some experiments that clarify the differences between compounds and mixtures.

Color the heading Elements, titles A, B, and C, and related structures of the iron and sulfur in both views in the upper section of the plate, along with their symbols. Natural sulfur is yellow, and iron is dark gray. (The symbol for iron, Fe, comes from its Latin name, *ferrum*.)

Iron and *sulfur* are elements. If you mix sulfur with iron filings or powdered iron, it doesn't matter how long or how thoroughly you mix, you will still see separate specks of iron among the particles of sulfur. You have made a *mixture,* but each element still retains its separate properties. Even if you absolutely pulverize the iron and sulfur so that the mixture looks completely homogeneous to your eye, the atoms of iron will remain independent of the atoms of sulfur. You can easily demonstrate this by bringing a magnet close to the mixture: the iron particles will be drawn out, leaving the sulfur behind.

Color the heading Chemical Reaction and titles C through E¹. Color the middle section of the plate. The actual color of iron sulfide (the compound shown) in nature is a dark tan.

If you heat a mixture of iron and sulfur over a flame, the iron and sulfur combine in what we call a chemical reaction. Each iron atom becomes attached to a sulfur atom, and neither of them has quite the same properties as before. If you now bring a magnet near, the iron does not separate from the sulfur, and there is no reaction to the magnet at all. A new substance, a *compound* called iron sulfide, has been formed. We symbolize this compound by the *empirical formula* FeS. This formula tells us that the two elements combine in a ratio of one atom of iron to one atom of sulfur. If we began with a mixture

containing more atoms of iron than of sulfur, there would be some iron left over. An initial surplus of sulfur would result in some sulfur being left over.

Color the heading Compounds and Elements and titles H through F. Color the hydrogen and oxygen inside the pressure bottle at the lower left. Color the explosion in the second pressure bottle and the water droplets inside the third pressure bottle.

If two parts of *hydrogen* and one part of *oxygen* are mixed in a sealed pressure bottle, a mixture is formed. If a spark is made inside the bottle, a powerful *explosion* occurs, which will result in the formation of a compound called *water.* (Don't try this experiment yourself; the resulting explosion can be very dangerous.) After the explosion, the hydrogen and oxygen no longer show their separate properties. In fact, they now form a liquid: water.

Color the oxygen, hydrogen, and water in the glass apparatus at the lower right, as well as the remaining title and representation.

Just as separate elements can be combined to form a compound, a compound can be broken down into its separate elements. In the case of water, this is done by passing an electric current through it. (This is another dangerous procedure you should NOT try for yourself.) If some glass tubing is put together as shown in this plate and a direct current (such as one obtained from a battery) is passed through the water, the water will break down into the two elements that make it up: hydrogen and oxygen. Hydrogen always bubbles off the negative wire and oxygen off the positive wire. The volume of hydrogen collected is always twice the volume of the oxygen. This constant ratio of elements is characteristic of compounds and is expressed in the *empirical formula* for water, H_2O. Two parts of hydrogen and one part of oxygen are required to form water, and water breaks down to release two parts of hydrogen and one part of oxygen.

COMPOUNDS AND MIXTURES.

MACROSCOPIC VIEW

"MICROSCOPIC VIEW"

Fe $_A$

S $_B$

ELEMENTS★
IRON$_A$
SULFUR$_B$
MIXTURE$_C$
CHEMICAL REACTION★
HEAT$_D$
COMPOUND$_E$

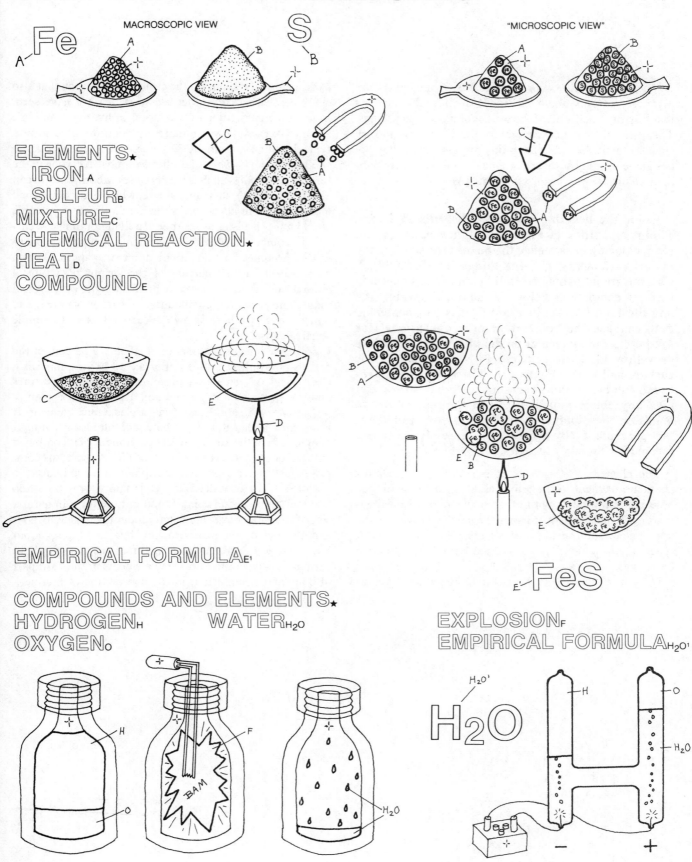

EMPIRICAL FORMULA$_{E'}$

$_{E'}$ FeS

COMPOUNDS AND ELEMENTS★
HYDROGEN$_H$ WATER$_{H_2O}$
OXYGEN$_O$

EXPLOSION$_F$
EMPIRICAL FORMULA$_{H_2O'}$

H_2O'

H$_2$O

6
ELEMENTS IN LIVING THINGS

Of the 92 naturally occuring elements, living things are composed of only about 26, and 6 of those 26 elements make up practically all of the weight of most living things. The other 20 elements essential for life are present in very small amounts, some in such tiny amounts that they are designated simply as "trace" elements. The six most abundant elements are visualized in this plate.

Color the heading Majority Elements in Living Things and titles H through B. Then color as you read, coloring in sequence the boxes representing the majority elements in living things (H, O, C, N, P, Ca), the atoms rising and falling out of the boxes and forming compounds below the boxes, the water (A), and the food (B). As in Plate 5, it is recommended that you use the colors that are conventional for textbooks and molecular model kits, which are white or yellow for hydrogen, red for oxygen, black for carbon, and blue for nitrogen. There are no conventional colors for phosphorus and calcium, so you can pick any colors you like. Fill in the representative bands on the rabbit and the bush as you read about the percentage (by weight) of these living things formed by the six elements.

The element *hydrogen* is a gas, and it is the most abundant element in the universe. Most of the hydrogen on earth is combined with other elements. Hydrogen gas is so light that it easily escapes the earth's gravity at the upper edges of the atmosphere and is lost into outer space. Plants and animals are about 10 percent hydrogen by weight.

Oxygen is also a gas, but it is 16 times as heavy as hydrogen, so it remains in the atmosphere, although it also combines with many other elements. As we have seen, *water* is a combination of two atoms of hydrogen and one atom of oxygen. Oxygen makes up about 63 percent of a typical animal and about 77 percent of a typical plant.

The element *carbon* is familiar to almost everyone in the form of charcoal, which is nearly pure carbon. Carbon can also form crystals such as graphite, which is used in pencil lead and in lubricants, or as diamond, which few people would guess is also pure carbon. In living organisms, carbon is combined with other elements to form carbohydrates, proteins, fats, and many other substances. Carbon comprises about 19 percent of the weight of a typical animal and about 12 percent of the weight of a typical plant. The particular combination of carbon, oxygen, and hydrogen shown here below the carbon box is the simple sugar glucose.

Nitrogen is a gas that makes up about 79 percent of the earth's atmosphere and is important to living things as a component of genes and proteins. Nitrogen constitutes about 4 percent of an animal and 1 percent of a plant.

Phosphorus is not found in nature as a pure element. It is so reactive that it will combine with almost anything it contacts, even the air. Yet a certain amount of it combined with other elements is necessary for life. Phosphorus comprises just under 1 percent of a plant or an animal.

Calcium is a grayish silver metal in pure form, but it too is rather reactive and is not found in pure form in nature. Very few plants have significant amounts of calcium, so it is not shown on the plant drawing, but it makes up about 2 percent of the weight of a mammal, such as a rabbit, because it is one of the principal elements of the hard part of bone and is important in the contraction of muscle.

ELEMENTS IN LIVING THINGS.

MAJORITY ELEMENTS IN LIVING THINGS.★
HYDROGEN$_H$ CARBON$_C$ PHOSPHORUS$_P$
OXYGEN$_O$ NITROGEN$_N$ CALCIUM$_{Ca}$

WATER$_A$
FOOD$_B$

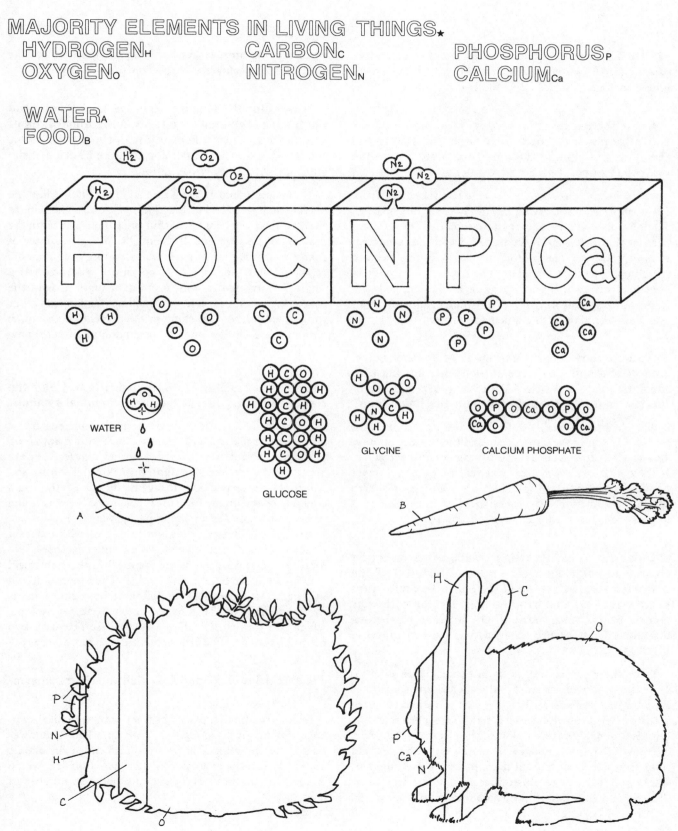

WATER

GLUCOSE

GLYCINE

CALCIUM PHOSPHATE

SUBATOMIC PARTICLES

The atom has been found to be composed of several dozen parts, called subatomic particles. We will discuss three important ones, illustrated in the hydrogen atom.

Color the headings Hydrogen Atom and Nucleus and titles p and e. Color the proton and electron in the hydrogen atom at the top of the plate. Leave the symbol H uncolored for now.

Every atom consists of a nucleus (central mass) with one or more *electrons* flying around it at very high speed. The nucleus contains one or more *protons.* The simplest possible atom is found in hydrogen, which has a nucleus consisting of only one proton and has only one electron traveling around that nucleus. The proton has a positive electric charge, which attracts the negatively charged electron, but the electron is traveling so fast that it is never pulled all the way in to contact the proton.

Select a dark and a light shade of another color; color title A and the blades of the stationary fan with the dark shade and title A¹ and the blurred image of the blades of the rotating fan with the light shade.

The electron moves so fast (close to the speed of light) that it is virtually everywhere around the nucleus at once. The best way to visualize this is to think of an electric fan. If the *blades* are *stationary,* you can easily pass a finger between them. But if the blades are *rotating* at a high speed, they are virtually everywhere at once, and it is impossible to pass a finger between them.

Color title e¹ with a light shade of the electron color, and color the electron cloud (orbital) of the hydrogen atom at the top of the plate. You may want to trace the path of a hypothetical electron with your pen or pencil, going around the nucleus at various distances from it until the entire electron cloud is filled in.

Since the electron travels so fast and moves in three dimensions, sometimes close to the nucleus and sometimes farther away, the space where the electron travels is often called an "electron cloud," although the modern scientific term is *orbital.* Like other clouds, the electron cloud does not have a sharp boundary; it merely gets thinner and thinner near the edges until there just isn't anything left. The traditional way to illustrate this boundary is to draw the part of the orbital within which the electron spends 90

percent of its time, since that is about the extent of the electron's real influence on other atoms nearby.

Now color the heading Relative Dimensions and the titles and related structures in the illustration of the stadium. Color the candy with the proton color, the fly with the electron color, and the interior of the stadium with the orbital color.

No diagram on paper can give an idea of how small the nucleus is in relation to the electron cloud. One way to think of relative distances within the atom is shown in the drawing of the *stadium.* If the nucleus is a small piece of *candy* about the size of a pea that is sitting in the middle of the 50-yard line, the electron can be visualized as a small *fly* attracted to and in constant flight around the candy, sometimes close to it, sometimes farther away, but somewhere within the stadium 90 percent of the time, although it even flies outside the stadium perimeter now and then.

Color the heading Isotopes and title n. Color the rest of the plate, except the three chemical symbols.

Neutrons are particles present in the nucleus of all atoms except the type of hydrogen atom shown at the top of the plate. Neutrons have no electrical charge. They are approximately equal in weight to protons. Protons and neutrons are responsible for virtually all of the mass (weight) of the atom. Electrons have only $1/1840$ as much mass, and hence their mass is negligible.

Atoms of a particular element can exist with different numbers of neutrons. Such atoms are called isotopes. Isotopes of the same element are identical in their chemical reactions but have different weights. Thus the isotope of hydrogen with one neutron (often called deuterium) is twice as heavy as the most common hydrogen isotope, which has no neutron. The hydrogen isotope with two neutrons (often called tritium) is three times as heavy.

Color titles H, B, and C and all the chemical symbols.

Numbers are often placed in front of the chemical symbols to distinguish one isotope from another. The lower number is the number of protons (also called the *atomic number*). The upper number is the *mass number,* which is equal to the number of protons plus the number of neutrons.

SUBATOMIC PARTICLES.

HYDROGEN ATOM.★
  NUCLEUS.★
    PROTON$_p$
  ELECTRON$_e$
   ORBITAL$_{e^1}$
STATIONARY FAN BLADES$_A$
ROTATING FAN BLADES$_{A^1}$

RELATIVE DIMENSIONS.★
CANDY$_{p^1}$
FLY$_{e^2}$
STADIUM$_{e^3}$
ISOTOPES.★   ATOMIC NUMBER$_B$
NEUTRONS$_n$ MASS NUMBER$_C$
SYMBOL$_H$

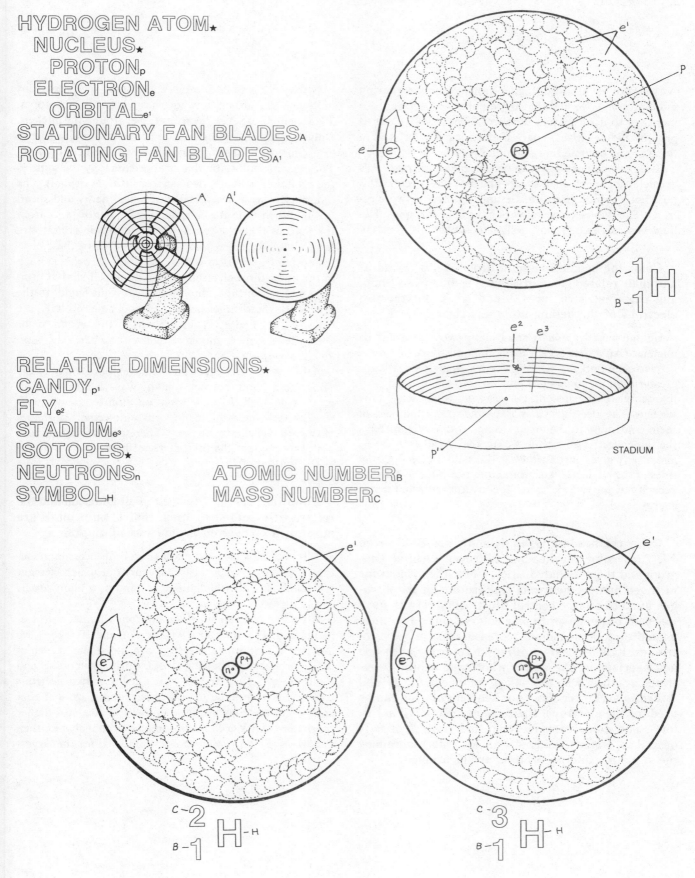

STADIUM

8
ATOMIC STRUCTURE

As we examine atoms of higher atomic numbers (having more protons), we find that the number of electrons also increases to assure equal numbers of positively charged protons and negatively charged electrons. The orbitals in these atoms become more complex.

Electrons can exist only in certain major energy levels or shells, and each shell except the first has several orbitals. Each orbital has a maximum capacity of two electrons. The two electrons always have opposite spins. Helium has only one orbital, designated as the "1s" orbital.

Color the heading Helium, titles He, p$^+$, and n^0 and their related structures, the heading First Shell, and titles and structures A and A^1. Color the two electrons of the helium atom and their orbital.

Helium has two *protons* and usually two *neutrons* in its nucleus. (All atoms except the common isotope of hydrogen have a number of neutrons approximately equal to or greater than the number of protons.) To balance the positive charges of the protons, two *electrons* travel around the nucleus. The electrons share the same space, and both of them contribute to the orbital. Since both electrons have the same electric charge, however, they repel each other, and at any given instant they will usually be on opposite sides of the nucleus. This orbital, the *1s orbital,* is filled to capacity by the two electrons. It constitutes the first major energy level.

Color the heading Orbitals and titles B^1 through E^1. Also color the heading Nitrogen Orbital Diagram and the associated titles. Color the symbol for nitrogen and the neutrons and protons of the nitrogen atom (the large central illustration). If you have five pens or pencils with pale colors, you can produce a good three-dimensional representation of the electron orbitals by using a different color for each one and overlapping the colors where one orbital penetrates another. Use the palest colors for the 1s and 2s orbitals. Darker shades or heavier coloring can be used for the electrons. Read the description that follows, and color each electron and orbital as it is described. Refer to and color the orbitals diagram as the s, p_x, p_y, and p_z orbitals are mentioned.

Nitrogen has seven protons and usually seven neutrons in its nucleus, although it, too, has a number of isotopes, all of them rare. Seven electrons travel around the nucleus. Only two of those electrons are at the first major energy level (innermost shell), which we saw in the helium atom. They occupy an orbital that is approximately the same as the orbital of helium's two electrons (the 1s orbital). The other electrons have a great deal more energy and spend much of their time at a considerably greater distance from the nucleus than the two 1s electrons. Their orbitals are larger but take more complex forms. Two of these electrons are in a spherical orbital called the *2s orbital.* Each of the other three electrons is in a dumbbell-shaped orbital, and each of those dumbbells is at right angles to the others in three-dimensional space. These orbitals are called p_x, p_y, and p_z. They are part of the second major energy level and are therefore referred to as $2p_x$, $2p_y$, and $2p_z$ *electrons.*

All of the 2s and 2p electrons have approximately the same energy, and they act in many ways as if they are all in a single shell. In fact, when an atom combines with some other atom or atoms to form a molecule, the shapes of the orbitals change anyway. Therefore, it is often more useful to diagram the orbitals as a single shell, as shown at the bottom of the plate.

Color the heading Nitrogen Shell Diagram and the related titles and structures. Individual orbitals are not drawn or colored in this sort of diagram.

Although this last sort of diagram is an oversimplification, it makes it much easier to figure out which element will combine with which other one and how many atoms of each will be involved.

If we continue up the scale of atomic numbers to elements number 8 (oxygen), 9 (fluorine), and 10 (neon), we find that the additional electrons are added one to each of the 2p orbitals until in neon all of the orbitals in the second shell (2s, $2p_x$, $2p_y$, and $2p_z$) are filled with two electrons each. As we continue to still larger atoms, we find that additional electrons are in a third shell at a still higher energy level with even more complex orbital shapes, then a fourth shell, and so on to a seventh shell for the largest atoms.

ATOMIC STRUCTURE.

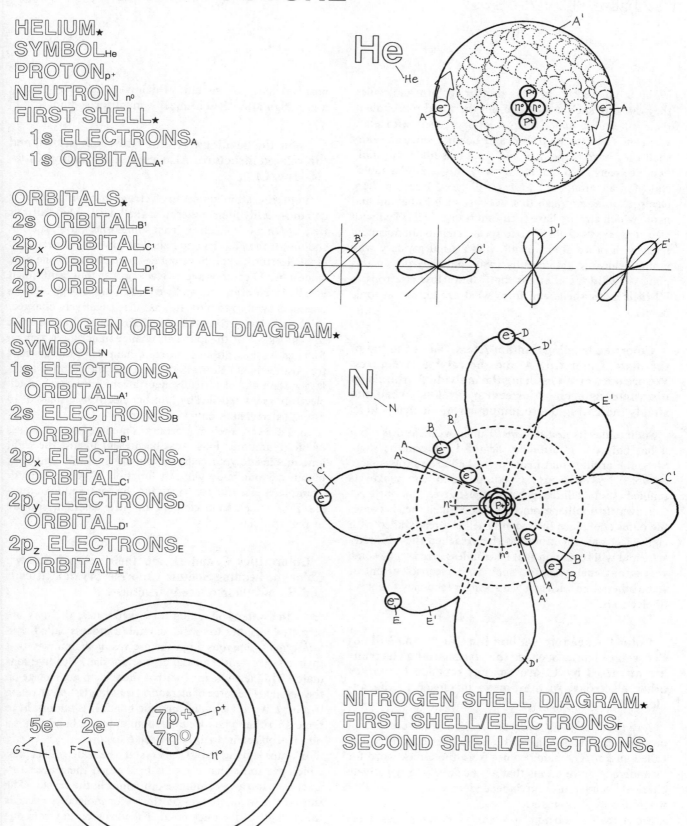

HELIUM★
SYMBOL$_{He}$
PROTON$_{p^+}$
NEUTRON $_{n^0}$
FIRST SHELL★
  1s ELECTRONS$_A$
  1s ORBITAL$_{A^1}$

ORBITALS★
2s ORBITAL$_{B^1}$
2p$_x$ ORBITAL$_{C^1}$
2p$_y$ ORBITAL$_{D^1}$
2p$_z$ ORBITAL$_{E^1}$

NITROGEN ORBITAL DIAGRAM★
SYMBOL$_N$
1s ELECTRONS$_A$
  ORBITAL$_{A^1}$
2s ELECTRONS$_B$
  ORBITAL$_{B^1}$
2p$_x$ ELECTRONS$_C$
  ORBITAL$_{C^1}$
2p$_y$ ELECTRONS$_D$
  ORBITAL$_{D^1}$
2p$_z$ ELECTRONS$_E$
  ORBITAL$_E$

He$_{He}$

N$_N$

NITROGEN SHELL DIAGRAM★
FIRST SHELL/ELECTRONS$_F$
SECOND SHELL/ELECTRONS$_G$

5e⁻ 2e⁻
G   F

7p$^+$ — P$^+$
7n^0
n^0

9
IONIC BONDS

When atoms react with one another to form molecules, they do so in such a way as to fill the s and p orbitals in their outermost major energy level or shell with eight electrons (two for hydrogen and helium, since the only shell they have is the first shell, and it is filled with only two electrons). No one knows why this is a "desirable" state for an atom, but atoms will gain, lose, or share electrons to accomplish this. Atoms, such as helium and neon, which already have their outermost shells filled with electrons, are "satisfied," so to speak, and do not normally react with other atoms at all. It is a total mystery why atoms behave in this peculiar way, but they do. In this plate we will look at atoms that gain or lose electrons to fill their outer shells and form what are known as ionic bonds.

Color the heading Sodium Atom/Na at the top of the plate. Color title A and the related structures. We are once again ignoring the individual orbitals by diagramming each major energy level as a shell and simply indicating the total number of electrons in it.

Sodium has its first two *shells* filled with *electrons,* but it has only one electron in the third (outermost) shell. Since the orbitals that make up a shell are simply spaces occupied by electrons, a sodium atom can make its unfilled outer shell disappear if it just gives that one electron away to another atom. The second shell then becomes the outer one. Since the second shell is already filled with eight electrons, the atom has then reached the "desired" state and will remain in that state unless acted on by some very strong energy source, such as an electric current or some other atom with an even stronger tendency to get rid of electrons.

Color the heading Sodium Ion/Na$^+$. You will notice a size change because the 10 remaining electrons are attracted by 11 protons but repelled by only 9 other electrons, so they move in closer to the nucleus.

As a result of giving away an electron, the sodium atom becomes electrically charged, and in that condition it is called an ion. The name comes from the Greek word for "wanderer," since atoms that are positively or negatively charged wander around independently of one another when dissolved in water.

Since the sodium ion now has 11 protons (positively charged) in its nucleus and only 10 electrons (negatively charged) to neutralize them, the atom as a whole has a net positive charge of one unit, which is indicated by placing a plus sign after the chemical symbol for sodium.

Color the heading Chlorine Atom/Cl, title B, and the related structure. Also color the heading Chloride Ion/Cl$^-$.

A chlorine atom has seven electrons in its *outer shell.* It can most easily fill its outermost shell with eight electrons by accepting one electron from some other atom, such as sodium. It then has its outer shell filled with the necessary eight electrons, but it has a net negative charge of one unit, since it has 17 protons and 18 electrons. In that condition it is called a chloride ion, and we place a minus sign after its chemical symbol to indicate that it is negatively charged. (Note the change from "chlorine" for the uncharged atom to "chloride" for the ion. This change in ending is typical for negative ions, although no such change in ending is used for positive ions.) You will observe that the chloride ion is larger than the electrically neutral chlorine atom. Each electron is now repelled by 17 other electrons instead of 16 but still attracted by only 17 protons.

Ions are extremely important in many of the processes of life. Numerous ions must be present in just the right amounts inside your cells, in your tissue fluids, and your blood, or your body will not function. Calcium ion, for example, is essential for triggering the contraction of muscle. If your heart muscle gets too little calcium, you are in real trouble.

Color titles C and D and their related arrows. Color the heading Sodium Chloride Crystal, titles E and F, and their related structures.

As the water containing ions evaporates, the *ions* are attracted together to form a crystal (a process called *crystallization*). The ions align themselves in such a way that their charges tend to balance one another. Positive ions match with negative ions so that there is no net charge on the crystal. The force of attraction that holds a positive ion to a negative ion is called an ionic bond. It is important to remember that when we talk about atoms, a bond is not a physical object; it is a force or a relationship.

If water is re-added to the crystal, the water molecules, which are somewhat electrically charged themselves, attract each ion away from the other ions in the crystal. The sum of the weak charges on the water molecules exceeds the strength of the ionic bond. The ions separate from one another in a process called *dissociation*. This is illustrated in Plate 11.

IONIC BONDS.

SODIUM ATOM/Na⋆
OUTER ELECTRON/SHELL_A

CHLORINE ATOM/Cl⋆
OUTER ELECTRON SHELL_B

$1e^-$
$8e^-$
$2e^-$
$11 p^+$
$11 n^o$
A

$7e^-$
$8e^-$
$17 p^+$
$18 n^o$
$2e^-$
B

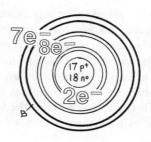

CHLORIDE ION/Cl⁻⋆

e⁻ A

A+B

SODIUM ION/Na⁺⋆

$8e^-$ $2e^-$
$11 p^+$
$11 n^o$

$8e^-$
$8e^-$
$2e^-$
$17 p^+$
$18 n^o$

C

D

CRYSTALLIZATION_C
DISSOCIATION_D

SODIUM CHLORIDE
CRYSTAL ⋆

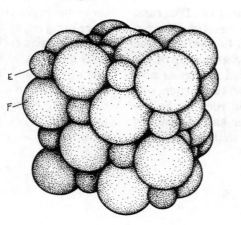

E
F

SODIUM ION_E
CHLORIDE ION_F

10
COVALENT COMPOUNDS

In Plate 9 you learned that many atoms gain or lose electrons to become ions and that these ions wander around separately in water or collect into crystals in the absence of water. The electrical attraction holding two ions together in a crystal is often called an ionic bond. Many elements, however, have atoms that are unable to form ions that will last for more than the tiniest fraction of a second. One such element is carbon.

Color the titles in the upper third of the plate. Choose contrasting colors for C and H. Color the protons, the neutrons, the electrons, and all the space between the electron shells and the nucleus as labeled. Color the symbols for the elements (H and C).

A neutral *carbon atom* has six *protons* and six *electrons*. It also has six, seven, or eight *neutrons*. If it were to gain four electrons to satisfy its "desire" to fill its outer shell to its capacity of eight, the atom would be in the rather unbalanced condition of having six positive charges and ten negative charges. Since it is only the attraction of opposite electrical charges that keeps the electrons around a nucleus, the outer electrons would not be held very strongly and would easily be removed by other atoms. If that same carbon atom were to lose four electrons to get rid of its second shell entirely so that the already filled first shell becomes the outer one, it would have six positive charges balanced by only two negative charges; the overabundance of positive charges would have a strong attraction for additional electrons, and the atom would attract electrons away from other atoms to become a neutral atom again. Instead, carbon almost always fills its outer shell by sharing electrons. One good example of this is the compound methane (CH_4), which is the principal component of the natural gas that is widely used for heating and cooking.

Color the heading Methane Electron Diagram. Use the C_e- and H_e- colors for the title Shared Electrons. Color the title Empirical Formula and the structures of the diagram as you did at the top of the plate. To emphasize that the shared electrons now travel around both the hydrogen and the carbon, one half of each shared electron should receive the hydrogen electron color and the other half the carbon electron color as labeled. Note that in methane

all of the electrons are shared except those of the inner shell of carbon. Color the symbol CH_4 as well.

While carbon needs four additional electrons to fill its outermost shell, hydrogen needs only one additional electron to fill its outermost shell, which has a capacity of only two electrons. So a carbon atom "makes a deal" with four hydrogen atoms. "You let me use one of your electrons, and I'll let you use one of mine." Thus one electron from carbon spends part of its time going around the hydrogen atom at the top of our diagram while the electron of that hydrogen atom spends part of its time going around the carbon atom. One other electron from the carbon atom spends part of its time going around the hydrogen atom on the left of our diagram while the electron of that hydrogen atom spends part of its time going around the carbon atom. The same occurs with the other two hydrogen atoms in the diagram. This satisfies carbon, since it now has eight electrons going around it in the outer shell, and it satisfies the hydrogen atoms, which each have two electrons going around them. Where atoms are concerned, it apparently doesn't matter whether *shared electrons* are present full-time or part-time. This sharing of electrons holds the atoms together to form a molecule, and each pair of shared electrons is called a covalent bond.

Color the remaining headings, title S, and all related structures.

The projection formula provides a simple way to represent the arrangement of the atoms in a molecule that is formed by covalent bonds. Each atom is represented by the letter of its name, and the bond, or *shared electron pair*, is represented by a single straight line. In an actual methane molecule the hydrogen atoms are not radiating out at 90 degrees from one another in a flat plane. They are actually separated by equal angles in three-dimensional space, as is shown in the ball-and-stick model, which shows the angles correctly but does not give an idea of the space occupied by the electrons. That space is more accurately represented by the space-filling model, which represents the space within which the electrons spend 90 percent of their time. The angles at which the various atoms project out from a molecule and the spaces they occupy have great importance to their functioning in a living thing, so biological scientists are generally most interested in the space-filling model of a molecule.

COVALENT COMPOUNDS.

CARBON ATOM$_c$
NUCLEUS★
  PROTON$_{Cp+}$
  NEUTRON$_{Cn^o}$
ELECTRON/SHELL$_{Ce-}$

HYDROGEN ATOM$_H$
NUCLEUS★
  PROTON$_{Hp+}$
  ELECTRON/
  SHELL$_{He-}$

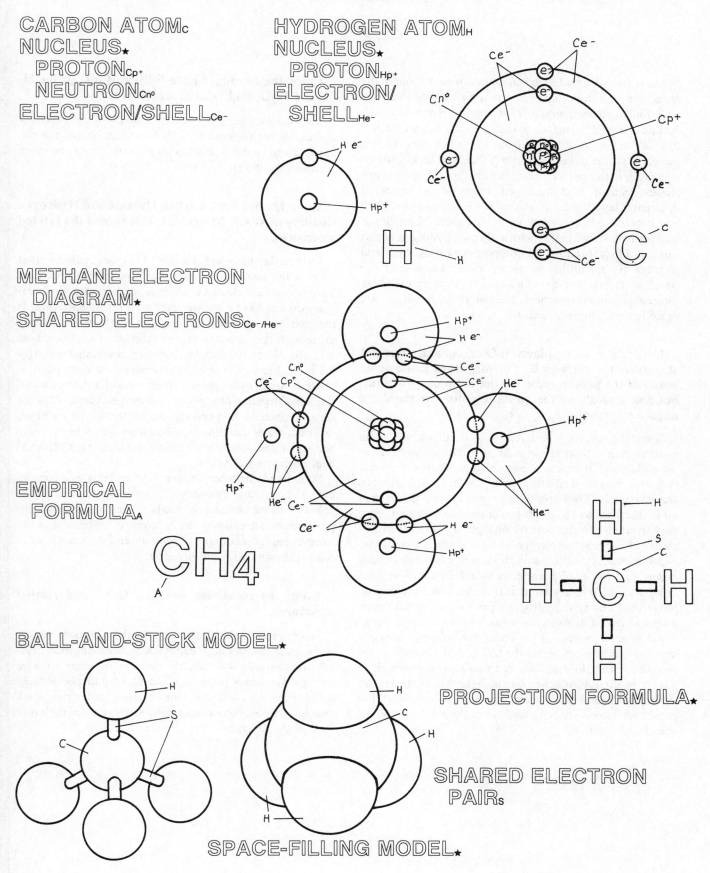

METHANE ELECTRON
  DIAGRAM★
SHARED ELECTRONS$_{Ce-/He-}$

EMPIRICAL
  FORMULA$_A$

BALL-AND-STICK MODEL★

PROJECTION FORMULA★

SHARED ELECTRON
  PAIR$_S$

SPACE-FILLING MODEL★

WATER MOLECULES

We have just seen that a carbon atom can contribute to the formation of a molecule of methane by sharing electrons with four hydrogen atoms. That sharing is nearly equal, and no part of the surface of the molecule has an electric charge that is any different from any other part. Such a molecule is said to be "nonpolar"; that is, it lacks distinctive poles with opposite electrical charges. When oxygen shares electrons with hydrogen, the result is somewhat different. Oxygen has six electrons in its outermost shell and it needs only two more to fill that shell. Therefore, it will form covalent bonds with only two hydrogen atoms instead of four, producing a compound with the empirical formula H_2O, familiar to us as water. Because of the peculiar orbitals that electrons "like" to occupy, the water molecule is not symmetrical. The two hydrogen atoms are more toward one side than the other.

Color the heading Electron Diagram and titles and structures p$^+$ through B. You may want to use some shade of the proton color for the delta positive symbol and a shade of the electron color for the delta negative symbol.

Since the oxygen atom has two more *protons* in its nucleus than carbon, it has a stronger attraction for *electrons* than the hydrogen atoms. As a result, the electrons that the oxygen is sharing with the hydrogen are not shared equally. They spend more time around the oxygen atom than around the hydrogen atom—enough more that they overbalance the positive charges of the oxygen nucleus. This gives the oxygen atom a distinct, but weak, *negative charge,* indicated by the Greek letter delta and a minus sign (δ^-). The electrons spend enough less time around the hydrogen atom that they don't quite neutralize the positive charge of hydrogen's protons, so the hydrogen atoms acquire a distinct, but weak, positive charge, designated as *delta positive* (δ^+). These delta charges are only about one-fifth as strong as the full unit of charge we saw on sodium and chloride ions, but that is strong enough to have a profound influence on the behavior of water and similar molecules. Such molecules are called "polar" because they have distinctive regions, or "poles," with opposite electric charges.

Color the heading Space-Filling Model, titles H, O, and H$_2$O, and related structures.

As with methane, the electron diagram is useful for discussing the formation of the covalent bonds, but the space-filling model is a better representation of the space occupied by the electrons.

Color the headings Carbon Dioxide and Hydrogen Bonding of Water Molecules, title C, and the related structures.

Carbon dioxide is an example of another molecule that is about the same size as water but is nonpolar. Since they have no surface electrical charges, carbon dioxide molecules do not have a strong attraction for one another. If they are released from the pressure bottle in which they are stored, they separate and expand into the atmosphere as a gas. Water molecules, however, are rather strongly held together by the attraction between the delta positive charges on the hydrogen atoms of one water molecule and the delta negative charges on the oxygen atoms of other water molecules. As a consequence of this attractive force, or bond, water remains as a liquid and does not rapidly become a gas, as carbon dioxide does, unless additional heat energy is put into it.

Such bonds between hydrogen atoms in a polar molecule and a negatively charged atom in some other molecule are called "hydrogen bonds." They are of great importance in determining the behavior of water and in the structure and functioning of proteins and the nucleic acids that determine our own heredity.

Color the remaining heading, titles, and related structures.

Since water molecules are polar, they are able to dissolve ionic compounds readily. A sodium chloride crystal, for example will quickly dissolve in water because four or five water molecules can fit around one *sodium* or *chloride ion* and the sum of their collective weak charges is enough to attract the ion away from the other ions in the crystal.

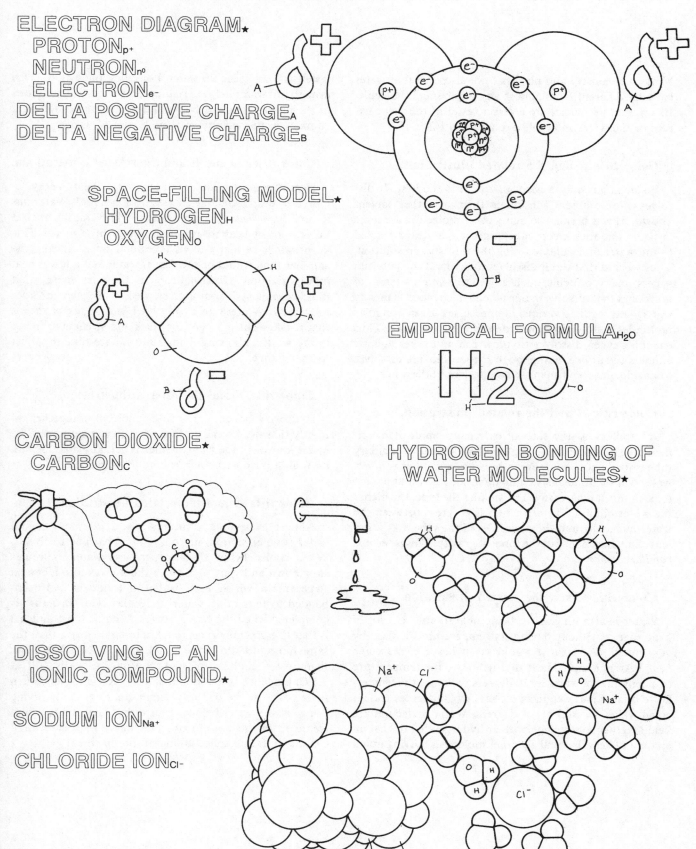

WATER MOLECULES.

ELECTRON DIAGRAM★
  PROTON$_{p+}$
  NEUTRON$_{n^0}$
  ELECTRON$_e$
  DELTA POSITIVE CHARGE$_A$
  DELTA NEGATIVE CHARGE$_B$

SPACE-FILLING MODEL★
  HYDROGEN$_H$
  OXYGEN$_O$

EMPIRICAL FORMULA$_{H_2O}$

$H2O$

CARBON DIOXIDE★
  CARBON$_C$

**HYDROGEN BONDING OF
WATER MOLECULES**★

**DISSOLVING OF AN
IONIC COMPOUND**★

SODIUM ION$_{Na^+}$

CHLORIDE ION$_{Cl^-}$

12
UNUSUAL PROPERTIES OF WATER

Modern chemistry and physics have disclosed that water is vastly different from almost any other small molecule. Its unusual properties are a direct result of the polar nature of the water molecule explained in Plate 11.

Color title A and the related illustration.

Water is not only a *good solvent,* it is the best. It dissolves more different substances than any other solvent known. This is because so many other molecules are ionic or polar, and their electrical charges make them attracted to the water molecules, causing them to stay in solution. Thus we find that water dissolves many kinds of salts and sugars, many proteins, such as gelatin, and a variety of hormones that dissolve in our blood (since blood is mostly water) and regulate various life processes. Even nonpolar molecules dissolve to some extent if they are small. Thus enough oxygen dissolves to allow fish and other aquatic animals to survive, and enough carbon dioxide dissolves to enable algae and many plants to live underwater.

Color title B and the related illustration.

A capillary is any tube of extremely small diameter, including the tiniest of our blood vessels. If a capillary tube is made of glass or any other substance that is polar, water will spontaneously climb up inside it without having to be pumped in any way. The smaller the tube, the higher the water climbs. The attraction is so great between the water molecules and the molecules of the tube that water will climb in defiance of the force of gravity. This is termed *capillary action.*

Color title C and the related illustration.

Water is also unusual in being able to absorb a lot of heat energy without having its temperature increase by very much. (A scientist would say it has a *high specific heat.*) An amount of heat that will raise the temperature of a container of water by 10 degrees will raise the temperature of an equal weight of alcohol by 20 degrees and an equal weight of iron by 94 degrees. Water molecules are held together so strongly by their hydrogen bonds that an amount of heat that will get other molecules moving much

faster will not speed up water molecules much at all. This property of water helps to reduce temperature fluctuations in the animal or plant body, and it also makes for mild climates in the vicinity of large bodies of water.

Color titles D and E and the related illustrations.

Heat of vaporization is the amount of heat energy required to evaporate a given weight of a liquid. Water has a very *high heat of vaporization,* which means that it takes a lot of heat to evaporate just a little water. This keeps water in many more lakes and ponds during the summer than would be the case if water had a lower heat of vaporization. Heat of fusion is the heat energy that must be removed from a given weight of water in order to freeze it. Water's relatively *high heat of fusion* means that it takes much longer for lakes and streams to freeze in the winter, allowing living things more time to adjust to the change.

Color title F and the related illustration.

Hydrogen bonds hold water molecules together so tightly that the water's surface acts like a membrane. The insect known as the water strider is actually able to walk on that surface without breaking through.

Color title G and the related illustration.

Almost everything contracts when it is cooled, and water is no exception, up to a point. That point is 3.8°C. When cooled below that temperature, water molecules slow down and start to arrange themselves into a crystal structure in which each water molecule is hydrogen-bonded to four other water molecules. This structure is completed when the water freezes. What is most unusual is that this crystal structure takes up more space than the same molecules did in the liquid state, so ice is less dense than water, and it floats. The water below is still at 3.8°C. Since ice is a good insulator, lakes and ponds can freeze over in the winter without freezing all the living things in the water below. If ice didn't float, lakes would freeze from the bottom up, and many of them would eventually freeze solid, killing all life in them.

UNUSUAL PROPERTIES OF WATER.

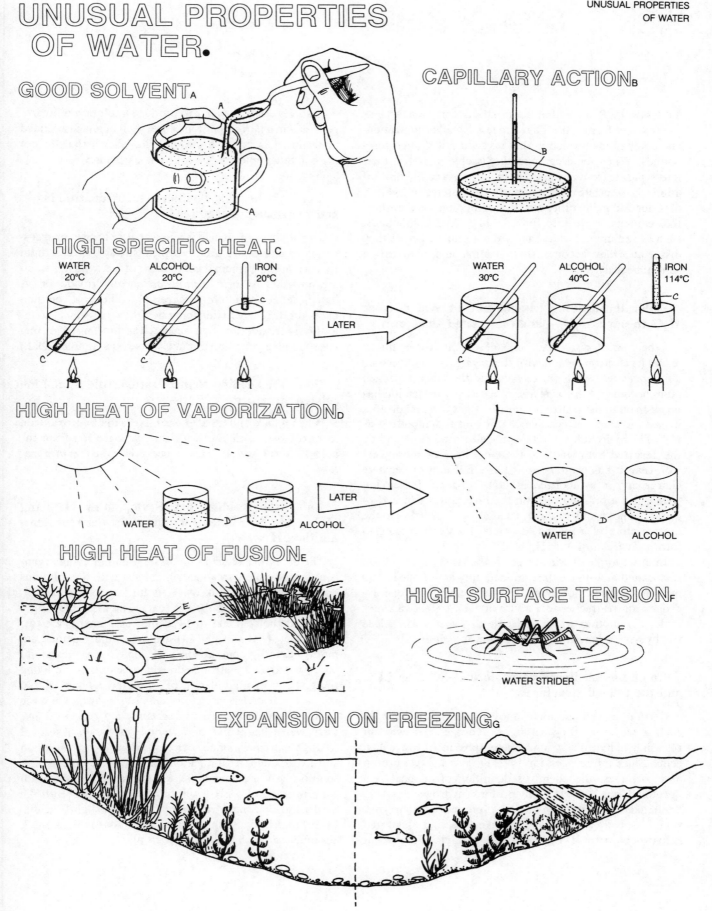

GOOD SOLVENT$_A$

CAPILLARY ACTION$_B$

HIGH SPECIFIC HEAT$_C$

WATER 20°C ALCOHOL 20°C IRON 20°C

LATER

WATER 30°C ALCOHOL 40°C IRON 114°C

HIGH HEAT OF VAPORIZATION$_D$

WATER ALCOHOL

LATER

WATER ALCOHOL

HIGH HEAT OF FUSION$_E$

HIGH SURFACE TENSION$_F$

WATER STRIDER

EXPANSION ON FREEZING$_G$

13
ACIDS AND BASES

You saw in Plate 9 that some atoms can gain or lose electrons to form ions. That process is called ionization, and compounds formed in that way are called ionic compounds. When ionic compounds dissolve in water, their ions separate from one another in a process called dissociation (the opposite of association). One interesting feature of water and many other covalent compounds is that they, too, can dissociate into ions. Unlike ionic compounds, such as sodium chloride, they are not ionized before they dissociate; they accomplish ionization and dissociation simultaneously.

Color the heading Dissociation of Water, titles H_2O through H_3O^+, and the related structures.

When *water dissociates,* one of the hydrogen nuclei leaves its electron behind with the oxygen atom to become a *hydrogen ion,* while the oxygen and the other hydrogen atom become a *hydroxide ion.* Since the hydrogen ion has no electron to neutralize the positive charge on its proton, it has a full unit of positive charge and is symbolized as H^+. The hydroxide ion retains the electron left behind by the departed hydrogen and therefore has one more electron than it has protons, so it has a full unit of negative charge and is symbolized as OH^-. The hydrogen ion (really just a proton except in the rare isotopes of hydrogen) does not wander long by itself before it attaches to the oxygen atom of a second, un-ionized water molecule to form a *hydronium ion* (H_3O^+).

In any sample of water, very few of the molecules are dissociated at any one time: in fact, only about one in 550 million. There is, however, a constant change; as one hydrogen ion reattaches to a hydroxide ion to form a water molecule, another water molecule dissociates to replace the hydrogen ion and hydroxide ion in solution.

Color the heading Hydrochloric Acid, title Cl^-, and the related structures.

Certain molecules, ionic and covalent, dissociate in such a way that they release a hydrogen ion without releasing a hydroxide ion. These substances are called acids. Since a hydrogen ion is really just a single proton in most cases, the chemist's definition of an acid is a "proton donor." If very many protons (hydrogen ions) are "donated," the effect can be very profound, burning your skin, dissolving a metal, and the like. The acid illustrated is hydrochloric acid. Pure hydrochloric acid is a gas, but it dissolves easily in water to produce a solution of hydrogen ion and *chloride ion.* Since nearly all of it is dissociated in water, it is called a strong acid. (Acids that do not dissociate completely are called weak acids.)

Color the heading Sodium Hydroxide, title Na^+, and the related structures.

The opposite of an acid is a base, better known in everyday language as an alkali. A typical strong base is sodium hydroxide, the principal component of lye, which dissociates in water to form *sodium ion* and hydroxide ion. A base is defined as a "proton acceptor." The most common bases produce hydroxide ion when they dissociate, and it is the hydroxide ion that accepts the proton. (A strong base can give your skin a much worse burn than an acid.)

Color the heading Neutralization, title B, and the related structures.

When a base and an acid are mixed, the hydroxide ion from the base combines with the hydrogen ion from the acid to form water. This process is called *neutralization.*

Color the heading pH Scale, titles H^+ and OH^-, and the related portions of the bar representing the pH scale.

The quantities of acids and bases found in living organisms are extremely small in comparison with the solutions normally used in chemistry laboratories. As a result, biologists have adopted the pH scale. The pH scale ranges from 0 at the acid end to 14 at the basic end. These two extremes (pH 0 and pH 14) are only mildly acidic and mildly basic by comparison with many other acids and bases, but they are strong enough to be lethal to a living thing. Each change of one pH unit indicates a tenfold increase or decrease in *hydrogen ion concentration* and the opposite tenfold change in *hydroxide ion concentration.* Pure water has a pH of 7, which means it has equal (though extremely small) concentrations of hydrogen ion and hydroxide ion (10^{-7} molar, if you are a chemist). A solution with a pH of 6 has ten times the hydrogen ion concentration of a pH 7 solution and only one-tenth the hydroxide ion concentration; it is therefore slightly acidic. Most fluids in living things have a pH not too far from 7, although stomach acid can get to pH 1.

ACIDS AND BASES.

DISSOCIATION OF WATER.★
WATER$_{H_2O}$　　　DISSOCIATION$_A$　　　HYDROXIDE ION$_{OH^-}$
HYDROGEN ION$_{H^+}$　　　HYDRONIUM ION$_{H_3O^+}$

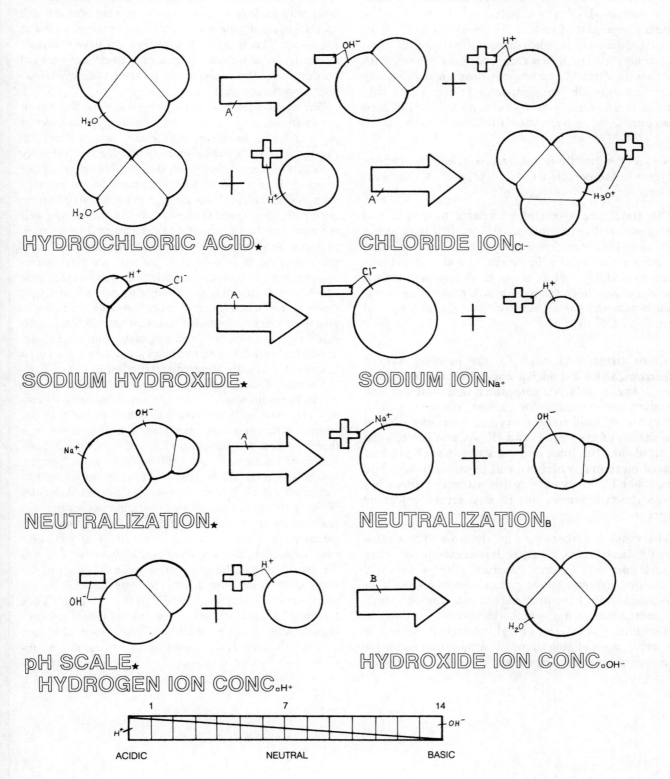

HYDROCHLORIC ACID.★　　　CHLORIDE ION$_{Cl^-}$

SODIUM HYDROXIDE.★　　　SODIUM ION$_{Na^+}$

NEUTRALIZATION.★　　　NEUTRALIZATION$_B$

pH SCALE.★
HYDROGEN ION CONC.$_{\square H^+}$　　　HYDROXIDE ION CONC.$_{\square OH^-}$

ACIDIC　　　NEUTRAL　　　BASIC

14

CARBOHYDRATES I

The processes of life are primarily the result of the chemistry of compounds of carbon. In fact, all except a very few simple compounds of carbon are called "organic" compounds. Because of carbon's tendency to form four covalent bonds in four different directions, as we saw with methane in Plate 10, carbon can form an unbelievably large number of different compounds of high complexity. This plate introduces you to one major category of such compounds, the carbohydrates (hydrates of carbon).

Color the headings L-Glycerose Isomer and D-Glycerose Isomer, title C+H+O, and its representation.

"Hydrate" means something combined with water, and carbohydrates have approximately two hydrogen atoms and one oxygen atom (amounting to one water molecule) for every carbon atom. The general formula for carbohydrates is $(CH_2O)_n$, where n can be almost any number. For glycerose, illustrated here, n is 3; multiplying everything within the parentheses by 3, we get an empirical formula of $C_3H_6O_3$.

Color titles A through O, the heading Shared Electrons, titles B and B_1, and all remaining structures. Again it is recommended that you use the standard colors: black for carbon, white or yellow for hydrogen, and red for oxygen. You may wish to use shades of gray for B and B^1. As you color, keep in mind that the lines and sticks represent pairs of shared electrons, which we call covalent bonds. The term "bond" refers only to the attractive force between the two atoms, not to any actual object or structure.

The simplest carbohydrate is glycerose. The ending "-ose" indicates that it is a sugar. It is an example of a class of molecules called monosaccharides ("single sugars"). Notice that glycerose has only three *carbon* atoms. Other monosaccharides have from four to nine carbon atoms. The most common monosaccharides have five or six.

Sometimes glycerose is called glyceraldehyde because of the arrangement of atoms around carbon atom number 1 (the uppermost one). Any carbon atom sharing *one pair*

of electrons with a *hydrogen atom* and *two pairs of electrons* with an *oxygen atom* constitutes what chemists call an *aldehyde group.* Several other important sugars are also aldehydes. This is the first compound we have discussed in which two atoms share two pairs of electrons instead of only one pair. This arrangement is called a "double bond" and is not uncommon.

The most common way of representing the three-dimensional structure of carbon compounds on a flat sheet of paper is the Fischer projection formula, named after Emil Fischer, a great German chemist of the nineteenth century who first proposed this method. The three-dimensional structure is arranged so that the carbon atoms are in a vertical line (although the upper and lower carbon atoms are farther away from the viewer than the middle one), and the atoms to the left and right of the center carbon atom project toward the viewer. The atoms are "projected" onto the paper, much as you might do with an overhead or opaque projector or with just a bright light to cast shadows. The ball-and-stick and space-filling models are arranged this way, and a projection of them onto a flat surface will give the projection formula illustrated here. If this is difficult for you to visualize, you can make your own three-dimensional models with toothpicks and different-colored gumdrops or other moderately soft candies.

Because of the arrangement of carbon's electron orbitals in three-dimensional space, many compounds of carbon can exist with the identical empirical formula but entirely different arrangements of the same atoms and therefore entirely different chemical properties. Such molecules are called isomers (Greek: *iso*, "same"; *meros*, "parts"). A complete discussion of isomers must be left to a chemistry book, but even a brief study of the molecules of living things introduces us to certain isomers that are exact mirror images of one another, technically known as enantiomers (Greek: *enantios*, "opposite"). If you compare the models of D-glycerose and L-glycerose, you will see that they have the same atoms joined together in the same combinations, but in arrangements that differ in the same way that your right hand differs from your left. They are equal but opposite and are said to be "mirror images" of each other. It is a peculiarity of living things that they use right-handed (D-) isomers of sugars almost exclusively. Left-handed (L-) sugars are extremely rare.

CARBOHYDRATES I.

L-GLYCEROSE ISOMER.★
EMPIRICAL FORMULA$_{C+H+O}$

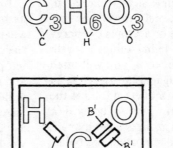

$C_3H_6O_3$
_C _H _O

D-GLYCEROSE ISOMER.★

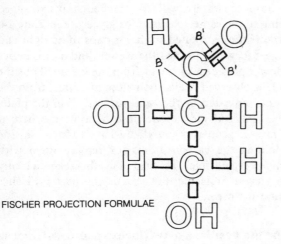

FISCHER PROJECTION FORMULAE

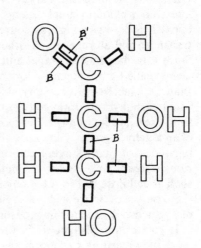

ALDEHYDE$_A$
CARBON$_C$
HYDROGEN$_H$
OXYGEN$_O$

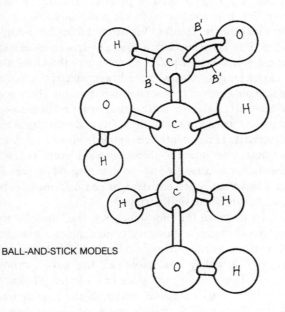

BALL-AND-STICK MODELS

SHARED
ELECTRONS.★
ONE PAIR$_B$
TWO PAIRS$_{B^1}$

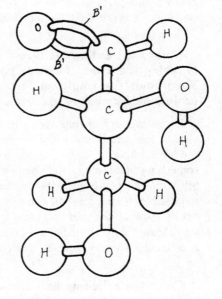

SPACE-FILLING MODELS

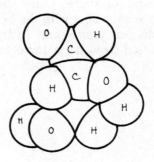

15
CARBOHYDRATES II

Glucose is the monosaccharide (simple sugar) that circulates in our blood to supply energy to cells throughout the body. It is used as an energy source or a means of energy transport and storage by all other living things as well. Since it is abundant in grapes and in corn, you will often hear it called grape sugar or corn sugar. Like glycerose in Plate 14, glucose has an empirical formula that follows the usual carbohydrate ratio of $(CH_2O)_n$ (with n being 6 in this case), and its number 1 carbon atom is a component of an aldehyde group. Glucose is also important as a building block of many larger molecules, called polysaccharides (Greek: *poly,* "many"), including some giant ones such as cellulose and starch. Only D-glucose is used for all these purposes, so we will not even consider the structure of L-glucose, its mirror-image isomer.

If you are interested only in the biological role of glucose, the preceding paragraph has told you most of what you need to know, and you may want to omit coloring this plate. On the other hand, if you are taking a class in which you are expected to learn some details of the structure of glucose, this plate will help you to understand some concepts that are not well explained in many biology books.

Color the heading D-Glucose and titles C, H, and O, using the same colors as you used in Plate 14. Then color the empirical formula and the Fischer projection formula of glucose.

If you ever try to construct a ball-and-stick model of a glucose molecule from a Fischer projection formula, your model is likely to come out all wrong because there is something that biology and chemistry books almost never tell you. The angles of carbon's electron orbitals make it impossible to arrange such a model so that a projection of it will show all the *carbon* atoms in a straight line unless you zigzag them—but that isn't the way chemists look at a molecule when they make a projection formula.

Color the rolled-up ball-and-stick model, title A, and the chemist's eyes in the four viewing positions around the model. Color also the straight-chain ball-and-stick model with the broken sticks at right. The hydrogen atoms and hydroxyl groups have been omitted from these for clarity.

In making a projection formula, a chemist views each carbon atom separately, with its attached atoms arranged according to the same rule used for glycerose in Plate 14: the carbon atom is rotated so that groups to its right and left project forward, toward the viewer, and atoms above and below it project back, away from the viewer. Thus the molecule is observed one carbon atom at a time, from the viewpoints indicated by the *"chemist's eyes"* in the plate. To get a ball-and-stick model into a position where it would project properly, you would have to break all the sticks joining the carbon atoms (or replace them with something very flexible). To most of us, this seems a funny way to view a molecule, but that is the way a Fischer projection is drawn.

Color the headings α-D-Glucose and β-D-Glucose and each of the remaining structural formulae as you come to them in the reading below.

Glucose in a living animal or plant is virtually always dissolved in water, and the molecule bends and folds as it collides with other molecules. Whenever it folds far enough to bring carbon 1 close to carbon 5, one hydrogen atom and several electrons change places to rearrange the molecule into a closed ring, as shown in the lower part of the plate. Depending on which of the two bonds is disrupted between carbon 1 and its oxygen atom, this ring structure of glucose can take two forms, known as alpha (α) and beta (β). Any single molecule is constantly changing from one form to another, and any solution of glucose in water consists of all three forms, approximately 36 percent α ring, 64 percent β ring, and less than 1 percent straight chain form. (Other monosaccharides form similar rings.)

When we diagram the ring structures, the "chair" formula is useful because it shows the bond angles accurately. The Haworth projection formula is less accurate but easier to draw, so it is widely used. Note that the carbon atoms in this projection are not represented by a letter. The user must remember that each intersection of the lines represents a carbon atom. (The chair formula can also be drawn with only intersections to represent carbon.)

The space-filling model shows approximately what we think the electron clouds (orbitals) would look like if we could see them.

CARBOHYDRATES II.

D-GLUCOSE★   CARBON$_C$ HYDROGEN$_H$ OXYGEN$_O$

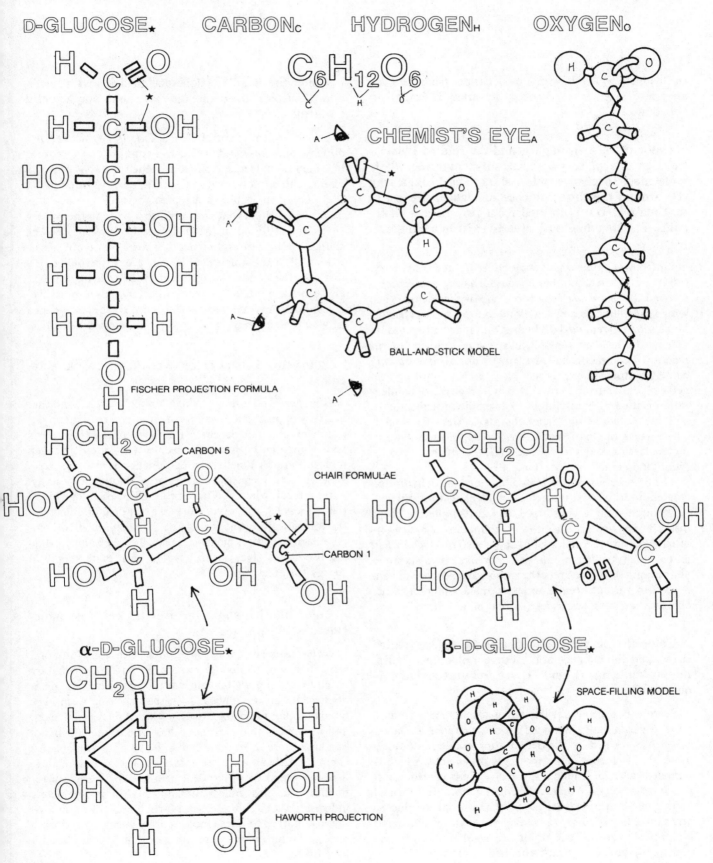

$C_6H_{12}O_6$

CHEMIST'S EYE$_A$

BALL-AND-STICK MODEL

FISCHER PROJECTION FORMULA

H CH$_2$OH

CARBON 5

CHAIR FORMULAE

CARBON 1

H CH$_2$OH

α-D-GLUCOSE★

β-D-GLUCOSE★

CH$_2$OH

SPACE-FILLING MODEL

HAWORTH PROJECTION

16

CARBOHYDRATES III

In this plate we see how monosaccharides (single sugars) are joined together to make a wide variety of larger carbohydrates.

Color titles A through E and the related illustration. In this plate you need only color the whole molecules, so choose pale colors. The broken ring (D) around the lower glucose molecules illustrates that maltose (D) is formed from two glucose molecules. The ring does not actually exist in a molecule.

A *disaccharide* is formed by the removal of a hydrogen atom from one *monosaccharide molecule* and a hydroxyl (OH) group from another monosaccharide molecule to form a *water molecule*. The two monosaccharides are then joined together by a covalent bond to form a disaccharide. Since *water* is removed (dehydration) and the two monosaccharides are "condensed" into a single molecule, this reaction is commonly called a *"dehydration condensation"* or "dehydration synthesis" (*synthesis,* "putting together"), indicated by arrow B. This same type of condensation is used by living things to assemble proteins, lipids, and nucleic acids from their subunits, as we will see later. The reverse of this is called *hydrolysis* because a water molecule is broken in the process (Greek: *hydro,* "water"; *lysis,* "loosening" or "breaking").

The first section of this plate illustrates the dehydration condensation of two molecules of *glucose,* the principal sugar circulating in your blood, to form a molecule of the disaccharide *maltose.* Maltose is abundant in malt (germinated grain used in brewing) and is also produced when an enzyme (ptyalin) in your saliva breaks down starch in your mouth. Glucose and other monosaccharides are condensed in various combinations to make different disaccharides or polysaccharides (Greek: *poly,* "many").

Color the heading Disaccharides, the associated titles, and the sucrose and lactose molecules. Again, the broken rings (F and H) are for illustrative purposes only.

Sucrose is a disaccharide that is familiar as common table sugar, usually obtained from sugarcane or sugar beets. It is formed by the dehydration condensation of a molecule of glucose and a molecule of *fructose,* which is a monosaccharide abundant in many kinds of fruit. *Lactose* is a disaccharide that is abundant in milk. It is formed by the dehydration condensation of a molecule of glucose and a molecule of *galactose.* Galactose has a structure that is identical to glucose except for a reversal of the hydrogen and hydroxyl (OH) groups attached to carbon 4.

Color the heading Polysaccharides and titles J and K. Color over the amylose and amylopectin molecules.

Starch is a mixture of two different polysaccharides, *amylose* and *amylopectin.* Each is composed of hundreds of glucose molecules joined together. Starch provides plants with a good means of storing energy. In amylose, the glucose molecules are joined together in one long, unbranched chain that coils up into a helix in the watery environment found in living things. In amylopectin, the chain is branched and sometimes forms a complex network. All of the glucose molecules are in the alpha ring form, so the bonds linking them together are known as alpha linkages. Humans and other animals are able to digest both amylose and amylopectin, so they serve as a good source of energy.

Color title L, and color over the glycogen molecule.

Glycogen is commonly called "animal starch," and that is a very appropriate name for it. It is exactly like amylopectin except that its chain branches at closer intervals and it apparently does not link up with other glycogen molecules to form networks. Its function is also the same. It serves as a means of storing large amounts of energy. After a meal, when you have more glucose in your blood than you need, your liver stores a great deal of glucose in the form of glycogen. Hours later, when your blood glucose begins to drop, the liver replenishes the supply. Muscle also stores glycogen to have an immediate supply of energy for emergencies.

Color title M, and color over the cellulose molecules.

Cellulose is very much like amylose in consisting of a straight chain of glucose molecules, but the chains are much longer (up to 4000 glucose molecules in some plant fibers), and all the glucose is in the beta ring form. The bonds joining them together are therefore beta linkages, and it happens that no animal has an enzyme capable of breaking those linkages to make the glucose available for energy. Cattle and similar animals have bacteria in their digestive tracts that digest the cellulose for them, but for humans and most other animals, cellulose is completely indigestible. Cellulose does not form a helix like amylose; it joins to other cellulose molecules by means of hydrogen bonds, making an immense complex that is totally insoluble in water.

CARBOHYDRATES III.

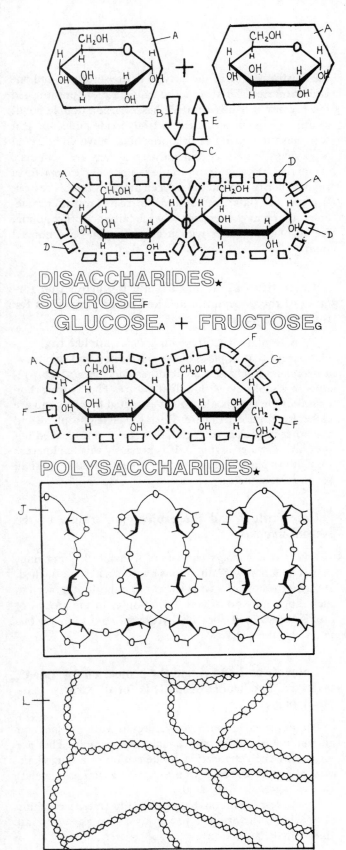

GLUCOSE
 (MONOSACCHARIDE)$_A$
DEHYDRATION
 CONDENSATION$_B$
WATER$_C$
MALTOSE
 (DISACCHARIDE)$_D$
HYDROLYSIS$_E$

DISACCHARIDES$_\star$
SUCROSE$_F$
 GLUCOSE$_A$ + FRUCTOSE$_G$

LACTOSE$_H$
 GALACTOSE$_I$ + GLUCOSE$_A$

POLYSACCHARIDES$_\star$

AMYLOSE$_J$
AMYLOPECTIN$_K$
GLYCOGEN$_L$
CELLULOSE$_M$

Lipids are fats, waxes, and similar molecules that do not dissolve well in water.

Color titles A, B, C, H, O, D, and D¹ and the Space-Filling Model headings. Color the projection formulae and the space-filling models of glycerol and the saturated and unsaturated fatty acids. It is recommended that you use the same colors as in previous plates for the carbon, hydrogen, and oxygen atoms. Choose light colors for A, B, and D.

Fats are composed of *glycerol* and *fatty acids.* Glycerol always has three *carbon atoms* and three hydroxyl (OH) groups, but there are several dozen kinds of fatty acids, ranging in size from 4 carbon atoms to 24. On one end of a fatty acid we find a carbon atom with a double bond to an oxygen atom and a single bond to a hydroxyl group. This entire group of four atoms, often written as —COOH, is called a carboxyl group and is able to ionize to release a hydrogen ion into solution, thus acting as an acid. (The ionized carboxyl group is symbolized as —COO⁻.) In any group of such molecules, only a few are ionized at any one time, so fatty acids are all weak acids. All the rest of a fatty acid molecule is pure hydrocarbon (hydrogen and carbon). Fatty acids are designated as *saturated* or *unsaturated* according to whether they are filled to capacity with hydrogen atoms or not. In a saturated fatty acid, all of the carbon atoms are joined to one another by single bonds, and each one (other than the carboxyl carbon) is bonded to at least two hydrogen atoms. (The one on the end has three.) In an unsaturated fatty acid, at least one pair of carbon atoms is joined by a *double bond,* so that each of those carbon atoms is bonded to only one hydrogen atom, leaving the fatty acid with at least two fewer hydrogen atoms than it would have if it were saturated. The double bond often throws a kink in the hydrocarbon chain as shown in the space-filling model here.

Color title E and the projection formula of the triglyceride. The broken ring (E) is to illustrate that the triglyceride is composed of glycerol, saturated fatty acids, and an unsaturated fatty acid.

A fat—chemically known as a *triglyceride*—consists of a molecule of glycerol joined to three fatty acid molecules by the same kind of dehydration condensation we saw in the formation of disaccharides and polysaccharides. The three fatty acids may be all the same or any combination of different ones. Note that in the triglycer-ide illustrated, two of the fatty acids are saturated and one is unsaturated. This would be called a monounsaturated fat, because it is unsaturated (has a carbon double bond) at only one point in the entire triglyceride molecule. If it were unsaturated at two or more points, it would be called a polyunsaturated fat. Since hydrocarbons are nonpolar, the entire triglyceride molecule is nonpolar except for a slight polarity around the oxygen atoms. For this reason, triglycerides (fats) are not much attracted to water molecules. If you have ever tried to wash butter or other animal fat off of your hands with just water, you have noticed that.

Color titles F, G, and H, and the projection formula of the phospholipid. The broken ring (F) is for illustrative purposes only.

In molecular structure *phospholipids* are like triglycerides except that in place of the third fatty acid they have a *phosphate group* and some other *polar group*. This results in a molecule with a dual nature. The hydrocarbon chains of the fatty acids are not attracted to water and are called hydrophobic ("water-fearing"). The phosphate and the other group are attracted to water and are called hydrophilic ("water-loving"). It is precisely this dual nature that allows phospholipids to form membranes, as we shall see in a later plate.

Color title I and the projection formula of the steroid nucleus.

The *steroid nucleus* consists of four interlocking rings of carbon atoms with numerous hydrogen atoms attached. It forms the core of a wide variety of important molecules including many hormones, which differ in the groups of atoms substituted for the hydrogen atoms at various points on the rings.

Color titles J and K and the projection formula of beeswax. The broken ring (J) is for illustrative purposes only.

Waxes provide protective coatings for various plant and animal tissues and for bees to make honeycombs. They are formed by the dehydration condensation of a long-chain *alcohol* (hydrocarbon with a hydroxyl group at one end) and a long-chain fatty acid.

Other, less common lipids (not illustrated) combine fatty acids with various other groups, such as sugars and amino acids.

LIPIDS.

GLYCEROL
PROJECTION FORMULA_A
SPACE-FILLING MODEL ★
CARBON_C
HYDROGEN_H
OXYGEN_O

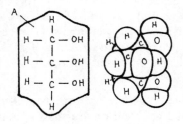

SATURATED FATTY ACID
PROJECTION FORMULA_B
SPACE-FILLING MODEL ★

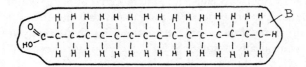

UNSATURATED FATTY ACID_D
CARBON DOUBLE BOND_D'

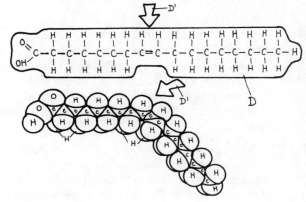

TRIGLYCERIDE_E

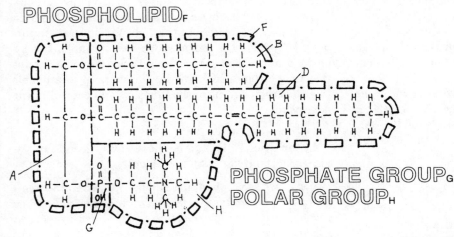

PHOSPHOLIPID_F

PHOSPHATE GROUP_G
POLAR GROUP_H

STEROID NUCLEUS_I

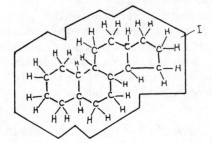

BEESWAX_J
ALCOHOL_K

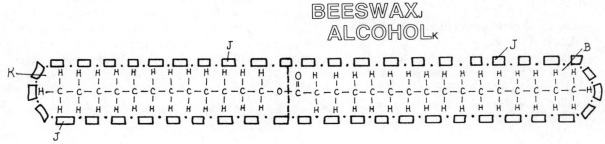

INTRODUCTION TO PROTEIN

Protein gets its name from a Greek word meaning "first" or "primary" because it is the material of primary importance in every process we associate with being alive. Virtually none of the chemical reactions in a living thing would occur at any useful speed if it were not for those specialized protein molecules called enzymes. Other proteins serve as transport molecules, carrying things in the blood or across membranes or transporting electrons that are removed or added in important chemical reactions. Proteins are the main structural components of claws, hooves, and hair as well as the tough surface layer of skin. Contraction of muscle and movement within living cells is accomplished by protein. Many of the chemical messengers we call hormones are proteins, as are the antibodies that protect us from disease.

Color the heading Amino Acid and titles C through N using the same colors as in previous plates. Now color the ball-and-stick model and the space-filling model at the top of the plate. The sticks that represent shared electrons (bonds) are to be colored gray. Leave the side groups uncolored for now.

Proteins are made up primarily (or exclusively, in some cases) of long chains of amino acids. Amino acids (here illustrated by L-alanine) consist of a two-carbon portion that is common to all amino acids and a side group, which varies from one amino acid to the next. *Carbon 1* is part of a carboxyl group (—COOH). The carboxyl group gives these molecules their acid properties by dissociating to release a hydrogen ion (proton). Carbon 2 has a nitrogen-containing amino group (NH_2).

In contrast to the situation in carbohydrates, where the right-handed or D-isomers are used for nearly everything, we find that proteins are made exclusively of the left-handed or L-isomers of amino acids. (D-amino acids are found in some antibiotics but not in proteins.)

Color the title ionization and the projection formulae of un-ionized alanine and the alanine zwitterion. Again, the gray bars are shared pairs of electrons.

Under the conditions found in living cells, nearly every molecule of any amino acid is doubly *ionized:* the carboxyl group releases a *hydrogen* ion and becomes negatively charged, while the amino group picks up a hydrogen ion and becomes positively charged. The resulting double ion is called a zwitterion (German: *zwitter,* "hybrid"). Note

that in the ionized carboxyl group we use one solid line and one broken line to join each oxygen atom to carbon 1. This is to indicate that each *oxygen* shares 3 electrons (1½ pairs) with the carbon atom. Thus the two oxygen atoms share the negative charge. These are true covalent bonds because a sharing of electrons is involved, but they are unusual in sharing one pair of electrons among one carbon atom and two oxygen atoms. (For convenience, projection formulae are often written as if all the negative charge were on one oxygen atom, but it isn't so).

Color the side groups of the top illustrations. Color the heading Peptide Formation and title S. The conventional color for sulfur is yellow. Then color the L-alanine and the L-cysteine in the bottom half of the plate.

More than 50 different amino acids have been discovered in living things, but only 20 of them are used to make proteins. Of the 20, 19 have exactly the same arrangement of atoms around carbon atoms 1 and 2 that you see here in L-alanine. (The twentieth is almost the same.) They differ in their side groups, which are the groups attached to carbon 2. They are called side groups because they stick out to the side of the long chain that is formed when numerous amino acids are joined together to make a protein. Alanine's side group consists of a carbon atom with three hydrogens attached (known as a methyl group). (Notice that in the lower drawing, the molecule has been flopped over 90 degrees to the right for illustration purposes.) Cysteine has a side group that is similar but includes a *sulfur* atom. Other amino acids have side groups that range from a single hydrogen atom up to a double ring of carbon and nitrogen atoms, as you can see in the next plate.

Color titles B, D, and E, the arrows (B and D), the dipeptide, and the water molecule.

Just as with carbohydrates and fats, proteins are assembled by a *dehydration condensation* of their subunits. The covalent bond joining two amino acids in this way is called a *peptide bond.* The resulting molecule is called a dipeptide. If we add one more amino acid to the chain in the same way, we will have a tripeptide. When we have many amino acids joined in a chain in this way, the molecule is called a polypeptide. A functional protein molecule may consist of a single polypeptide or a number of polypeptides joined together. It may also include some nonpolypeptide portions.

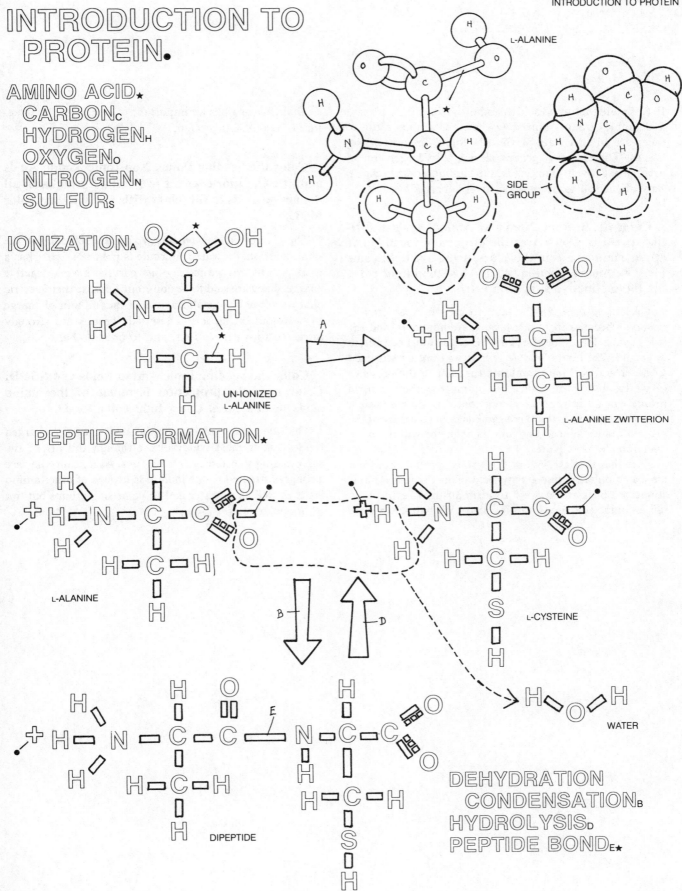

INTRODUCTION TO PROTEIN.

AMINO ACID ★
CARBON ꜱᴜʙ C
HYDROGEN ꜱᴜʙ H
OXYGEN ꜱᴜʙ O
NITROGEN ꜱᴜʙ N
SULFUR ꜱᴜʙ S

IONIZATION ꜱᴜʙ A

L-ALANINE

SIDE GROUP

UN-IONIZED L-ALANINE

L-ALANINE ZWITTERION

PEPTIDE FORMATION ★

L-ALANINE

L-CYSTEINE

WATER

DIPEPTIDE

**DEHYDRATION
CONDENSATION** ꜱᴜʙ B
HYDROLYSIS ꜱᴜʙ D
PEPTIDE BOND ꜱᴜʙ E ★

In this plate the amino acids are shown with their carboxyl and amino groups ionized (zwitterions), since most of them are in this state at the pH found in most living tissues. The carboxyl group releases a hydrogen ion to become negatively charged, and the amino group picks up a hydrogen ion to become positively charged.

Color the heading Nonpolar Amino Acids and titles A and B. Color over the projection formulae of all the amino acids in this class, using one light color for the common portion (A) and a second light color for the distinctive side group (B).

The *amino acids* in the first group are designated as *nonpolar* because their side groups are nonpolar, and the side groups determine the behavior of the finished protein molecule. The nonpolar *side groups* will not be attracted to water and will tend to clump together in the presence of water, just as oil molecules clump together to form droplets of oil in the watery portion of a salad dressing. How this influences the three-dimensional structure of the protein molecule and therefore its behavior will be introduced in the next plate.

Note that the last acid in this group, proline, does not have a complete amino group because its nitrogen atom is connected back to carbon 5 to form a ring. (Technically, it is an imino acid, not an amino acid.) We will see in Plate 21 that this also has an important influence on the structure of completed protein.

Color the heading Polar, Non-Ionic Amino Acids and title C. Color over the projection formulae of all the amino acids in this class with A and a light color for C.

These amino acids have *side groups* that are *polar* in the same sense that a water molecule is polar: one part has a weak positive charge and another part has a weak negative charge. They are said to be non-ionic because their electric charges are only a small fraction of the full unit of charge to be found on an ion. These side groups are strongly attracted to water, to ions, and to one another.

Color the heading Ionic Amino Acids and title D. Color over the projection formulae of the amino acids in this class. Use a light color for D.

The *side groups* of these amino acids are fully charged *(ionic)* under most conditions, although the pH of the surrounding solution can change this. As a result, they are strongly attracted to the ionic side groups of other amino acids as well as to water and to polar side groups but not to nonpolar ones.

AMINO ACIDS.

NONPOLAR AMINO ACIDS.
COMMON PORTION_A
NONPOLAR SIDE GROUP_B.

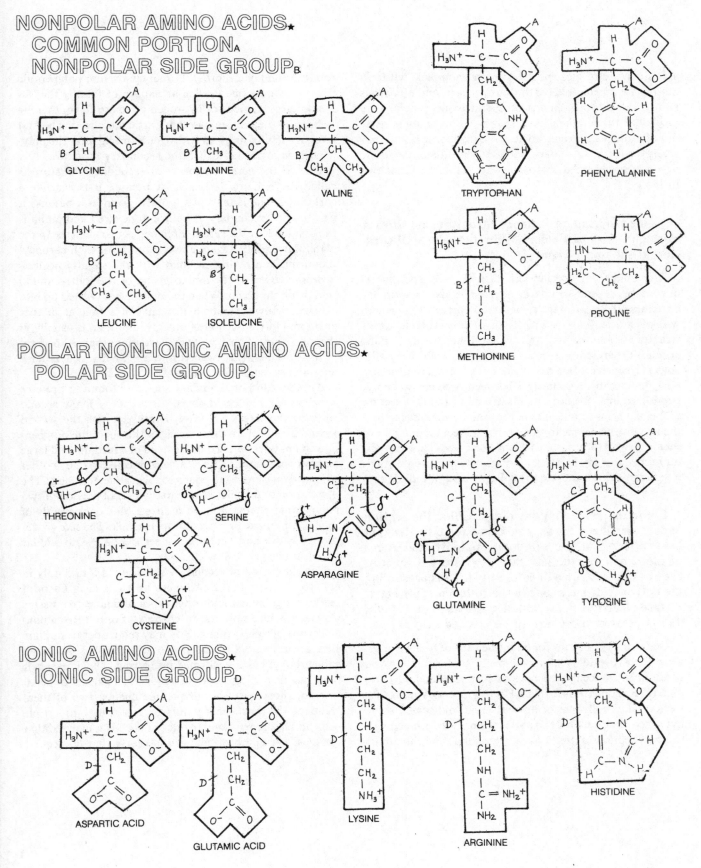

GLYCINE

ALANINE

VALINE

TRYPTOPHAN

PHENYLALANINE

LEUCINE

ISOLEUCINE

METHIONINE

PROLINE

POLAR NON-IONIC AMINO ACIDS.
POLAR SIDE GROUP_C

THREONINE

SERINE

ASPARAGINE

GLUTAMINE

TYROSINE

CYSTEINE

IONIC AMINO ACIDS.
IONIC SIDE GROUP_D

ASPARTIC ACID

GLUTAMIC ACID

LYSINE

ARGININE

HISTIDINE

The structure of most proteins is very complex, with numerous twists and folds of the various parts and numerous additional chemical bonds to stabilize the protein in a particular three-dimensional shape. To make some sense out of protein structure, we customarily divide it into four levels: primary, secondary, tertiary, and quaternary. The first two levels are discussed in this plate, the latter two in Plate 21.

Color the heading Primary Structure and titles A through F with pale colors. Color the related illustration at the upper right.

The linkage of specific amino acids in a specific sequence by peptide bonds is referred to as the primary structure. If the primary structure is correct, the protein will twist and fold itself into all of the other levels of structure spontaneously, provided that the conditions around it (temperature, pH, concentration of other molecules) are correct. In a few proteins the primary structure also includes disulfide bridges between two separate polypeptide chains. We see this illustrated in the first section of the plate. Two molecules of the amino acid *cysteine* lose their hydrogen atoms to some other molecules and join to each other by forming a covalent bond between their sulfur atoms, thereby forming a bridge between two polypeptide chains. Such a bridge is called a *disulfide bridge*.

Color the remaining headings. Color the section under each heading as you come to it in the text below. Note that you will color the individual atoms of the α helix at the left and the β pleated sheet at the bottom, but you will color only lines representing the polypeptide backbone in the collagen triple helix. Shared electrons, represented by sticks in the α helix and β pleated sheet, are to be colored gray.

Under the conditions found in the cells of living things, a new polypeptide chain will begin to form hydrogen bonds with adjacent molecules or between different parts of itself as soon as it is formed. Hydrogen bonds are rather weak, and as collisions occur with other molecules, some hydrogen bonds are broken and new ones are formed, with the polypeptide taking a variety of shapes as the atoms rotate around one another. For any particular polypeptide there is usually one particular pattern of folding that allows a large number of hydrogen bonds to form, thereby stabilizing the molecule in that particular shape. The first level of such folding and hydrogen bonding after the polypeptide is formed is called the secondary structure.

One of the common forms of secondary structure is called the α helix. It is a helix because it twists like a corkscrew. The letter alpha (α) was given to it because it was the first secondary structure discovered and alpha is the first letter in the Greek alphabet. As you can see in the plate, the slightly negative *oxygen atom* of each carboxyl group forms a hydrogen bond with the slightly positive *hydrogen atom* of the amino group of the fourth amino acid down the chain. When the molecule is twisted up into a helix, many separate hydrogen bonds hold it in that position. Only a portion of any given protein is usually in the helix form. Some proteins have more than 90 percent of their amino acids twisted into an α helix; others have only a very small percentage.

Another kind of secondary structure found in proteins is called the β pleated sheet, after beta (β), the second letter in the Greek alphabet, because it was the second secondary structure discovered. In this form the polypeptide chains are stretched out instead of coiled up, and large numbers of them lie side by side connected to one another by numerous hydrogen bonds to form a pleated sheet. The illustration in this plate shows only a small portion of two polypeptide chains in such a sheet. Many thousands of such chains make up a protein such as silk. The *side groups* project above and below the sheet to connect to similar sheets above and below this one.

The third kind of secondary structure is found only in collagen, the protein that makes up the strong fibers of tendons, ligaments, and other types of connective tissue. Collagen is made up almost entirely of only three amino acids, one of them proline. You may remember from Plate 19 that proline is the one amino acid that has its amino and carboxyl portions formed into a ring. This prevents it from turning at the correct angle to form an α helix. Instead, three separate polypeptide chains (two of them identical, the third slightly different) twist up into a triple helix and hydrogen-bond to one another. It is claimed that the tensile strength of collagen is nearly that of steel.

PROTEIN STRUCTURE I.

PRIMARY STRUCTURE.*
GLYCINE_A
SERINE_B
CYSTEINE_D
ALANINE_E
DISULFIDE BRIDGE_F
SECONDARY STRUCTURE.*

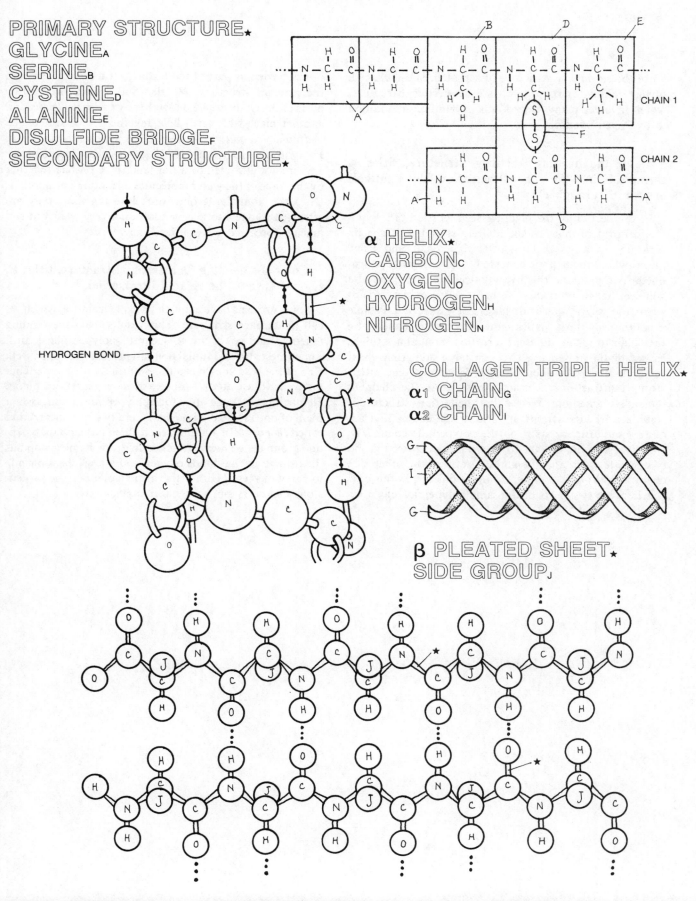

α HELIX.*
CARBON_C
OXYGEN_O
HYDROGEN_H
NITROGEN_N

COLLAGEN TRIPLE HELIX.*
α1 CHAIN_G
α2 CHAIN_I

β PLEATED SHEET.*
SIDE GROUP_J

HYDROGEN BOND

CHAIN 1

CHAIN 2

PROTEIN STRUCTURE II

Once the secondary structure of a protein is established, further folding or rotating constitutes tertiary structure, which involves many other kinds of bonding in addition to hydrogen bonding.

Color the heading Tertiary Structure, titles A through E, and the upper illustration. Use light colors for B through E.

In this section of the plate we see a portion of a hypothetical protein that has six different stretches of α helix separated by a few sections of straight chain and a few places where the helix is distorted. You will observe proline at two points in this hypothetical molecule, and you will see that it interrupts the helix and provides a point where the polypeptide chain takes a new direction. Proline is not the only thing that can interrupt the helix and cause the chain to take a different direction. Sometimes this is caused by the presence of two very large side groups next to each other; at other times it is simply that other, stronger attractions at other points cause the chain to remain stretched out. We see that the tertiary structure is stabilized by interactions of various side chains that are quite some distance apart on the polypeptide chain. The *ionic bridge* is an ionic bond between the carboxyl group of one side chain and the amino group of the other side chain. We also find another *disulfide* linkage; in this case it is between two parts of the same polypeptide chain, so

it is regarded as part of the tertiary structure rather than the primary structure. We also find a point where two parts of the chain are held together by *hydrogen bonds* and another place where it is held together by what is called *hydrophobic bonding* between large hydrocarbon groups. In hydrophobic bonding, side groups that are nonpolar and are not attracted to water tend to be pushed together by the force of the water molecules attracting one another. Once the nonpolar groups are close together, they are attracted to one another by additional very weak but significant forces called van der Waals forces.

Color the heading Quaternary Structure, titles F, G, and H, and the related illustration.

Some proteins have a fourth level of structure, which we call quaternary structure. This results from the binding together of two or more separate polypeptide chains, sometimes with additional nonpolypeptide groups as well. One of the best-known examples of this is the hemoglobin molecule (illustrated), which carries oxygen in our blood. Hemoglobin is made up of four polypeptide chains, two of them of one kind called α *chains* and two of another kind called β *chains*. In addition, each chain has a nonpolypeptide group called *heme* attached to it; the heme group has the iron atom that binds oxygen and allows the molecule to carry oxygen. Neither the heme alone nor the protein alone can carry out this function efficiently.

PROTEIN STRUCTURE II.

TERTIARY STRUCTURE ★
POLYPEPTIDE BACKBONE A
IONIC BRIDGE B
DISULFIDE BRIDGE C
HYDROGEN BONDING D
HYDROPHOBIC BONDING E

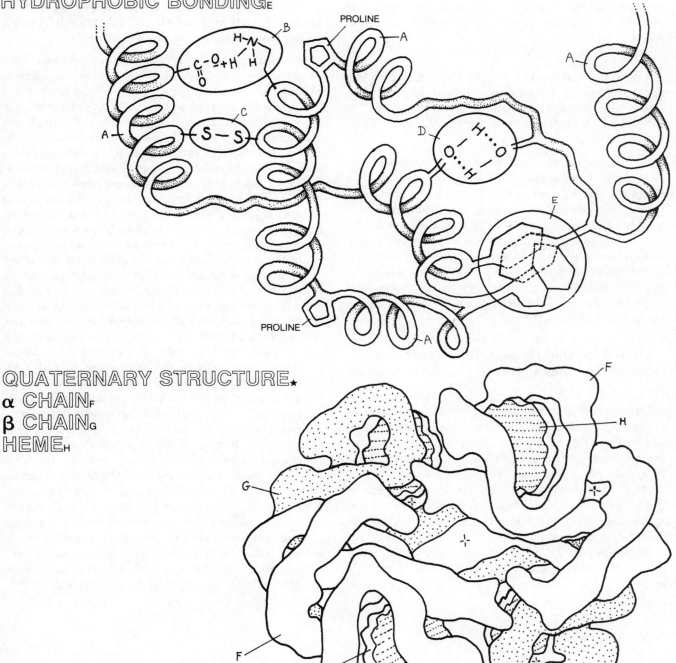

QUATERNARY STRUCTURE ★
α CHAIN F
β CHAIN G
HEME H

Our final major category of biological molecules is the class known as nucleotides, which received this name because they were first discovered in the nuclei of cells. We will deal with them in more detail later, but any discussion of the molecules of life would be incomplete without at least an introduction to this important group.

Color the headings Nitrogenous Bases, Purines, and Pyrimidines. Using pale colors so as not to obscure the structural formulae, color titles A through E and the five nitrogenous bases illustrated.

A nucleotide consists of at least three components: a nitrogenous base, a pentose (five-carbon sugar), and a phosphate ion. The nitrogenous bases receive their name because they all contain nitrogen and chemically act as weak bases (proton acceptors). The five shown here are the most important ones, although there are others with lesser roles. The two purines (adenine and guanine) consist of double rings of carbon and nitrogen atoms, while the three pyrimidines (cytosine, thymine, and uracil) consist of a single ring. They differ from one another in the groups of atoms attached to the rings.

Color the heading Pentoses, titles F and G, and the related illustrations. Color title H and the phosphate ion.

Two different pentoses are found in nucleotides. Most nucleotides contain ribose, but some contain deoxyribose, which is exactly like ribose except that it is missing the oxygen atom from the hydroxyl (OH) group on carbon 2. Phosphate ion is the negative ion of phosphoric acid (H_3PO_4). It is shown here with a single negative charge where its proton (hydrogen ion) has dissociated away, but each of the hydroxyl groups remaining can also dissociate a proton, producing phosphate ions with two or three negative charges. (In any group of phosphate ions, a few have two or three negative charges, but most have only one.)

Color the heading Ribonucleotide and the adenine, ribose, and phosphate components of the ribonucleotide, using the colors you used in the top part of the plate. Then color title I and the broken ring labeled I (which is for illustrative purposes only). Color the heading Functions in a Cell, titles J through M using light colors, and the related diagrams.

We see at the bottom of this plate one particular kind of nucleotide called adenosine monophosphate. (Ribose plus a base, without the phosphate, is called a nucleoside.) Adenosine monophosphate is classified as a ribonucleotide because it has ribose as part of its structure. The ribose has a molecule of adenine and a phosphate ion attached to it. Like polysaccharides, proteins, and lipids, the nucleotides are assembled from their subunits by dehydration condensation. In this case it occurs in several steps too complicated to go into here, but the net result is that a molecule of water is formed when a base and a pentose are joined, and another water molecule results when a pentose and a phosphate are joined. As we will see in various later plates, nucleotides can perform a number of functions in a living cell: they can be used as chemical messengers to regulate cell activities; they can function as coenzymes, essential parts of certain enzymes; with one or two additional phosphates added, they can serve to carry energy from one part of a cell to another; or they can become components of a nucleic acid. Nucleic acids are of two types: deoxyribonucleic acid (DNA) and ribonucleic acid (RNA). Both consist of very long chains of nucleotides, usually double-stranded in DNA and single-stranded in RNA. DNA is made up of nucleotides containing deoxyribose combined with one of the four bases, adenine, guanine, cytosine, and thymine. RNA nucleotides contain ribose combined with the bases, with uracil substituted for thymine. The sequence of the bases in DNA forms the hereditary code, which is passed from generation to generation. RNA molecules occur in several types (to be covered in later plates) and translate the hereditary code into specific proteins that determine how the cell (or a large organism made of large numbers of cells) will function under various conditions. There is also some experimental evidence to suggest that RNA may play a role in memory.

NUCLEOTIDES.

NITROGENOUS BASES.★
PURINES.★
  ADENINE_A
  GUANINE_B

PYRIMIDINES.★
  CYTOSINE_C
  THYMINE_D
  URACIL_E

PENTOSES.★
RIBOSE_F
DEOXYRIBOSE_G

PHOSPHATE ION_H

RIBONUCLEOTIDE.★
ADENOSINE MONOPHOSPHATE_I
FUNCTIONS IN A CELL.★
CHEMICAL MESSENGER_J
COENZYME_K
ENERGY CARRIER_L
NUCLEIC ACID_M

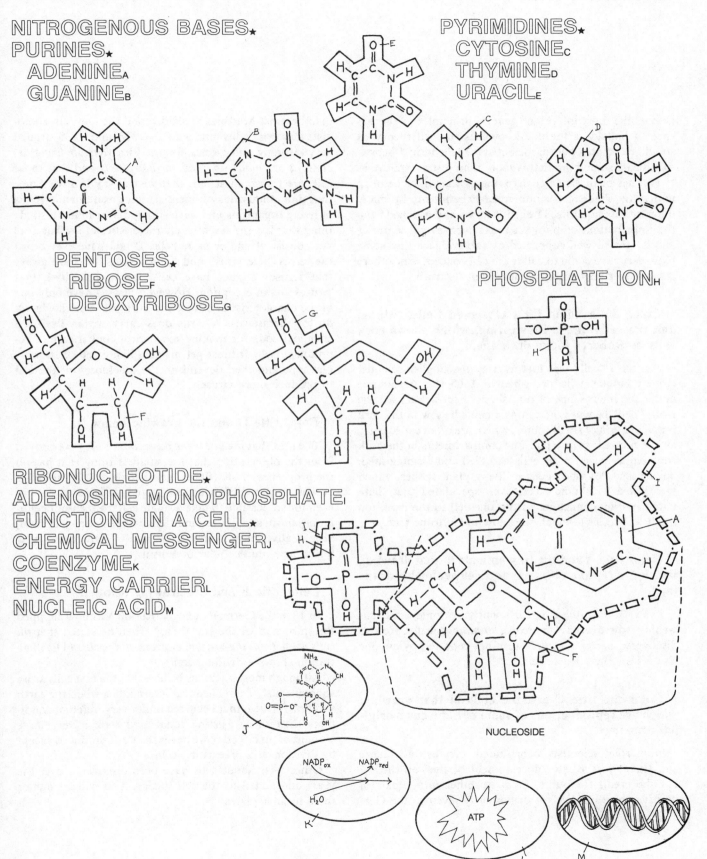

NUCLEOSIDE

NADP_ox   NADP_red

H_2O

ATP

23
CELL THEORY

One of the most important general principles in the science of biology is the one known as the cell theory. It should be pointed out that scientists use the term "theory" to indicate something that is more than just a hypothesis. It is a set of interrelated hypotheses that have held up under the challenge of numerous experiments by many different investigators. The cell theory was developed from the contributions of many scientists over several centuries and is now so well supported by the evidence that some biologists have come to call it the cell concept, since there is no longer any serious doubt about its truth.

Color the heading First Observed Cells, title A, and the cells in the illustration, which shows cork cells as Robert Hooke drew them.

The term "cell" was first used by the English scientist Robert Hooke, who described in 1665 the appearance under the microscope of thin slices of *cork*. He used the word "cell" because the compartments he saw in the cork reminded him of small rooms, which are often called cells, as in a monastery or a jail. The compartments in the cork were empty because the cells had died and disintegrated, but he also described cells in living plant tissues, which were filled with fluids. (He even speculated that there might be tiny connections from one cell to the next, too small to be seen—which we know today to be true.)

Color title A and the tiny animals in that drawing, which is a reproduction of van Leeuwenhoek's drawings.

In 1675 the Dutch amateur scientist Antonie van Leeuwenhoek discovered *microscopic animals* in water. He also discovered bacteria, which were not reported by anyone else for another two centuries.

Color the title C and the cells in that drawing, which are reproductions of some of Schwann's original drawings.

Numerous scientists contributed various bits, large and small, to the later development of the cell theory, but the credit for pulling it all together into a pair of positive generalizations is customarily given to the Ger-

man botanist Matthias Schleiden and the German zoologist Theodor Schwann, who in 1838 and 1839 argued very strongly and clearly for the idea that *all living organisms are made up of one or more cells* and that those cells are the smallest unit that we can say is alive. Since that time, thousands of scientists have examined millions of living organisms and have simply never found a single thing that has the essential characteristics of life but does not consist of one or more cells. If you have ever heard the terms "live virus" and "killed virus" and are aware that viruses do not have cells, you might think that viruses are an exception. However, viruses have only one of the characteristics of life, and even that one they can't do for themselves. A virus does carry within itself the chemical code for making more virus, but it cannot reproduce itself. It must get inside a living cell of the correct type and then depend on the machinery of the cell to produce more viruses.

Color title D and the drawing below it.

The idea that *the cell is the basic unit of life* was derived from the observation that the smallest thing that has all the properties of life (Plate 1) is a single cell. If the cell is broken open, the life processes stop. In this century it has been found possible to take a single cell from a multicellular organism and, if the conditions can be gotten just right, keep it alive, have it reproduce, and in some cases grow into a new multicellular organism.

Color title E and the drawing below it.

In 1858 the German biologist Rudolf Virchow supplied the third part of the cell theory when he stated that *all cells come from the division of preexisting cells.* (The illustration shows a dividing cell.)

Although most scientists believe that the first cells arose spontaneously from chemical interactions when the earth was first formed, that occurred under very different conditions than those existing today and took a very large amount of time. Today we never see a cell produced except by division of a preexisting cell.

Other facts about cells have been discovered over the years and added to the cell theory. You will encounter them in later plates.

CELL THEORY.

FIRST OBSERVED CELLS.★
CORK CELLS_A

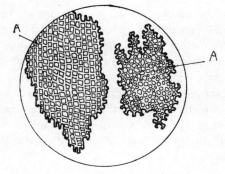

MICROSCOPIC ANIMALS_B

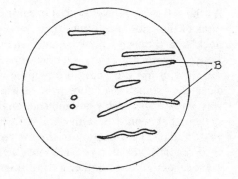

ALL LIVING THINGS ARE MADE OF ONE OR MORE CELLS_C

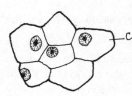 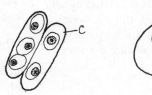

THE CELL IS THE BASIC UNIT OF LIFE_D

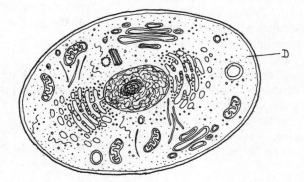

ALL CELLS COME FROM PREEXISTING CELLS_E

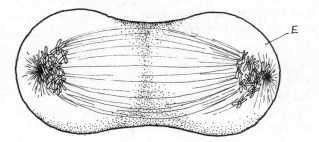

SIZES OF THINGS

As the cell theory developed, it became clear that descriptions of the sizes of cells and their parts such as Leeuwenhoek's "a thousand times smaller than the eye of a big Louse," were not really satisfactory. Such descriptions were useful up to a point, but as time went on it became clear that something more precise was needed. Science was already becoming an international enterprise, so a system that would be universally accepted would be of great value. The invention of the metric system just after the French Revolution provided just the thing. With all its units related to one another as multiples of 10, it was also extremely easy to use. Changes to larger or smaller units could be accomplished simply by moving the decimal point. As chemists and biologists dealt with smaller and smaller sizes, new prefixes were added to the system to create smaller units. Today the system is regulated by an international commission and is called the SI system (from the French, *Système international*). In this plate you will learn the units of length measurement that are useful in discussing cells.

Color titles A through C and the yardstick and meterstick; then color the portion of the meterstick and the centimeter in the inset.

The basic unit of the metric system is the meter. It is about 10 percent longer than the English yard. The *meter* is divided into 10 decimeters, a decimeter into 10 *centimeters,* and a centimeter into 10 millimeters. If something is 1 decimeter, 3 centimeters, and 7 millimeters long, we don't have to write all that out as we would if they were feet, inches, and fractions of inches. Since the units all differ by multiples of 10, we can just write 13.7 centimeters, 1.37 decimeters, or 137 millimeters. In actual practice, decimeters are rarely used; most sizes in this range are specified in centimeters or millimeters.

Color title D and the section of the meterstick representing 1 millimeter (D), the bar representing the magnified millimeter below (also D), and the symbol for 1 millimeter to the right. Use the meter color (B[1]) to color the exponential expression for the fraction of a meter that 1 millimeter is equal to. Color the scale of relative sizes as you read each section.

A *millimeter* is one-thousandth of a meter. This fraction is often symbolized by a 10 with a negative exponent of 3 (10^{-3}). If you are not familiar with exponents, you can convert them to common fractions by writing a 1 for the numerator and writing a denominator of 1 followed by the number of zeros indicated in the exponent (the small superscript number). Thus 10^{-3} is equal to $1/1000$. Millimeters are the smallest units that are commonly marked on a metric ruler. Fractions of a millimeter are always expressed in tenths, hundredths, or thousandths, never as halves, eighths, sixteenths, etc. One-tenth of a millimeter is about the size of the smallest object the human eye can see without the aid of magnification. A human egg cell is about one-tenth of a millimeter in diameter.

Color title E and the micrometer shown in the magnified millimeter, the bar below representing the magnified micrometer, and the abbreviation for one micrometer to the right. As before, color the exponential expression with the meter color.

A *micrometer* is one-thousandth of a millimeter, which is equal to one-millionth of a meter. (10^{-6} is the exponential form for one-millionth.) The strange letter in the symbol for micrometers is the Greek letter mu (μ) (pronounced "myoo"). Wherever you see it used in science, it means "one-millionth" and represents the prefix "micro-." A number of common bacteria have diameters of about one micrometer. (In older textbooks you may find a unit called a micron, symbolized by the letter μ alone. That is simply the old name for micrometer and should no longer be used.)

Color title F and the nanometer section as you did the previous sections.

A *nanometer* is one-billionth of a meter (10^{-9}). Its symbol is *nm,* so you have to read carefully to avoid confusing it with *mm* for millimeters. The α helix of a protein is about 1 nanometer in diameter.

Color title G and the angstrom section as you did the previous sections.

The *Angstrom* is not an official unit of the SI system but is often used. It is unusual in that it is only one-tenth of the next higher unit rather than one-thousandth. It is therefore equal to one ten-billionth of a meter (10^{-10}). It is also unusual in using a capital letter (A) for its symbol. You will sometimes see it with a tiny "o" over the A, since the unit is named after a Swedish physicist of the nineteenth century whose name in Swedish is spelled "Ångstrom." A hydrogen atom has a diameter of about one Angstrom.

SIZES OF THINGS.

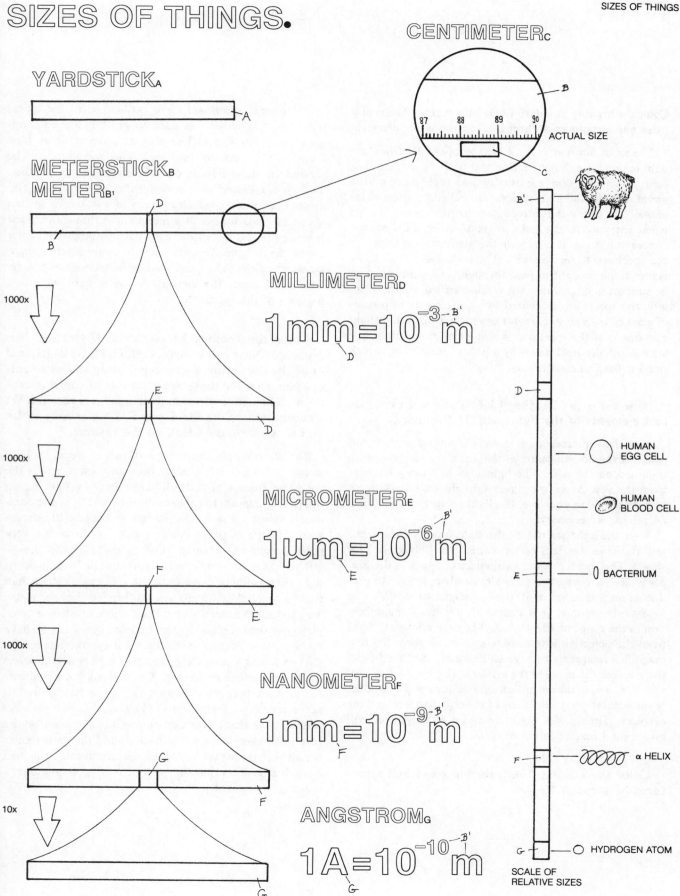

YARDSTICK_A

CENTIMETER_C

ACTUAL SIZE

METERSTICK_B
METER_B'

1000x

MILLIMETER_D

$$1mm = 10^{-3}m$$

1000x

MICROMETER_E

$$1\mu m = 10^{-6}m$$

1000x

NANOMETER_F

$$1nm = 10^{-9}m$$

10x

ANGSTROM_G

$$1A = 10^{-10}m$$

HUMAN
EGG CELL

HUMAN
BLOOD CELL

BACTERIUM

α HELIX

HYDROGEN ATOM

SCALE OF
RELATIVE SIZES

25
STUDENT MICROSCOPES

Color the heading Standard Light Microscope. Then color titles and structures A through G in the upper drawing.

The most familiar type of microscope is the standard light microscope shown in the upper half of this plate. The *base, pillar,* and *arm* are usually one single piece of cast metal, but the different regions are called by these three names. The arm is the correct place to grip the microscope when carrying it; the palm of your other hand should support the base. The *stage* is the platform that supports the specimen to be observed. The stage has a hole in its center to allow light to pass through, so specimens must be supported on a thin piece of glass called a *slide.* Usually, the specimen is covered by an extremely thin piece of glass called a *cover glass* (or cover slip). Since any slight movement of the specimen is magnified many times, the slide is usually held down by a pair of *stage clips,* which press it flat against the stage.

Now color the heading Light Path and the titles and elements of the light path, H through L.

A student microscope is usually equipped with a *concave mirror* to concentrate the light on the specimen mounted on the slide. The light may be from a lamp or from the sky. *Never* use direct sunlight unless you want to burn a hole in your eye. If a light is built into the base, no mirror is necessary.

After the light has passed through the specimen, it enters the objective lens (often called just "objective" for short). The shorter of the two objective lenses is the *low-power objective lens,* which is almost always made to produce a magnification of 10 times, designated as 10X. The *high-power objective lens* nearly always has a magnification in the range of 40X to 45X. The *tube* allows the light from the objective lens to pass upward to form the first magnified image; that image is then magnified further by the *eyepiece* (also called the *ocular*). The eyepiece is usually 10X. The total magnification obtained is the product of the separate magnifications of the objective lens and the eyepiece. Thus a 40X objective and a 10X eyepiece will give a total magnification of 400X.

Color the heading Controls and titles and structures M through R.

The *coarse adjustment knob* is often called the coarse focus knob because it is used to get the specimen approximately in focus. The *fine adjustment knob* (fine focus) gets it exactly in focus. The *nosepiece* can be rotated to change from one objective lens to another. The *iris diaphragm lever* operates the *iris diaphragm* (indicated open here), which consists of a dozen or so thin sheets of brass ingeniously mounted so that moving the lever opens and closes the light path through the center of the diaphragm. This increases or decreases the angular width of the light beam passing up through the stage, which influences the contrast between light and dark portions of the specimen.

Color the heading Stereoscopic/Dissecting Microscope. Next color parts A, B, C, D, G, K, M, and O of the dissecting microscope, using the same colors you used for those parts on the standard microscope. Then color titles and structures S through W. Choose light colors for S and T. W is usually white on one surface and black on the reverse.

The stereoscopic microscope (Greek: *stereos,* "solid"; *scopic,* "vision") is so called because each eye sees the specimen from a slightly different angle, producing an image that appears to be three-dimensional. It is also commonly called a dissecting microscope, because the stereoscopic image is particularly useful for dissecting very small specimens. Magnification is always rather low—10X to 60X total—and observation is usually with incident light (light shining down onto the specimen) rather than transmitted light (light passed up through the specimen). To produce the stereoscopic effect there must be a completely separate optical system for each eye. Thus we have *two low-power lenses* (one for the right eye and one for the left) in a single *mounting* and *two high-power objective lenses* in another *mounting.* The *binocular head* allows the distance between the *two eyepieces* to be adjusted to accommodate different users. The *stage* usually has a *reversible plate* that is white on one surface and black on the reverse to allow choice of a background that will render the subject most visible. Stage clips are usually attached in such a position that they can be used or swung out of the way, as desired.

STUDENT MICROSCOPES.

STANDARD LIGHT MICROSCOPE★
BASEA
PILLARB
ARMC
STAGED
SLIDEE
COVER GLASSF
STAGE CLIPSG
LIGHT PATH★
CONCAVE MIRRORH
LOW-POWER OBJECTIVE LENSI
HIGH-POWER OBJECTIVE LENSJ
TUBEK
EYEPIECE/OCULARL
CONTROLS★
COARSE ADJUSTMENT KNOBM
FINE ADJUSTMENT KNOBN
NOSEPIECEO
IRIS DIAPHRAGM LEVERP
IRIS DIAPHRAGM OPENQ/
   CLOSEDR

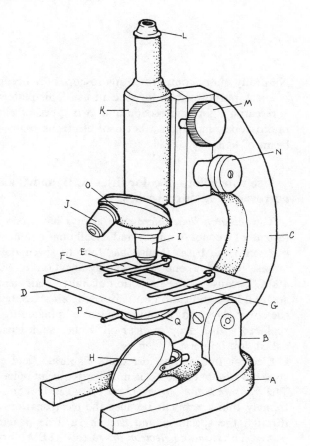

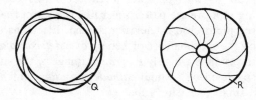

STEREOSCOPIC/DISSECTING
   MICROSCOPE★
PAIRED LOW-POWER
   OBJECTIVE LENSESS/
   MOUNTINGS¹
PAIRED HIGH-POWER
   OBJECTIVE LENSEST/
   MOUNTINGT¹
BINOCULAR HEADU
PAIRED EYEPIECES/OCULARSV
REVERSIBLE STAGE PLATEW

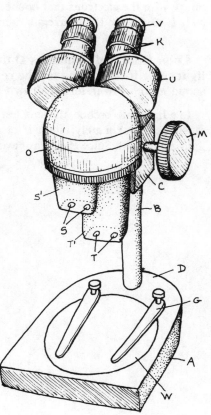

Naturally, more complicated microscopes are needed for research purposes than for student use. This plate shows a research light microscope and two types of electron microscopes, in which a beam of electrons replaces the beam of light.

Use pale colors to color titles A, B, and C and the corresponding microscopes.

The *research light microscope* always has a condenser instead of a concave mirror and usually has a light source built into the base. It also has four to seven objective lenses, including one with a magnification of 90X or 100X. Because of the refraction of light by air, lenses of such high magnification must have the air space between the cover glass and the objective lens eliminated by being replaced with a drop of clear oil. Hence such lenses are called "oil immersion" lenses.

Electron microscopes are quite elaborate (and expensive) devices that take up as much space as an office desk. The *transmission electron microscope* (TEM) requires extremely thin specimens because the electrons must pass through the specimen and they have little penetrating power. The *scanning electron microscope* (SEM) forms an image with the electrons that bounce off of the specimen, so it is useful for thick, irregular specimens.

Color titles and structures D through P. Choose a light color for F. Note that the research light microscope at the lower left is drawn upside down.

In a light microscope, the smallest detail we can distinguish is approximately one-half of a wavelength of the light being used. Since electrons have a wavelength that is much shorter than that of light rays, we can see much smaller detail if a beam of electrons is used in place of light. In place of glass lenses, which would absorb electrons instead of focusing them, electron microscopes focus the electron beam by means of magnetic fields, which are produced by running electricity through coils of wire. These coils of wire, then, are the "lenses" of electron microscopes. The magnification can be easily varied by changing the current running through the coils. Since electrons are easily absorbed even by air, the specimen and the entire path of the electron beam must be in a vacuum chamber.

In the transmission electron microscope, the path of the *electron beam* is almost identical to the path of the *light beam* in a light microscope, except that it is upside down. There is a *condenser lens* to concentrate the beam on the *specimen,* an *objective lens,* and the equivalent of an *ocular lens,* which is called a *projector lens* in this case since the electrons must be projected onto a *phosphor screen* to produce a visible image. Windows allow us to look into the vacuum chamber to see the image. A pair of *short-focal-length binoculars,* placed in front of one of the windows, allows a still more magnified view of that image. The phosphor screen can be folded out of the way to record the image on *photographic film* just below it.

In the scanning electron microscope, a very thin electron beam passes through a *scanning coil* that causes the *beam* to sweep back and forth across the specimen according to a precise pattern just like the beam that traces out the picture on a television screen. Electrons that bounce off the specimen are picked up by a *detector plate* and used to produce a highly magnified image of the surface of the specimen on a small *cathode ray tube,* which is just like a small television tube.

RESEARCH MICROSCOPES.

RESEARCH LIGHT MICROSOPE A
TRANSMISSION ELECTRON MICROSCOPE B
SCANNING ELECTRON MICROSCOPE C

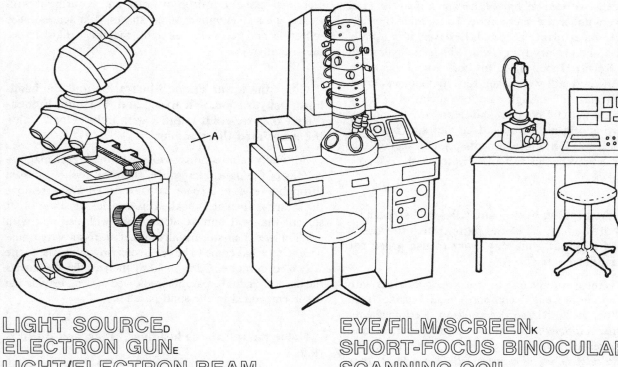

LIGHT SOURCE D
ELECTRON GUN E
LIGHT/ELECTRON BEAM F
CONDENSER LENS G
SPECIMEN H
OBJECTIVE LENS I
OCULAR LENS J
PROJECTOR LENS J'

EYE/FILM/SCREEN K
SHORT-FOCUS BINOCULARS L
SCANNING COIL M
SCANNING BEAM N
DETECTOR PLATE O
CATHODE RAY TUBE DISPLAY P

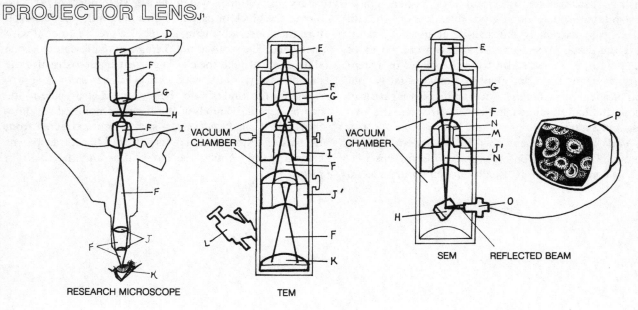

RESEARCH MICROSCOPE

TEM

VACUUM CHAMBER

VACUUM CHAMBER

SEM   REFLECTED BEAM

CELL PREPARATION
FOR LIGHT MICROSCOPES

To look at microscopic organisms consisting of one or a very few cells, we need only place them on a glass slide and look at them under the microscope. To examine the cells of a multicellular plant or animal, however, it is usually necessary to make very thin slices, which biologists call sections. Robert Hooke sliced his cork with a penknife, and that method still works well for any plant materials with cell walls stiff enough to allow slicing. But some materials are not stiff enough, and in any event the contents of living cells are semiliquid, so they spill out when the cells are cut. Certain treatments prior to slicing (sectioning) can prevent this and allow us to see the cell contents in place.

Color the heading Fixing and titles and structures A and B at the top left of the plate. Since the tissue illustrated is a leaf, you may want to use green for A. Use a light color for B.

Fixing is the name given to the treatment of *tissues* with chemicals that prevent decomposition and make them firmer so that they can be sectioned evenly. Certain chemicals *(fixative solutions),* such as formaldehyde, are used to create numerous cross-linkages among the molecules within the *cells.* This, of course, kills the cells, if they aren't already dead.

Color the headings Supporting and Hand-Sectioning and Wet Mount, titles C through G, and the associated illustrations.

For most tissues, fixing is helpful but is not enough to make good, uniform sectioning possible. Sections that do not need to be extremely thin can be obtained by sandwiching the tissue to be sectioned between two layers of *carrot,* potato, or the soft pith from the center of a stem of cottonwood or a similar plant. To make a fairly thin section of a leaf, for example, you can just cut a carrot in half longitudinally, sandwich the leaf between the two halves, and section the carrot and leaf together with a *razor blade.* The carrot can then be discarded and you can mount the leaf section on a *glass slide* for observation. This sort of section is usually for temporary use, so

it is merely made into a wet mount by immersing the sectioned tissue in a drop of *water* and covering it with a *cover glass* to eliminate all air spaces. (Air spaces cause reflection and refraction of light, making it hard to see the specimen.)

Color the lower Fixing illustration and the headings Dehydration, Clearing, and Paraffin Embedding. Color titles H, I, and J with light colors. Color the associated illustrations.

For really thin sections usable with the higher magnifications of the light microscope, the fixed tissue is passed through a series of *alcohol* baths to dehydrate it (remove the water). Then it is soaked in *toluene* or some other solvent that will remove alcohol but will also mix with paraffin wax. This step usually makes the tissue very transparent, so it has come to be called clearing. Next the tissue is soaked in molten *paraffin* until the paraffin penetrates the tissues. Then the paraffin is allowed to cool, leaving the tissue embedded in the solid paraffin.

Color the remaining headings, titles, and illustrations.

The block of paraffin with the embedded tissue is normally sliced in a special slicing machine called a *microtome* (Greek: *micro,* "small"; *tomos,* "a cutting"), which can produce sections as thin as 2 micrometers ($2/1000$ of a millimeter or $1/12,500$ of an inch). There are several kinds of microtomes, but the most common is the rotary microtome, which has a stationary *blade* and moves the block of paraffin down across the blade as you turn the crank, advancing the block a precise amount with each slice. The sections are then glued to glass microscope slides, put through the clearing agent again to dissolve the paraffin away, and stained with whatever *stain* is appropriate for the kind of study being done. The prepared slide can be kept indefinitely if the tissue is mounted in something like Canada *balsam,* a sticky resin extracted from the balsam fir tree, which dries to form a hard, transparent cement, firmly gluing the cover glass and the tissue in place.

CELL PREPARATION FOR LIGHT MICROSCOPES.

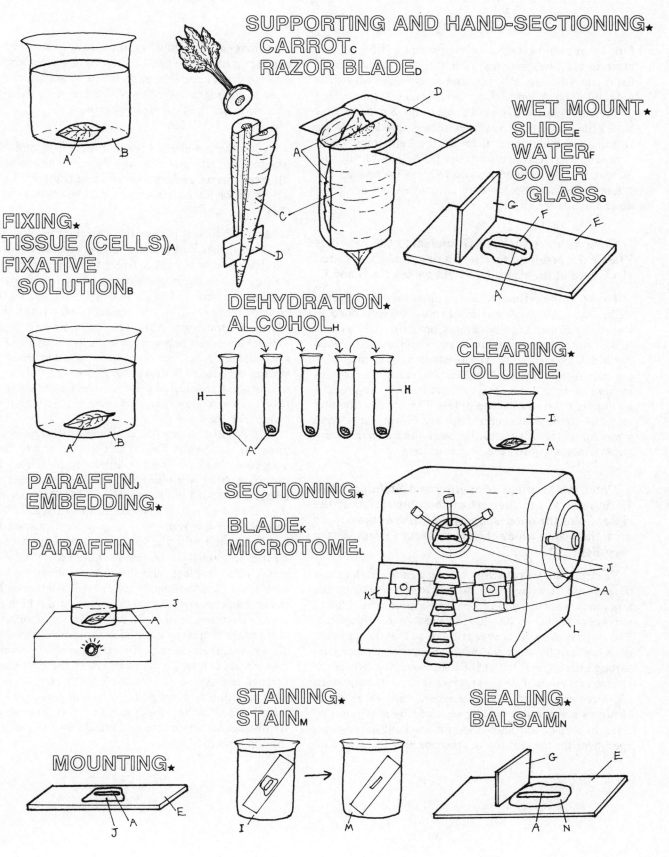

SUPPORTING AND HAND-SECTIONING★
CARROT_C
RAZOR BLADE_D

WET MOUNT★
SLIDE_E
WATER_F
COVER
GLASS_G

FIXING.★
TISSUE (CELLS)_A
FIXATIVE
SOLUTION_B

DEHYDRATION.★
ALCOHOL_H

CLEARING.★
TOLUENE_I

PARAFFIN_J
EMBEDDING.★

PARAFFIN

SECTIONING.★

BLADE_K
MICROTOME_L

STAINING.★
STAIN_M

SEALING.★
BALSAM_N

MOUNTING.★

28
CELL PREPARATION FOR TRANSMISSION ELECTRON MICROSCOPES

For the transmission electron microscope (TEM), tissues must usually be prepared with the same processes of fixing, dehydrating, clearing, and so on required for the light microscope, but with certain differences necessitated by the use of electrons instead of light rays. Since electrons have little penetrating power, extremely thin sections are required—much thinner than for the light microscope. (The scanning electron microscope [SEM] shows only surface features, so sectioning is generally not necessary.) The "stains" used need not have any color at all, but they must absorb or scatter electrons.

Color titles A through G, including the headings Plastic Embedding and Sectioning, and the related structures at the right. Use light colors for B and C.

To obtain the extremely thin sections necessary for the TEM, *cleared, dehydrated tissue* must be embedded in something firmer than paraffin. Generally, the tissue is soaked for a time in a liquid *plastic resin,* which is then *hardened* by a catalyst. The plastic with the embedded tissue is then sectioned by a machine called an *ultramicrotome,* which uses a glass or diamond knife to produce sections of 0.05 micrometers or less. These are so thin that they have to be floated on the surface of water in a small *water trough* right next to the *blade,* and a *stereomicroscope* is necessary to observe the process.

Color the headings Mounting and Staining, titles H through J, and the rest of the illustration at the right. Color the outer supporting frame of the copper grid; then color along the crosshatched wires. Use a very light color for I.

Electrons do not pass through any significant thickness of glass, so specimens for the electron microscope are mounted on small *grids* of very fine copper wire, 8 to 12 wires per millimeter (200–300 per inch). A very thin *plastic film* is sprayed on the copper grid first so that the specimen does not sag between the wires. Electrons cannot pass through the copper, but at the high magnifications of the electron microscope, the portions of the specimen between the wires amount to quite a large area to look at. Since electrons are not absorbed or scattered by much of anything in cells or the stains used for the light microscope, specimens for the electron microscope are "stained" with

such things as lead citrate or uranium acetate, which contain atoms of *heavy metals* to absorb or scatter electrons, creating dark areas in the image. Plate 37 is a drawing of an electron micrograph (photograph taken with an electron microscope) of this type of preparation.

Color the headings Freeze Fracturing and Metal Shadowing, titles K through N, and the associated illustrations as each is discussed in the text. Choose a light color for L, and color over the small gold wire (M).

Metal shadowing is a method useful for examining isolated organelles, viruses, and protein molecules or cells that have been subjected to freeze fracturing. Freeze fracturing involves freezing the specimen at a very low temperature (about −100°C) and allowing the ultramicrotome blade to strike it rather hard to fracture it instead of slicing through smoothly. The fracture planes generally follow boundaries between parts or divide layers of lipid, since lipids do not freeze as hard as water and water-soluble materials. Because *membranes* contain a great deal of lipid, this technique often peels a membrane into two layers and allows us to see some of the internal structure of the membrane.

The isolated organelles or freeze-fractured cells are mounted on a plastic film on a copper grid and placed in a *vacuum chamber.* There, a small *wire* of a heavy metal (usually gold or platinum) is mounted off to one side and electrically heated until it vaporizes. The *metal atoms* are then deposited on all exposed surfaces. Since they approach the tissue from one side, they are deposited on the high spots and miss the low spots, creating *"shadows."* Next a uniform layer of carbon is applied to cement the metal atoms in place, and the cells are dissolved away.

When the carbon-metal replica is photographed in the electron microscope, the metal atoms on the high places scatter electrons, so they do not strike the photographic film and the film is not exposed in those areas (not shown). Those areas then appear light when the film is developed. The low spots have no metal atoms, so the electrons pass through and expose the film, making it dark in those areas when the film is developed. Surface features are readily distinguished in the developed film in the same way that we distinguish mountains and craters on the moon from the shadows they cast.

CELL PREPARATION FOR TRANSMISSION ELECTRON MICROSCOPES.

CLEARED, DEHYDRATED
  TISSUE$_A$
PLASTIC EMBEDDING.★
RESIN$_B$
HARDENER$_C$
SECTIONING.★
ULTRAMICROTOME$_D$
BLADE$_E$
WATER TROUGH$_F$
STEREOMICROSCOPE$_G$
MOUNTING.★
COPPER GRID$_H$
PLASTIC FILM$_I$
STAINING.★
HEAVY METAL COMPOUND$_J$

FREEZE FRACTURING.★
MEMBRANE$_K$

METAL SHADOWING.★
VACUUM CHAMBER$_L$
GOLD WIRE/ATOMS$_M$
SHADOW$_N$

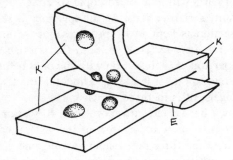

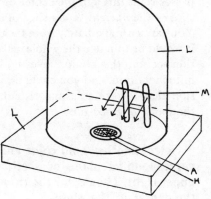

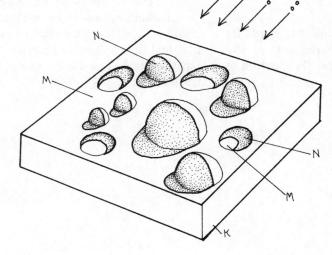

INTERPRETING THIN SECTIONS

Whether you become involved in the actual use of the microscope to observe cells or your study of biology is entirely confined to the illustrations in books, you will sooner or later face the problem of correctly interpreting thin sections of cells. It's not a terribly difficult problem, but some serious misconceptions can occur if you forget what you are looking at. To avoid these misconceptions, you must form the habit of reminding yourself constantly that you are looking at a thin section that includes only a fraction of the entire specimen. As an example, let's look at thin sections taken at various angles through a greatly oversimplified cell with a curved section of endoplasmic reticulum, a nucleus, and a mitochondrion.

Color the heading Cell and titles and structures A through D in the cell at the upper right, using relatively subdued colors. Color the Planes of Sections and Thin Sections headings; then use a bold color to color title E and its plane of section. Using the same colors, color the longitudinal thin section at center left.

A plane of section is an imaginary flat surface through which a blade passes in cutting a section. The thin section is the actual slice of tissue or cell that results. Any section made parallel to the longest axis is called a *longitudinal section*. The particular longitudinal section shown here passes right through the center of the cell and includes a slice of the *nucleus* and a slice of *endoplasmic reticulum*. You can see immediately from the thin section how foolish it would be to judge the whole cell on the basis of a single thin section: this section gives no hint that the cell has a mitochondrion, and you could easily get the false impression that the endoplasmic reticulum is spherical rather than an elongated tube.

Select another bold color and color title F and its corresponding plane of section in the cell at the upper right. Then color the transverse thin section in the center of the plate.

A *transverse section* is one made perpendicular to the longest axis, and you can see that this particular transverse section misses the nucleus and the mitochondrion entirely while cutting the endoplasmic reticulum in two places. If you had only the thin section to examine, you would have no hint that the two circular slices of the endoplasmic reticulum are part of one continuous tube.

With a third bold color, color title G and its corresponding plane of section in the original cell at the upper right. Then color the oblique thin section below it.

An *oblique section* is one cut at any angle that is neither parallel nor perpendicular to the longest axis. The particular oblique section shown here includes a slice of the nucleus but nothing of the endoplasmic reticulum or the mitochondrion. Once again, the single thin section gives us a very incomplete picture of the cell as a whole.

Color the heading Serial Sections and the remainder of the plate using the same color as before.

For accurate understanding of the three-dimensional arrangement of parts within a microscopic specimen, it is necessary to study serial sections. As the specimen is sectioned, each section is kept in its proper sequence and mounted on a slide or a series of slides in that sequence. A careful comparison of each section with the ones before it and after it allows one to build up a mental image of what the arrangement is in three dimensions.

The cell shown here is cut into more than 40 sections, but there is room to show only a few of them. You can see, however, how the three-dimensional arrangement of the endoplasmic reticulum can be deduced by observing the gradual changes from one section to the next. In complex cases, a scale model can be made of each section, and the models can then be stacked up to form an accurate three-dimensional model of whatever is being studied.

INTERPRETING THIN SECTIONS.

CELL★
NUCLEUS_A
ENDOPLASMIC RETICULUM_B
MITOCHONDRION_C
CELL MEMBRANE_D
PLANES OF SECTIONS★
LONGITUDINAL_E
TRANSVERSE_F
OBLIQUE_G

THIN SECTIONS★

SERIAL
SECTIONS★

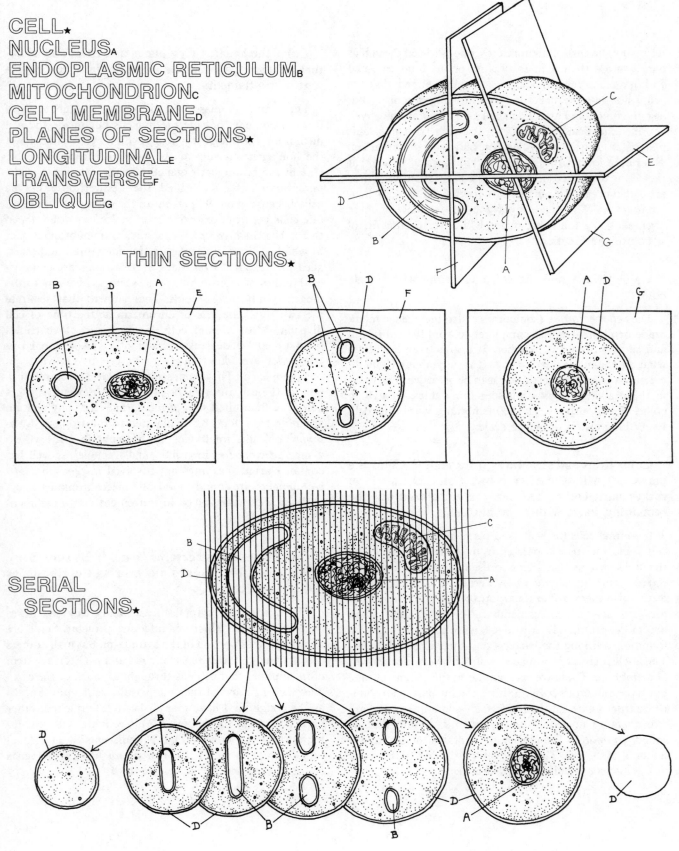

Although the earliest light microscopes showed the cell as hardly more than a mass of amorphous fluid enclosed within a membrane, modern research has shown that the cell is not only the structural unit of living organisms but also the functional unit. Each cell carries out all the physical and chemical reactions we associate with life. This plate is an artist's reconstruction of a typical animal cell as it might look with its upper half cut away. We see that the cell is organized into many distinct structures. These are called organelles, and each is specialized for a particular function. This plate gives you an overview of these organelles; the following plates will cover the details of structure and function.

Color title A and the cell membrane with a pale color.

The *cell membrane* (sometimes called the plasma membrane or plasmalemma) completely covers the entire cell and serves to hold it together. It also actively regulates what enters and leaves the cell. It is only about 10 nanometers thick, so its thickness has to be greatly exaggerated in the drawing to give you something thick enough to color. This is also true of the membranes within the cell. Everything else is drawn to scale.

Color titles and structures B through F. Color the pores (C) with a darker color. Color D and E in rather dark colors, and leave F uncolored (all the remaining space within the nucleus).

In animal cells (as well as in plant, protist, and fungus cells), the nucleus is separated from the rest of the cell by the *nuclear envelope.* Such cells are called eukaryotic (Greek: *eu,* "true"; *karyon,* "kernel" or "nucleus") to distinguish them from prokaryotic cells (Greek: *pro,* "before"), which lack a true membrane-enclosed nucleus and are more primitively organized. (Prokaryotic cells are found only among the bacteria and their close relatives.) The nuclear envelope is made up of two layers (not shown) of membrane. These are very similar to the cell membrane but have numerous *pores.* Within the nucleus is a prominent structure called the *nucleolus*—sometimes there are two or more nucleoli—and a network of thin threads called *chromatin.* The chromatin contains the hereditary material of the cell. The fluid that fills the rest of the space in the nucleus is called the *nuclear sap.*

Color the heading Cytoplasm and titles and structures G through N. Color over the lines that represent microfilaments.

The term "cytoplasm" is still used to designate all of the cell contents outside the nucleus but inside the cell membrane, although we realize that cytoplasm is not the homogeneous substance it was once thought to be. One of the prominent organelles in the cytoplasm is the *mitochondrion,* often called the "powerhouse of the cell" because about 90 percent of the energy that eukaryotic cells get from oxidizing food molecules is developed there. The *Golgi complex* is a stack of membranous sacs in which various molecules are manufactured and packaged for "export" from the cell. *Centrioles* are cylindrical bundles of microtubules that seem to give rise to the longer spindle microtubules (not shown) that separate the two duplicate sets of chromatin at the time of cell division. Most animal cells have a pair of centrioles lined up at 90 degrees to each other. Additional *microtubules* are found singly or in groups elsewhere in the cytoplasm. They appear to provide structural support to the cell and may be involved in movement. *Vacuoles* are fluid-filled sacs of membrane that may contain anything from food being digested to oil droplets. *Lysosomes* look like small vacuoles but contain digestive enzymes. *Microbodies* look like small vacuoles as well but contain various enzymes not involved in digestion. *Microfilaments* are found in various places around the cytoplasm and are involved in movement and attachment to other cells.

Color titles and structures O and P. Be sure to use a pale color for P to avoid obscuring the ribosomes (O). Do not color Q.

Throughout the cytoplasm are many tiny structures called *ribosomes,* which manufacture proteins. Some are free in the fluid portion of the cytoplasm, but many others are attached to the *endoplasmic reticulum* (ER), a system of membranes that extends throughout much of the cytoplasm. Some parts of the endoplasmic reticulum (known as the rough ER) have many ribosomes attached; other parts (known as the smooth ER) have none. The remaining portion of the cytoplasm, which seems to be a structureless fluid, is called the *hyaloplasm.* (Some biologists call it the cell sap or the cell matrix.)

ANIMAL CELL.

CELL MEMBRANE_A
NUCLEUS.★
NUCLEAR ENVELOPE_B
NUCLEAR PORE_C
NUCLEOLUS_D
CHROMATIN_E
NUCLEAR SAP_F÷
CYTOPLASM.★
MITOCHONDRION_G
GOLGI COMPLEX_H

CENTRIOLE_I
MICROTUBULE_J
VACUOLE_K
LYSOSOME_L
MICROBODY_M
MICROFILAMENT_N
RIBOSOME_O
ENDOPLASMIC RETICULUM_P
HYALOPLASM_Q÷

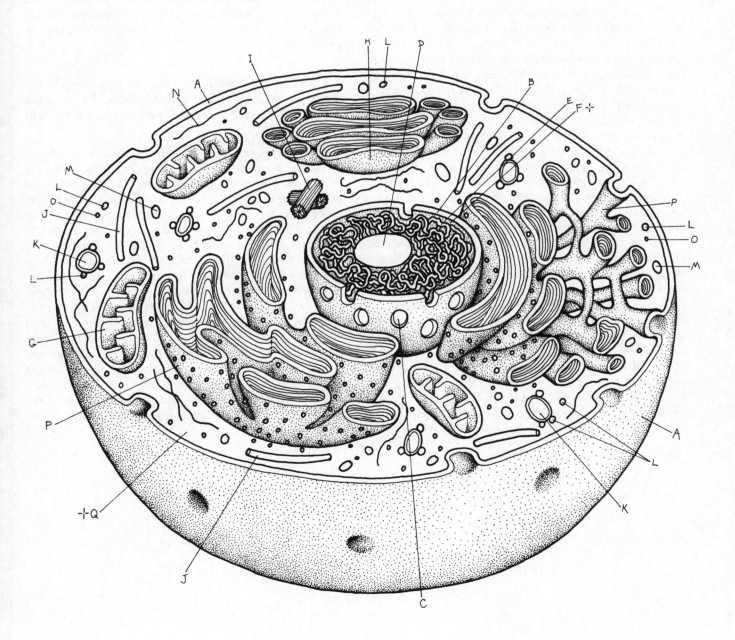

In a typical plant cell we see virtually all of the structures found in animal cells except for centrioles and certain protrusions used for locomotion or absorption. On the other hand, plant cells contain certain structures not found in animal cells at all.

Color titles and structures A through H, including the heading Plastids. Use green for F and a light color for D and D¹. In nature, G is white. Color H with any bright color you wish. (If this were a ripening tomato, H would be red.)

The plant *cell membrane* is essentially the same as an animal cell membrane. Immediately outside the plant cell membrane, however, is a *cell wall* consisting mostly of fibers of cellulose, although other kinds of molecules also become incorporated into it. When a cell is first formed by division of its parent cell, the cell wall is relatively elastic and is called a primary cell wall. As the cell grows, the wall is made thicker and more rigid and becomes known as a secondary cell wall (shown in this plate). The cell wall is perforated by numerous small pores called *plasmodesmata* (singular, plasmodesma), which appear to allow a direct bridge of cytoplasm from one cell to the next.

Although animal cells often contain some small vacuoles, plant cells usually contain one or a few very large ones. As plant cells mature, the vacuoles tend to get larger and usually fuse to form a single very large *vacuole* that may comprise up to 90 percent of the cell's volume. These large vacuoles are sometimes called "water vacuoles" because they contain large quantities of water. However, they also contain a wide variety of dissolved substances, including nutrients stored for later use and toxic substances, which may be broken down into harmless subunits in the vacuole. It is because of the dissolved substances that water flows into the vacuole and creates osmotic pressure, which is responsible for the rigidity ("turgor") of plants. When water is in short supply, the vacuoles lose their osmotic pressure, and the plant wilts (see Plate 34). Sometimes substances are stored in vac-

uoles as solid *crystals,* and many flowers receive their coloring from the pigments dissolved or crystallized in their vacuoles. The membrane of the vacuole is often called the *tonoplast.*

Plants are also colored by their plastids, but *chloroplasts,* which are green, have a much more important function than merely making plants green. They trap light energy and convert it to chemical energy for the manufacture of food in the process called photosynthesis. *Leucoplasts* are whitish in color and serve to store starch, lipid, or protein. *Chromoplasts* are plastids that produce and store other pigments that impart color to particular parts of a plant, as when fruit ripens or leaves turn color in the fall. They are formed by modification of chloroplasts or leucoplasts.

Color all the remaining titles and structures in the plate, including the heading Nucleus. Use a dark color for J, light colors for K and R, and a very light color for Q.

Golgi complexes in plant cells are usually called dictyosomes. They are very much like the Golgi complexes in animal cells except that they are usually smaller and more numerous. In addition to synthesizing various complex molecules needed within the cell, they appear to be responsible for manufacturing the components of the cell wall, which animal cells never have. All the remaining structures are virtually identical to those found in animal cells: *ribosomes* synthesize proteins and are found attached to the *endoplasmic reticulum* and free in the cytoplasm; *mitochondria* provide energy by oxidizing the carbohydrate made in the chloroplasts; *microtubules* and *microfilaments* seem to provide support and produce movement; *lysosomes* and *microbodies* contain enzymes; and the apparently structureless fluid making up the rest of the cytoplasm is called the *hyaloplasm.* The nucleus, too, is virtually the same; for that reason, this plate shows only the exterior of the *nuclear envelope* with its numerous *pores.*

PLANT CELL.

CELL MEMBRANE_A
CELL WALL_B
PLASMODESMA_C
VACUOLE_D
  TONOPLAST_D¹
  CRYSTAL_E
PLASTIDS_★
  CHLOROPLAST_F
  LEUCOPLAST_G
  CHROMOPLAST_H
GOLGI COMPLEX_I
RIBOSOME_J

ENDOPLASMIC RETICULUM_K
MITOCHONDRION_L
MICROTUBULE_M
MICROFILAMENT_N
LYSOSOME_O
MICROBODY_P
HYALOPLASM_Q
NUCLEUS_★
  NUCLEAR ENVELOPE_R
  NUCLEAR PORE_S

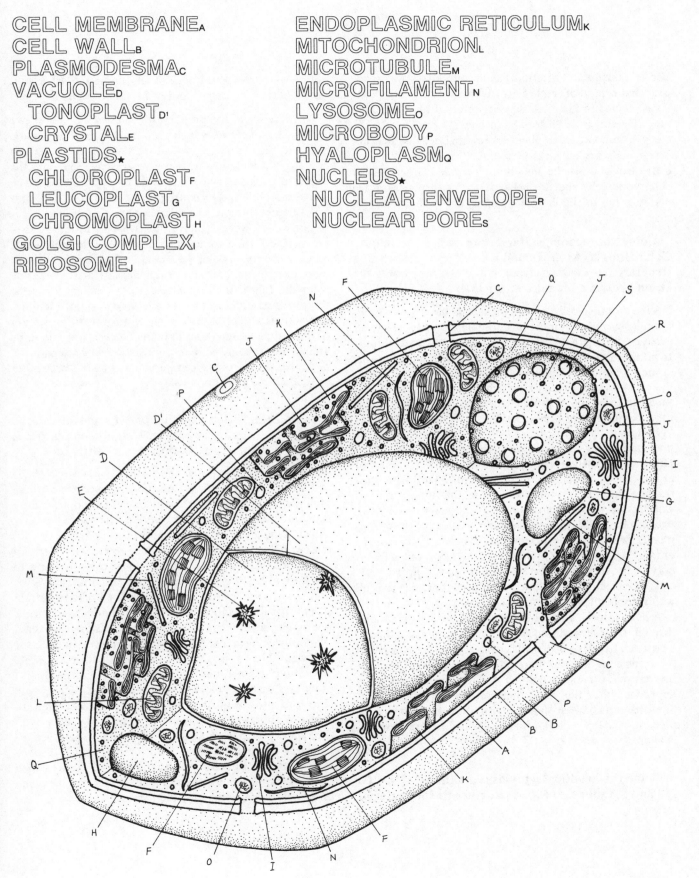

Bacteria (singular, bacterium) and cyanophytes are organisms that consist of single cells lacking a nuclear envelope. As explained in Plate 30, they are referred to as prokaryotic (Greek: *pro,* "before"; *karyon,* "kernel" or "nucleus"), in contrast to plant and animal cells, which are designated as eukaryotic (Greek: *eu,* "true"). Prokaryotic cells also lack all the membrane-bounded organelles of eukaryotic cells and are believed to retain the primitive organization of the first cells to appear on the earth.

Color the headings Bacterium and Cyanophyte. Color titles A, A¹, B, D, and E and the corresponding structures in the bacterium. Choose a pale color for D, and reserve blue-green for later.

Like all other cells, prokaryotic cells have a *cell membrane* (plasma membrane or plasmalemma). In nearly all prokaryotes, there is a *cell wall* outside the cell membrane. In addition to fibers of cellulose, found in plant cell walls, numerous other complex carbohydrate molecules, cross-linked to one another by short chains of amino acids, are found in the walls of prokaryotic cells. For nearly a century, bacteria have been classified as "gram-positive" or "gram-negative" according to whether they are stained or not stained by a standard staining technique called Gram's stain. Modern research has shown that these two groups have distinctly different cell walls. This plate illustrates a gram-positive bacterium; in this group the cell wall has no visible separation into layers. Gram-negative bacteria have cell walls very similar to those of the cyanophytes, with multiple layers and a second "outer membrane." An outer membrane outside the cell wall is found only in prokaryotic cells. Many, but not all, prokaryotic cells have a jellylike *capsule* or slime layer outside the cell wall (often called a sheath in cyanophytes).

A *pilus* (plural, pili) is a projection of the cell membrane through the cell wall that serves for attachment, either to a host cell that is to be parasitized or to another bacterium for transfer of genetic material. Some bacteria have numerous pili; others have none. A *flagellum* is a whiplike appendage that the bacterium can use to propel itself. Some bacteria have no flagella, while others have one, two, four, or as many as 100. Pili and flagella are not found in cyanophytes.

Color the heading Cytoplasm and titles F, F¹, G, H, and L. Color the related structures in the bacterial cell. Color over the fibers of the nucleoid (F) with a pale color. Use a pale color for L as well.

The interior of the prokaryotic cell is divided into two regions: the *nucleoid,* which contains the hereditary material, and the cytoplasm, which includes all the other cell contents. Note, however, that no membrane separates the nucleoid from the cytoplasm. The nucleoid consists of a single molecule of DNA, frequently described as "circular" because the end is doubled back and connected to the beginning. However, it has to be twisted and folded back on itself to form a compact mass because the molecule is about 1,000 times as long as the cell. A *plasmid* is a small piece of circular DNA (genetic material) that is separate from the DNA of the nucleoid. As in eukaryotic cells, there are numerous *ribosomes;* the endoplasmic reticulum, however, is absent. A *mesosome* is an inward projection of the cell membrane that in many bacteria appears to be attached to the nucleoid and may assist in the separation of duplicate nucleoids when the cell divides. As with other cells, the fluid portion is called the *hyaloplasm.*

Now color titles C, I, J, and K and all of the structures in the cyanophyte. Use blue-green for I. Choose a color for K that contrasts with L.

The cyanophytes receive their name from *cyan,* a shade of blue-green, and *-phyte,* meaning "plant," since these organisms are often blue-green in color and trap light energy to make food just as plants do. They were formerly called "blue-green algae," but it is clear that they are not closely related to any of the other algae, which are all eukaryotic. A few biologists call them cyanobacteria to emphasize their close relation to the bacteria.

A cyanophyte cell is similar to the cell of a bacterium except that it lacks mesosomes and pili and has numerous internal layers of membrane. Like gram-negative bacteria, cyanophytes have a second, *outer membrane* outside the cell wall. The internal membranes are called *thylakoid membranes* (sometimes lamellae) and contain the chlorophyll and carotene molecules that trap light energy and convert it into chemical energy in the process of photosynthesis. Within the *thylakoid compartments* of cyanophytes are dense particles called *phycobilisomes,* which are attached to the thylakoid membranes and contain an assortment of additional light-trapping molecules called phycobilins, which are often blue-green but may be many other colors, depending on the species.

PROKARYOTIC CELLS.

CELL MEMBRANE_A
PILUS_{A¹}
CELL WALL_B
OUTER MEMBRANE_C
CAPSULE_D
FLAGELLUM_E
NUCLEOID_F
PLASMID_{F¹}

CYTOPLASM★
RIBOSOME_G
MESOSOME_H
PHYCOBILISOME_I
THYLAKOID MEMBRANE_J
THYLAKOID COMPARTMENT_K
HYALOPLASM_L

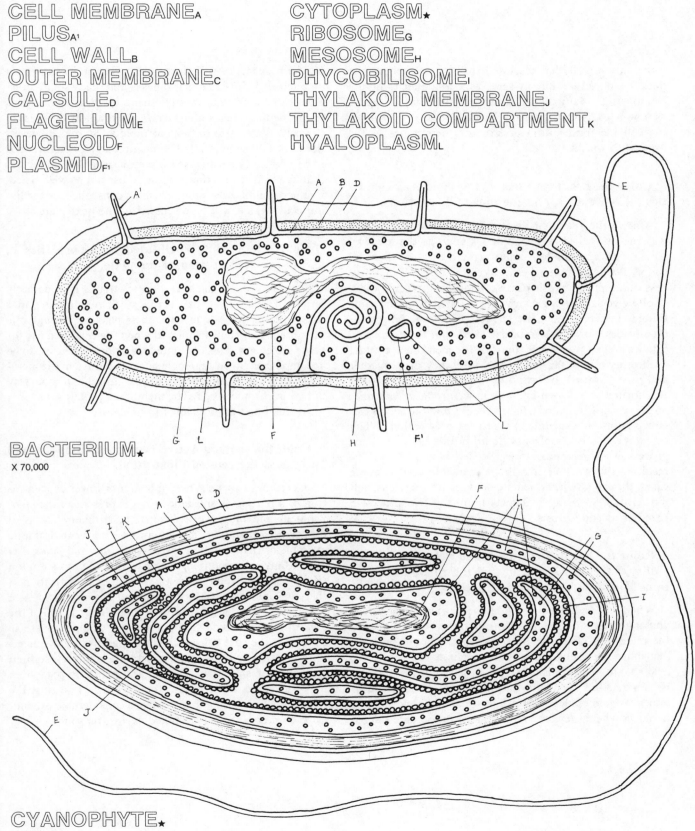

BACTERIUM★
X 70,000

CYANOPHYTE★
X 70,000

MEMBRANE PROPERTIES

The study of cells has disclosed that the cell membrane does far more than merely serve as a container to hold a cell together. Many of the activities of cells that we regard as unique to living organisms turn out to be properties of the cell membrane itself and of the membranes of the organelles within the cell.

Color the headings Phagocytosis and Amoeba, titles A, A¹, and B, and the upper illustrations.

Among the very first discoveries with the microscope was the one-celled water animal known as the amoeba, which captures smaller organisms and devours them by simply flowing around them and engulfing them in a process called phagocytosis (Greek: *phagein,* "to eat;" *kytos,* "hollow vessel"). (Our own white blood cells do the same thing to invading bacteria.) The *cell membrane* bulges out on all sides of the *prey* to be captured, and the cytoplasm follows. When the two portions of the membrane contact each other on the far side of the prey, they fuse together and pinch inward to form a vacuole. A portion of what was formerly cell membrane is now *vacuole membrane.* Such a vacuole is called a food vacuole, and the cell passes enzymes into the vacuole to digest the prey within. When cells take in tiny droplets of liquid in this same way, the process is called pinocytosis ("cell drinking"), and the resulting tiny vacuole is called a pinocytic vesicle. Some cases, though, are intermediate between phagocytosis and pinocytosis, so many biologists lump the two processes together as endocytosis.

Color the heading Diffusion and Selective Permeability, titles C, D, and E, and the related illustrations. Use a light color for C.

When you fry bacon in the kitchen, it isn't long before you can detect the aroma all through the house. Some of the molecules of the bacon spread out spontaneously to distribute themselves evenly throughout all the space available to them. This process is called diffusion and can be summarized by saying that molecules always tend to move from any place where they are highly concentrated to places where they are in low concentration unless some

barrier prevents them from doing so. One of the surprising characteristics of cell membranes is that many small molecules can diffuse right through them. If you add *fatty acids* to the medium surrounding a *cell,* they rapidly diffuse into the cell. Water also diffuses in rapidly. *Starch,* however, does not diffuse in at all. The membrane is therefore said to be selectively permeable (or semipermeable). Molecules that can diffuse into the cell can diffuse out as well. Selective permeability has some peculiar consequences for living cells, which will be discussed in the next plate.

Color the heading Facilitated Transport, title F, and the related portion of the plate.

Some molecules, such as *glucose,* that are moderately large and do not easily pass through cell membranes under most circumstances can have their passage through the membrane helped or facilitated when the cell has a use for those molecules. This facilitated transport is believed to be accomplished by protein molecules embedded in the membrane that help the molecules to pass through in some way not entirely understood. The membrane proteins that perform this function are called permeases.

Color the heading Active Transport, titles K^+ and Na^+, and the related illustration.

In certain cases, cells need a higher or lower concentration of a particular substance than is present in the environment and must work against the ordinary forces of diffusion to maintain that situation. The most dramatic example of that situation involves *sodium* and *potassium ions.* Many cells require a concentration of potassium ion (K^+) inside the cell that is 100 times the concentration in the environment outside the cell and a concentration of sodium ion (Na^+) inside the cell that is only 1/100 of the concentration outside. Sodium and potassium ions are so small that they are constantly diffusing through the membrane, so the cell has to be constantly pumping sodium ions out as fast as they come in and pumping potassium ions back in as fast as they escape. To work against the forces of diffusion like this requires a considerable expenditure of energy, so this is called active transport.

MEMBRANE PROPERTIES.

PHAGOCYTOSIS.★
  AMOEBA.★
  AMOEBA CELL MEMBRANE_A
  VACUOLE MEMBRANE_{A^1}
  PREY_B

NUCLEUS

DIFFUSION AND SELECTIVE
  PERMEABILITY.★
CELL_C
FATTY ACID SOLUTION_D
STARCH SOLUTION_E

FACILITATED TRANSPORT.★
GLUCOSE MOLECULE_F

ACTIVE TRANSPORT.★
POTASSIUM ION_{K^+}
SODIUM ION_{Na^+}

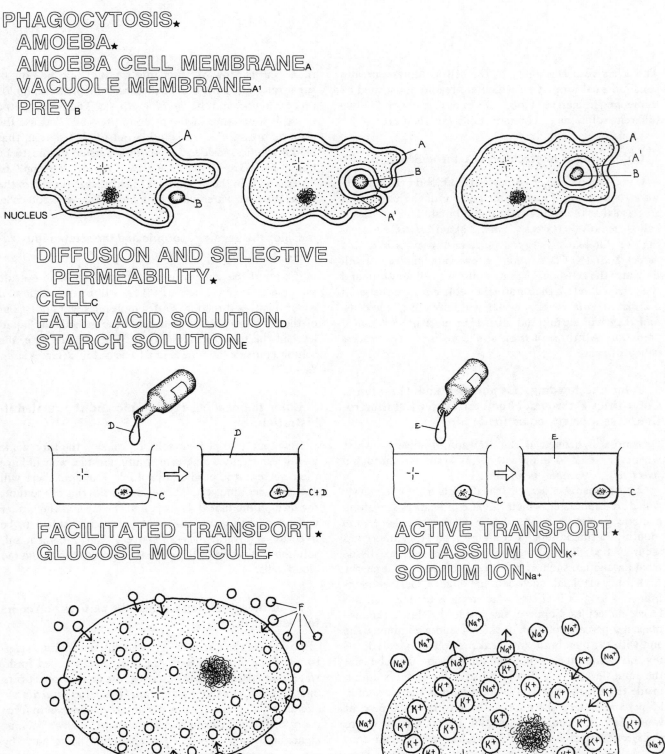

The selective permeability of the cell membrane creates some unusual properties, which are readily illustrated by immersing human red blood cells in pure water and in two different solutions of common table salt (NaCl).

Color titles and structures A through D.

If you place some *erythrocytes* (red blood cells) in *pure water,* water will flow into the cells until the pressure is so great that the cells swell up and burst. This process is called hemolysis (Greek: *hemo,* "blood"; *lysis,* "loosening" or "breaking"). If you place erythrocytes in a *0.85 percent salt* (NaCl) *solution,* a few water molecules will flow into the cells, but the same number will flow out, and the cells will retain their normal double-concave shape. In a *2 percent salt solution,* water will flow out of the cells, and they will shrink and shrivel up in a process called crenation. All three of these situations involve a process called osmosis.

Color the headings Osmometer and Hypotonic. Color titles E through H and the center left illustration. Use a bright color for H.

Osmosis is defined as the diffusion of solvent—water whenever we are dealing with living systems—through a *selectively permeable membrane.*

Osmosis can be demonstrated with a simple device called an osmometer, which will also indicate the resulting osmotic pressure. A membrane is stretched over the mouth of a thistle tube (a common laboratory item) and securely tied there. The membrane can be some tissue from an animal, such as intestine or bladder, or it can be something artificial, such as a cellophane dialysis membrane, as long as it allows water to pass through but not larger molecules. Suppose that inside the thistle tube we place a 5 percent solution of sucrose and we immerse the end with the membrane in a beaker of pure water, adjusting the thistle tube so that the liquid inside the tube is at the same level as the liquid outside. Since the solution inside the thistle tube is 5% sucrose, it is therefore only 95 percent water. The pure water outside is 100 percent water. Since molecules always diffuse from regions where their concentration is higher to regions where their concentration is lower, the *water molecules* will diffuse through the membrane into the thistle tube. Some water molecules will also diffuse out, but not nearly as many as will diffuse in. The *sugar molecules* will tend to diffuse from inside, where their concentration is 5 percent, to outside, where their concentration is 0 percent, but the membrane will not allow them to pass through. Since

there is a net flow of water into the tube and no flow of sugar out of it, a pressure, called osmotic pressure, builds up in the tube, and the *liquid level* rises. The height of the liquid is a measure of the osmotic pressure. Because the osmotic pressure of the solution outside is less than that of the solution inside, the solution outside is said to be hypotonic (Greek: *hypo,* "below"; *tonus,* "tension" or "pressure"). In the experiment illustrated at the top of the plate, the pure water was hypotonic to the red blood cells.

Color the heading Isotonic and the related illustration.

If we put the same 5 percent sucrose solution outside the tube as we have inside it, there will be no net flow of water in either direction. Water molecules will diffuse out of the tube just as fast as they diffuse in, and the liquid level in the tube will remain the same. In this case, the solution outside the tube is said to be isotonic (Greek: *isos,* "same").

Color the heading Hypertonic and the related illustration.

If we put a 10 percent solution outside the tube while we still have only a 5 percent solution inside, we will have the reverse of the original hypotonic situation: there will now be a net diffusion of water molecules out of the tube, the level in the tube will drop, and the level in the beaker will rise. In this case the solution outside is said to be hypertonic (Greek: *hyper,* "above"). The 2 percent salt solution at the top of the plate was hypertonic to the red blood cells.

Color the remaining parts of the plate as you come to them in the text.

To survive the forces of osmosis, different living organisms have adopted various strategies. Single-celled freshwater animals, such as the amoeba, use active transport to pump excess incoming *water* into special *contractile vacuoles,* which collect that water and then contract to force it out of the cell through a tiny opening. Cells of plants, algae, and fungi have *cell walls* to resist being burst by osmotic pressure. Plants actually depend on osmotic pressure to keep them erect. If you let plant cells lose water (or make them lose water by placing them in a hypertonic solution), they shrink away from their cell walls, and the plant wilts. That is why supermarkets spray their vegetables frequently and smart cooks keep vegetables damp in a humidifier drawer in the refrigerator.

OSMOSIS.

ERYTHROCYTE_A
PURE WATER_B
0.85% SALT SOLUTION_C
2% SALT SOLUTION_D

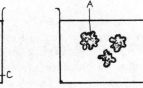

OSMOMETER★
SELECTIVELY PERMEABLE
 MEMBRANE_E
WATER MOLECULE_F
SUGAR MOLECULE_G
TUBE SOLUTION LEVEL_H

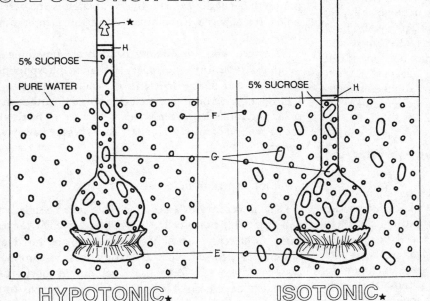
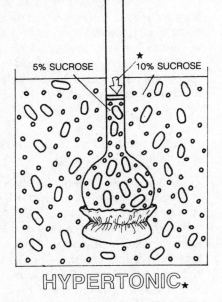

HYPOTONIC★   ISOTONIC★   HYPERTONIC★

AMOEBA★
WATER_I
CONTRACTILE VACUOLE_J

WILTING PLANT CELL★
CELL WALL_K
AIR SPACE_L
SHRUNKEN VACUOLE_M
NUCLEUS_N
CHLOROPLAST_O
HYALOPLASM_P

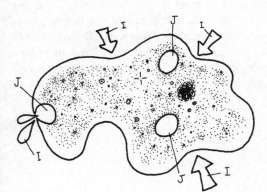
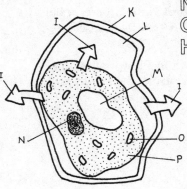

EVIDENCE OF MEMBRANE ULTRASTRUCTURE

The exact ultrastructure of cell membrane is not yet known. However, a hypothetical model has been formulated that fits the known facts and gives a possible explanation for most membrane functions. This plate and the following plate should be worked together.

Color the heading Electron Microscope Image and titles and structures A, B, and C.

Membranes are so thin (7.5–10 nanometers) that even the best electron microscopes today show them only as two fuzzy, *densely staining lines* sandwiching a *lightly staining area* between them. When we isolate membranes from the rest of the cell and analyze them chemically, we find that they are composed primarily of protein and lipid with small amounts of carbohydrates. The kinds and amounts of protein vary considerably from one type of cell to another. The kinds of lipid also vary, but most of them are phospholipids, and the amount is just about enough to make a double layer around the cell if the molecules are lined up in a particular way.

Color the heading Salad Dressing and titles and structures D and E.

Our hypotheses of how the lipids are lined up in membranes are largely derived from what we know about the structure and the properties of lipids (Plate 17) and their behavior when mixed with water, since cells are largely made of water and exist in watery environments. If you have ever shaken a bottle of *oil*-and-*vinegar* salad dressing and watched it for a few minutes, you know that even before the oil floats to the top, it gathers together in droplets. The water molecules (which make up most of the vinegar) are polar (Plates 11 and 12) and have such a strong attraction for one another and so little attraction for the oil, which is largely nonpolar, that the oil molecules are excluded from the water and gather together into droplets with other oil molecules. Sophisticated experiments have shown that once the oil molecules are closely packed together, a weak but effective attractive force develops between them and tends to keep them together.

Color the headings Micelle and Phospholipid and titles and structures F and G, using light colors.

As we saw in Plate 17, phospholipids have a *nonpolar* or *hydrophobic* ("water-fearing") *portion,* composed of two long hydrocarbon chains, and a *polar* or *hydrophilic* ("water-loving") *portion* consisting of a phosphate group plus some other polar group. If placed in water, phospholipids will become spontaneously arranged in such a way that the hydrophobic portions of the molecules pack closely together and avoid contact with the water molecules while the hydrophilic portions face outward to form hydrogen bonds with the water molecules. Depending on the conditions, this results either in ultramicroscopic spheres called micelles or a double layer of phospholipid, generally called a bilayer, which is the way the lipids are believed to be arranged in the membranes of living cells.

Whenever you use a detergent to clean something oily or greasy, you form molecules similar to phospholipids, since one end of the detergent molecule attaches to the hydrophobic oil or grease molecule while the other end is polar and forms hydrogen bonds with water. The result is the formation of micelles that are easily washed away with water.

Color the remainder of the plate.

Cells infected with a peculiar virus called Sendai virus will fuse with other cells they contact, even if they are from a different species. Mouse cells can be fused with human cells, for example. (No, this does not produce a two-legged mouse 6 feet tall; the cell dies in short order.) Biologists have been able to follow what happens to the membrane proteins after cell fusion by using stains that fluoresce under ultraviolet light. The *proteins of the mouse membrane* are stained with a substance that fluoresces in one color, and the *proteins of the human membrane* are stained with a substance that fluoresces in a different color. The cells are then caused to fuse and observed with a fluorescence microscope, which uses an ultraviolet light source. Immediately after fusion, the proteins are observed to be still confined to the portion of the membrane they originally came with. But in 30 to 60 minutes they migrate around until each kind of protein is evenly distributed over all of the newly formed cell. Some proteins do not migrate like this, but most do.

EVIDENCE OF MEMBRANE ULTRASTRUCTURE.

ELECTRON MICROSCOPE IMAGE.★
DENSELY STAINING PORTION_A
LIGHTLY STAINING PORTION_B
CYTOPLASM_C

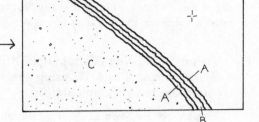

SALAD DRESSING.★
OIL_D
VINEGAR_E

MICELLE.★
PHOSPHOLIPID.★
NONPOLAR (HYDROPHOBIC) PORTION_F
POLAR (HYDROPHILIC) PORTION_G

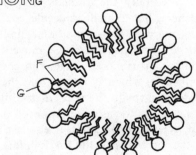

MICELLE

PHOSPHOLIPID BILAYER

CELL FUSION.★
MOUSE PROTEIN_H
HUMAN PROTEIN_I

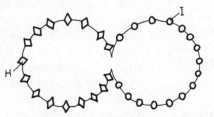

IMMEDIATELY AFTER FUSION

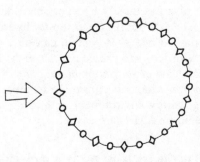

30–60 MINUTES LATER

THE FLUID MOSAIC MODEL

Although we are still not certain exactly how the molecules comprising a membrane are actually put together, the model shown in this plate is the best integration of the known evidence that anyone has come up with so far.

Color the heading Phospholipid Bilayer and titles and structures A and B, using the light F and G colors from the previous plate.

This model is known as the fluid mosaic model because it hypothesizes that the membranes of a cell consist of phospholipid bilayers in which proteins and carbohydrates are embedded like the stones in a mosaic. Other lipids (not shown) are also present. The phospholipid bilayer appears to be in a somewhat fluid state, allowing most of the proteins to migrate within it. (Recall the cell fusion portion of Plate 35.)

Color the headings Protein Molecule and Glycocalyx and titles C, D, and E. Color the hydrophobic and hydrophilic portions of the three protein molecules in the center and right portions of the central drawing, saving the leftmost protein (with the arrow) for later. Also color the carbohydrate.

Some of the proteins in a membrane are exposed only to the exterior of the cell or only to the interior, while others extend all the way through the membrane. Attached to some of the proteins on the exterior surface are certain carbohydrate groups (collectively called the *glycocalyx*), some of which are believed to assist cells in responding to environmental changes and in "recognizing" one another and interacting in appropriate ways.

The positioning of the proteins within the lipid bilayer is easily understood if you remember (from Plate 19) that some of the amino acids making up proteins have hydrophobic ("water-fearing": nonpolar) side groups while other have hydrophilic ("water-loving": polar or ionic) side groups. If a particular protein is constructed so that one portion of it has mostly hydrophobic side groups on the surface while the rest of it has mostly hydrophilic side groups on the surface, that protein can only attach to a membrane with its *hydrophobic portion* embedded in the *hydrophobic region of the phospholipid bilayer* and its *hydrophilic portion* protruding from the membrane into the surrounding watery environment or into the cytoplasm, both of which are *hydrophilic*.

Now color the leftmost protein molecule, titles F and G, and the molecule and arrow.

One very satisfying feature of the fluid mosaic model is that it offers a possible explanation for the rapid *diffusion* of *small polar molecules* across the membrane. We need only assume that some of the proteins have pores through their centers lined with hydrophilic side groups that attract small hydrophilic (polar) molecules and allow them to pass through.

Color the remaining sections of the plate as you come to them in the following text. For illustrative purposes, the individual polypeptide chains of proteins are not shown, and only the hydrophobic and hydrophilic regions are distinguished. The bilayer is shown only in the first pair of drawings.

The fluid mosaic model also offers several possible explanations for facilitated transport and active transport. We know from studies of the proteins of muscular contraction and of enzymes, which are proteins, that many proteins change their tertiary and quaternary structures when ions or small molecules attach to them or detach from them. If a membrane protein has a certain part that attracts a specific ion or small molecule, that part might move in some way to transport that ion or molecule through the membrane.

There are four different hypotheses for such transport. To simplify the discussion, only an *ion* is mentioned here, but a small molecule could be transported in the same way. One hypothesis stipulates that a *movable arm* binds the ion to be moved, which causes a shift in the distribution of electron energies in the protein molecule, which in turn causes the movable arm to move down through the pore, taking the ion with it and releasing it on the opposite side of the membrane. A second hypothesis assumes a sequence of *binding sites:* when the ion binds to the first binding site, that causes changes that make the second binding site have a stronger attraction for the ion, so it moves farther in; that change causes the third binding site to develop a stronger attraction, so the ion moves in farther, and so on until it is all the way through. The third idea proposes a traplike movement: the ion is attracted into a cavity in the protein, but as soon as it is in the cavity, the protein changes shape to close behind the ion and open in front of it so that the ion is essentially on the other side of the membrane already. A fourth hypothesis (regarded today as not really likely) assumes that the binding of the ion causes changes that make the protein flip over within the lipid bilayer, dragging the ion to the opposite side.

THE FLUID MOSAIC MODEL.

PHOSPHOLIPID BILAYER.★
NONPOLAR (HYDROPHOBIC)
  PORTION_A
POLAR (HYDROPHILIC)
  PORTION_B

PROTEIN MOLECULE.★
HYDROPHOBIC PORTION_C
HYDROPHILIC PORTION_D
GLYCOCALYX.★
  CARBOHYDRATE_E
ION/SMALL POLAR MOLECULE_F
DIFFUSION_G

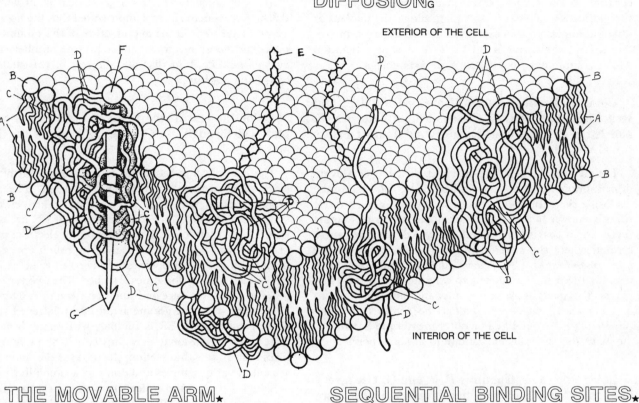

EXTERIOR OF THE CELL

INTERIOR OF THE CELL

THE MOVABLE ARM.★
MOVABLE ARM_D1

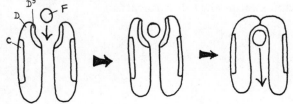

THE TRAP.★
BINDING SITE_D3

SEQUENTIAL BINDING SITES.★
BINDING SITE_D2

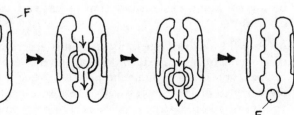

THE FLIPOVER.★

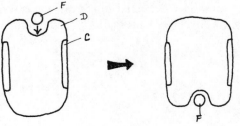

NUCLEUS AND ENDOPLASMIC RETICULUM

The nucleus of a cell contains virtually all of the hereditary material (DNA) that determines whether the cell will be a free-living single cell like an amoeba or will grow into a carrot, a rabbit, or some other kind of living thing. If the nucleus is removed by microsurgery, the cell will die. In contrast to the earlier plates on cells, which portrayed three-dimensional shapes, this plate shows the nucleus as it would appear in thin section under the electron microscope. Also shown is part of the endoplasmic reticulum, which is intimately associated with the nucleus.

Color the heading Nuclear Envelope and titles and structures A, B, and C. Choose a light color for C, and leave D uncolored. Retain your lightest colors for G and N.

The nucleus of eukaryotic cells is enclosed by two layers of membrane that are very similar to the cell membrane enclosing the entire cell. These two membranes comprise what is known as the nuclear envelope. At various points there are holes in this envelope called *nuclear pores,* created where the *outer membrane* turns in to join the *inner membrane.* These pores allow certain molecules manufactured in the nucleus to migrate out to the cytoplasm. The space between the inner and outer membranes of the nuclear envelope is called the *perinuclear space* (Greek: *peri,* "around"). The outer membrane is continuous with the endoplasmic reticulum at some points.

Color titles and structures E, F, and G. Use a very light color for G. Be aware that if the chromatin were drawn to scale, its diameter would be only about one-tenth of what is shown, and its total length would be several hundred thousand times what you are coloring here.

If a cell is stained, much of the stain is absorbed by a tangle of threads called *chromatin* (Greek: *chromos,* "color") spread throughout the nucleus. When a cell is preparing to divide into two cells, these threads coil up tightly into compact structures called chromosomes (Plate 65), but for most of the cell's life the chromatin is spread through the nucleus as shown here. At various points the chromatin is attached to the inner membrane of the nuclear envelope. Chromatin is made up of DNA, protein,

and a very small amount of RNA. The most prominent structure in the nucleus is the *nucleolus,* which is not separated from the rest of the nucleus by a membrane but consists of fibers and granules so densely packed that it is easily distinguished from the surrounding chromatin. It consists of some DNA and a great deal of protein and RNA. Experiments have demonstrated that the fibers and granules are steps in the manufacture of the subunits that make the ribosomes found in such large numbers in the cytoplasm. The fluid that fills the rest of the nucleus is called the *nuclear sap.*

Color the rest of the titles and structures on the plate (H through N). Choose light colors for J, J¹, and N.

The endoplasmic reticulum (ER) is an extensive network of membranes within the cytoplasm. The membranes enclose numerous passageways called *cisternae* (singular, cisterna). Two different types of ER are distinguished by their overall appearance in electron micrographs: *rough ER,* which receives its rough appearance from the numerous ribosomes attached to it, and *smooth ER,* which has no ribosomes attached. The two types also differ in their structure and function. Rough ER tends to have much of its membrane arranged in flattened sheets, while most smooth ER is tubular, with many branches. (The three-dimensional views in Plate 30 show these features best. In this thin section, the tubes of the smooth ER are cut at various angles and don't look much like tubes.)

Rough ER serves to collect the proteins synthesized by the *attached ribosomes* and allow them to be transported within the cell. Most of those proteins are for export from the cell, and as they collect, the ER pinches off to form *vesicles* filled with protein. These vesicles then migrate either to the Golgi complex (see Plate 39) or to the cell membrane, where the proteins are released to the outside. *Free ribosomes* (not attached to the ER) synthesize proteins for use within the cell. The smooth ER synthesizes steroids and, in the liver, breaks down toxins and converts glycogen to glucose to maintain the proper level in the blood. In muscle, the smooth ER is more flattened and stores calcium ion, which it releases to produce muscular contraction. The fluid filling the rest of the space in the cytoplasm is called the *hyaloplasm.*

NUCLEUS AND ENDOPLASMIC RETICULUM.

NUCLEAR ENVELOPE ★
 OUTER MEMBRANE A
 INNER MEMBRANE B
 PERINUCLEAR SPACE C
 NUCLEAR PORE D ÷
CHROMATIN E
NUCLEOLUS F
NUCLEAR SAP G
SMOOTH ENDOPLASMIC
 RETICULUM H

ROUGH ENDOPLASMIC
 RETICULUM I
CISTERNA J
ATTACHED RIBOSOME K
FREE RIBOSOME L
VESICLE M
 CONTENTS J1
HYALOPLASM N

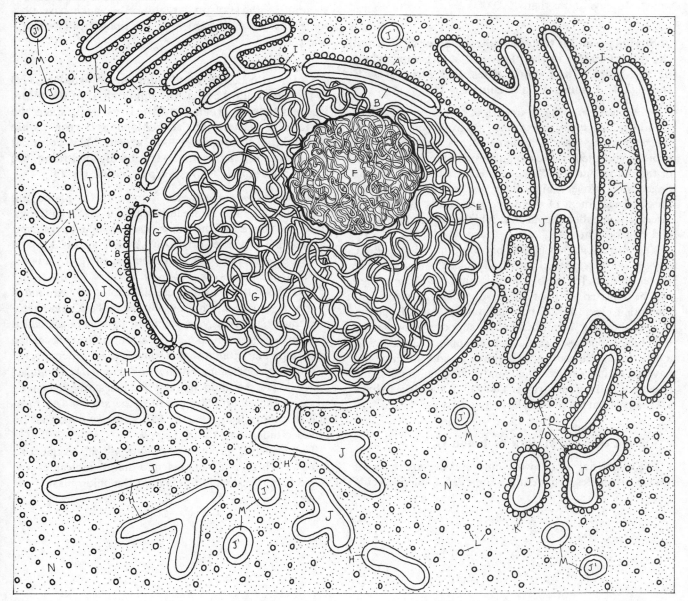

MITOCHONDRION AND CHLOROPLAST

The mitochondrion (plural, mitochondria) is the organelle that combines oxygen with food molecules to obtain energy for the cell. Mitochondria are so tiny that they are difficult to see with the light microscope, even with special staining, but they are very visible with the electron microscope. They take a wide variety of shapes—including ovoid, spherical, branching, pear-shaped, and threadlike —and in the living cell actually change shape constantly. The number of mitochondria in any given cell depends on the metabolic activity of that cell. One known kind of cell has only one mitochondrion; liver cells, which are very active, usually have more than 1000; and a few cells are known to have more than 100,000.

Saving green for the chloroplasts, color title A and the mitochondria in the plant cell at the upper right of the plate. Then color titles and structures B through E. Choose a light color for E.

A *mitochondrion* has a smooth *outer membrane* and a greatly folded *inner membrane,* separated by a distinct *intermembrane space.* The inner membrane sends many projections called cristae (singular, crista) into the interior of the mitochondrion. Some of these are tubular, like the fingers of a glove, while others are sheetlike folds. The inner compartment of the mitochondrion is filled with a viscous fluid called the *matrix,* which is about half water and half protein.

The mitochondrion is the only place in the cell where oxygen can be combined with food molecules to release the energy in them for use by the cell. Although some energy can be released without the involvement of oxygen, it amounts to less than 10 percent in most cases. To get an idea of the value of mitochondria, suppose your average light lunch consists of a sandwich, an apple, a cupcake, and a soft drink. Without mitochondria, a light lunch would have to include at least ten sandwiches, ten apples, ten cupcakes, and ten soft drinks. You would also have to have a digestive tract ten times as large as your present one.

Color titles and structures F and G.

Mitochondria, along with chloroplasts, are unusual among organelles in containing some *DNA,* the molecule that carries the hereditary code, as well as their own *ribosomes* and the RNA molecules necessary to make their own proteins. Although they are dependent on the rest of the cell for some of their proteins, it is clear that they make many of their own. Both mitochondria and chloroplasts reproduce themselves as the cell grows without waiting for the cell to divide. This has led to the speculation that perhaps they were originally free-living prokaryotic cells that somehow became engulfed by eukaryotic cells and established a mutually beneficial relationship. They are, after all, about the size of prokaryotic cells, and their ribosomes are the same size and are composed of the same subunits as those of prokaryotic cells, which have ribosomes distinctly different from the ribosomes of eukaryotic cells.

Color title H green and color the chloroplasts (H) in the small plant cell at the upper right of the plate. Use a shade of green for title L as well, since the thylakoids (L) contain the chlorophyll responsible for making chloroplasts look green. Then color all the remaining titles and structures, including the heading Granum. Choose a pale color for Q.

Although *chloroplasts* are found only in plants and algae, they are of immense importance to all living things, since virtually all of the energy used by living organisms in their life processes originally came from the sun and was trapped and converted to chemical energy by chloroplasts in the process of photosynthesis.

Like the mitochondrion, the chloroplast is surrounded by two *membranes.* Chlorophyll, the molecule that actually traps light energy, is found in flattened sacs called *thylakoids.* These thylakoids are arranged in stacks called grana (singular, granum), and some of them are connected to others by extensions called *stromal lamellae* (singular, lamella). The semifluid material that fills all the remaining space in the chloroplast is called the *stroma.* As mentioned, chloroplasts contain *DNA* and *ribosomes* and reproduce independently of the cell.

The principal product of photosynthesis is glucose, some of which is assembled into starch molecules and collected into *starch grains* within the chloroplast.

MITOCHONDRION
AND CHLOROPLAST.

MITOCHONDRION_A
  OUTER MEMBRANE_B
  INTERMEMBRANE SPACE_C
  INNER MEMBRANE_D
  MATRIX_E
  RIBOSOME_F
  DNA_G

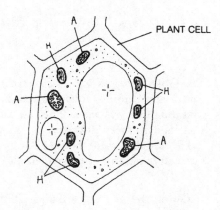

PLANT CELL

CRISTA

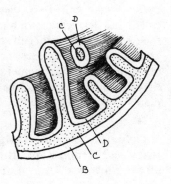

CHLOROPLAST_H
OUTER MEMBRANE_I
INTERMEMBRANE SPACE_J
INNER MEMBRANE_K
GRANUM_★
  THYLAKOID_L
STROMAL LAMELLA_M
DNA_N
RIBOSOME_O
STARCH GRAIN_P
STROMA_Q

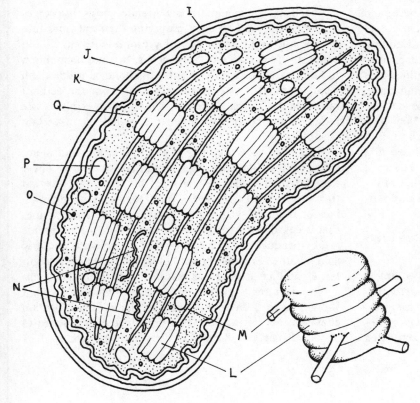

GOLGI COMPLEX, LYSOSOMES, MICROBODIES

The Golgi complex was discovered in the late nineteenth century by Camillo Golgi, who used a complex silver stain that didn't always work. This led some scientists to question its existence. However, the modern electron microscope has disclosed that the Golgi complex really is a distinctive organelle, present in all eukaryotic cells.

Color the headings Golgi Complex and Saccule/Vesicle and titles and structures A through B[1] at the upper right. Choose a pale color for B and B[1].

A Golgi complex consists of a stack of flattened membranous sacs, called saccules, and a number of tiny sacs called vesicles, which are budded off from the saccules at their edges. An animal cell will generally have between 10 and 20 Golgi complexes; a plant cell will usually have many more than that, sometimes as many as 200, although they are much smaller. These small Golgi complexes in plants are usually called dictyosomes.

Color the heading Golgi Complex in Action and titles and structures C through I of the animal cell at the bottom of the plate.

The function of the Golgi complex is to assemble simple molecules into complex ones and package them for use elsewhere. Complex carbohydrates are made there, and the assembly of complex proteins is completed. The *amino acids* making up complex proteins are first assembled into *polypeptide chains* (see Plate 18) by *ribosomes* attached to the *rough endoplasmic reticulum*. The polypeptides pass into the *cisternae* of the rough ER as they are being made. Sometimes glucose and a few other sugars are added to the polypeptides while they are in the cisternae. They then move to the *smooth ER*, where they are pinched off into vesicles at the end of a *cisterna*, becoming enclosed in a small piece of ER membrane. The resulting vesicle is called a *transition vesicle* because the polypeptide molecules are in transition from the ER to the Golgi complex.

Color the Golgi saccules (A and B + G) and titles and structures J and K.

When the transition vesicle reaches the Golgi complex, the vesicle membrane fuses with the *Golgi membrane,* emptying the polypeptides into the *Golgi saccule compartment.* There the polypeptides have numerous sugars, lipids, or other molecules attached to them. The completed *protein complexes* then migrate to the end of the Golgi saccule and are pinched off within a piece of membrane to form a Golgi vesicle. If that particular protein is for "export" (secretion), the Golgi saccule, called a *secretion vesicle,* migrates to the surface of the cell, where it will fuse with the *cell membrane,* emptying its contents to the exterior in a process called exocytosis (the reverse of endocytosis, explained in Plate 33).

Color the remaining parts of the plate.

The ER and the Golgi complex also cooperate to manufacture many enzymes (which are also *protein complexes*). Enzymes are capable of hydrolyzing ("breaking down"; see Plate 16) all kinds of large molecules. These are also pinched off into vesicles, but such vesicles are called *lysosomes* (Greek: *lysis,* "loosening" or "breaking"; *soma,* "body") because the enzymes they contain cause complex molecules to come apart into their components. It is not known how the lysosome membrane avoids being itself hydrolyzed by the enzymes, but if lysosomes are broken open, they may digest the entire cell. This occurrence is a normal part of the development of an embryo, as when a human embryo reabsorbs the tail (that is present in the early stages) and the webs between the fingers and toes.

In cells that take in materials by endocytosis, lysosomes fuse with the *food vacuole,* emptying their enzymes into the vacuole to digest the *food.* Lysosomes also engulf and break down worn-out cell organelles. It is not uncommon to find lysosomes with parts of mitochondria inside them being digested. Lysosomes are absent from the cells of most plants. Instead, the same hydrolytic enzymes are found in the central vacuole or occasionally in certain plastids.

Cells have been found to contain other kinds of membrane-enclosed vesicles very similar to lysosomes. These were originally called microbodies, but it is now clear that there are at least two types. They, too, contain enzymes, but for different purposes. *Peroxisomes* contain enzymes that break down hydrogen peroxide. Hydrogen peroxide is a by-product of the breakdown of amino acids and uric acid by other enzymes present in the peroxisome. A regular, crystalline structure, presumed to be a crystal of enzymes, is often seen in the center. Glyoxysomes (not shown) have been found only in plants. They contain enzymes that extract energy from glucose in a series of chemical reactions known as the glyoxylate cycle.

GOLGI COMPLEX, LYSOSOMES, MICROBODIES.

GOLGI COMPLEX★
SACCULE/VESICLE★
   MEMBRANE$_{A,A^1}$
    COMPARTMENT$_{B,B^1}$
GOLGI COMPLEX IN ACTION★
AMINO ACID MOLECULES$_C$
CELL MEMBRANE$_D$
RIBOSOME$_E$
ROUGH ER MEMBRANE$_F$
   CISTERNA$_{F^1}$
POLYPEPTIDE CHAINS$_G$
SMOOTH ER MEMBRANE$_H$
   CISTERNA$_{H^1}$
TRANSITION VESICLE$_I$

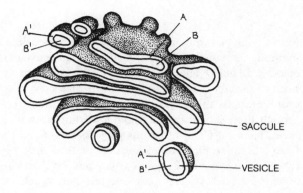

SACCULE

VESICLE

SECRETION VESICLE$_J$
   PROTEIN COMPLEX$_K$
LYSOSOME$_L$
FOOD VACUOLE$_M$
   FOOD$_N$
MICROBODY★
   PEROXISOME$_O$

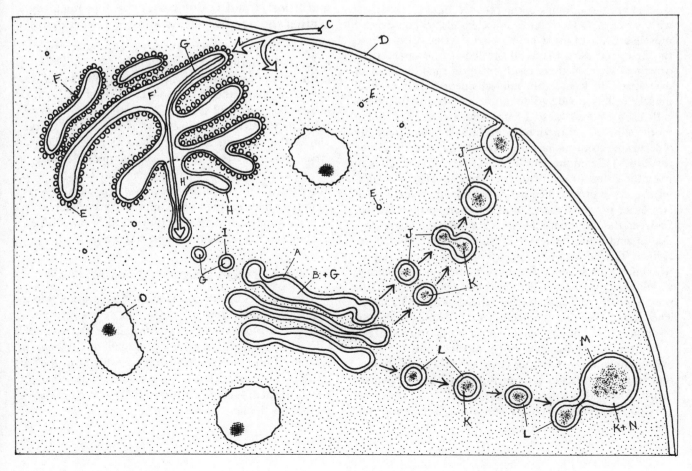

MICROFILAMENTS AND MICROTUBULES

Microfilaments and microtubules are responsible for maintaining cell shape and for virtually all movement in eukaryotic cells. The more common kinds of microfilaments are only 5 to 7 nanometers in diameter, and microtubules are only 20 to 25 nanometers in diameter. Although there is still considerable uncertainty, movement seems to result from numerous projections that reach out from one microfilament to another or from one microtubule to another and then shorten, pulling the microfilaments or microtubules past one another, much as the crew of a sailboat will pull in the anchor by standing in a line and pulling in the anchor rope hand over hand. The proteins comprising microfilaments are, however, entirely different from those comprising microtubules, and the two apparently never interact.

Color the headings Relaxed Sarcomere and Contracted Sarcomere and titles and structures A through E. Use a light color for A.

The microfilaments that have been most extensively studied are those found in the *skeletal muscle* of animals with backbones. This type of muscle is also called striated muscle because of the striations (stripes) that divide it into the light and dark bands illustrated in the upper right drawing. The striations result from alternating areas of overlap of thick and thin microfilaments. Each skeletal muscle cell is divided into numerous contractile units called sarcomeres, 1.5 to 2.5 nanometers long, depending on the state of contraction or relaxation. Each sarcomere is separated from the next by a partition called the *Z-line* (or Z-disk). Extending toward the center of the sarcomere from the Z-line at each end are numerous *thin microfilaments* of a protein called actin, with smaller amounts of two other proteins, troponin and tropomyosin, attached. Between the thin microfilaments are *thick microfilaments* that extend from the center of the sarcomere toward the Z-lines. The thick microfilaments are composed of a protein known as myosin and have numerous projections, which are commonly called *cross-bridges,* since they appear to cross over and attach to the thin microfilaments. The current "sliding filament theory" of muscular contraction is based on the belief that the cross-bridges attach to the thin microfilaments and pull them past the thick microfilaments so that the sarcomere shortens. Notice that in the contracted sarcomere, cross-bridges have attached to the thin microfilaments.

When the sarcomere is fully shortened, the thin microfil-

aments overlap somewhat in the center, and the Z-lines are drawn in until they almost touch the ends of the thick microfilaments. If enough sarcomeres shorten, the entire muscle shortens.

Color the heading Ratchet Action, titles F and G, and the associated illustration.

The movement of the cross-bridges is often compared to the action of a ratchet. Each cross-bridge apparently reaches out and attaches to a thin microfilament at an angle of 90 degrees, then moves (in the power stroke) to an angle of 45 degrees, forcing the thin microfilament to move a short distance; then it detaches to return (in the recovery stroke) to the 90-degree position for a new attachment. With numerous cross-bridges going through this same process, the microfilaments are made to slide past one another, and the sarcomere shortens.

Color the heading Cytoskeleton/Cytomusculature and titles H and I. Color over the fine lines representing the microfilaments and the microtubules.

Microfilaments are found in most cells, but in much less organized patterns than in muscle cells. In some cases they seem merely to provide a support to hold the cell in a particular shape and have come to be regarded as a sort of cytoskeleton ("cell skeleton"). In other cases they seem to be involved in the movement of the cell or of parts within the cell and are referred to as cytomusculature ("cell muscle system"). They are composed mostly of actin, but myosin is often present, particularly where movement is involved. *Microtubules* are generally more rigid than microfilaments and appear to have a larger role in determining cell shape, but they are also responsible for certain kinds of movement within the cell.

Color titles and structures J and K.

Microtubules do not consist of the same proteins found in microfilaments. Instead they are made up of two other proteins, α-tubulin and β-tubulin. These are assembled into dimers (double molecules), which in turn become assembled into left-handed spirals. In this form they are useful as structural elements for the cell; with other proteins attached, they can also produce movement, as you will see in the next plate.

MICROFILAMENTS AND MICROTUBULES.

SKELETAL MUSCLE CELL_A
Z-LINE_B
THIN MICROFILAMENT_C
THICK MICROFILAMENT_D
CROSS-BRIDGE_E

RELAXED SARCOMERE.★

CONTRACTED SARCOMERE.★

RATCHET ACTION.★
CROSS-BRIDGE IN POWER STROKE_F
CROSS-BRIDGE IN RECOVERY STROKE_G

CYTOSKELETON/CYTOMUSCULATURE.★

MICROFILAMENTS_H

MICROTUBULES_I

α-TUBULIN_J
β-TUBULIN_K

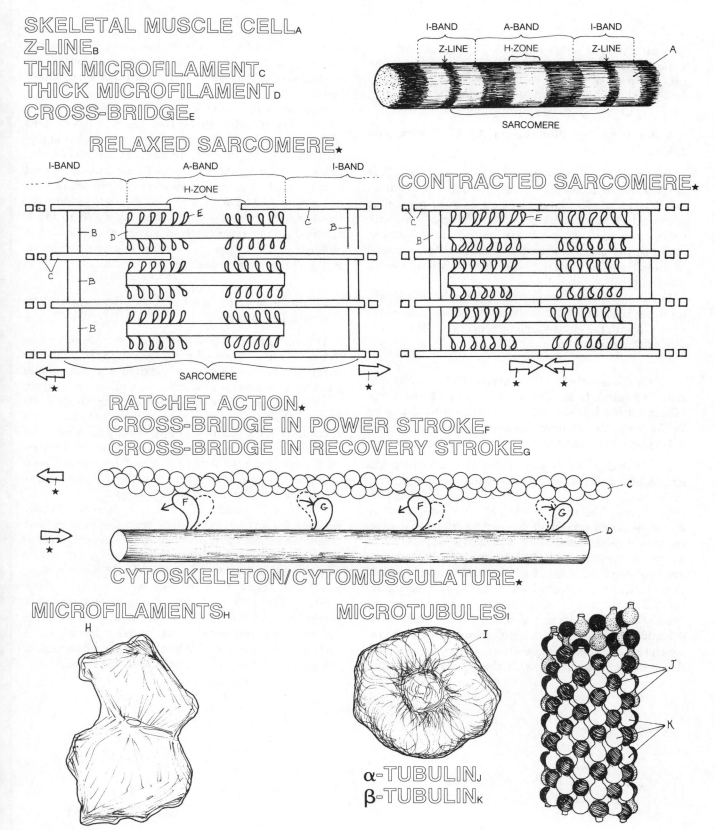

The earliest discoveries with the microscope included some single-celled creatures that propel themselves by cilia or flagella (Latin: *cilium,* "eyelash"; *flagellum,* "whip").

Color titles and illustrations A and B with pale colors.

Paramecium has about 2000 cilia on its single cell. Cilia are also used for locomotion in small multicellular animals and in large ones for movement of materials in tubes of the reproductive and respiratory tracts. Whereas cilia are short and numerous, flagella are long and usually single, as in *Euglena* or in the male reproductive cell (sperm cell) of a multicellular animal. Some algae cells have two or more flagella.

For all the differences apparent in the light microscope, however, cilia and flagella appear to be identical when observed with the electron microscope, except for the obvious difference in length.

Color the headings Microtubules and Doublet and titles C through K. Color each part of the large cilium on the left side of the plate as it is discussed in the text. Use contrasting colors for C and D and a very bright color for J.

In the electron microscope it becomes clear that cilia and flagella are composed of microtubules arranged in the same precise pattern. In the movable portion, which extends beyond the body of the cell, there are nine microtubule doublets around the perimeter. Each doublet consists of two subtubules, an *A subtubule* (slightly closer to the center) and an attached *B subtubule.* In the center are two single microtubules, commonly called *singlets.* (This pattern is commonly called the "9 + 2" configuration.) The base of the cilium or flagellum (known as the basal body), which anchors it within the cell, has a "9 + 0" configuration: nine microtubule triplets around the perimeter and no central singlets. Each triplet consists of an A subtubule (distinctly closer to the center) and attached

B and *C subtubules.* The C subtubules do not extend past the body of the cell, but the A and B subtubules are continuous from the basal body to the tip of the cilium or flagellum.

Detailed studies with the electron microscope have disclosed filaments called *spokes* that join each A subtubule to *sheath elements* surrounding the central singlets at regular intervals. In addition, filaments of a protein called *nexin* join each A subtubule to the B subtubule of the next doublet. Each A subtubule also has two projecting arms of a protein called *dynein,* which is believed to produce the movement by reaching up or down to attach to the B subtubule of the next doublet at a different level and then pulling the tubules past one another in a manner similar to the sliding of microfilaments in muscle.

Color the heading Mechanism of Bending and the related illustrations.

This section shows how a cilium or flagellum is made to bend. Suppose that a cilium is to bend to the right. The dynein arms of the doublets on the right side of the cilium reach up and attach to the adjacent doublets at a higher level. At the same time, the dynein arms of the doublets on the left side (not shown) reach down and attach at a lower level. Then the dynein arms contract, causing the doublets on the right to move up relative to the doublets on the left. Since the doublets are anchored in the basal body, the only way they can move relative to one another is to force the cilium to bend to the right.

The "9 + 0" pattern of the basal body is also exactly what we find in centrioles, which were discussed in Plate 30 and will come up again when we discuss cell division in animals (Plate 65). Originally it was a great mystery why these two different organelles should have identical structures. Today we know that when a cell needs to produce more cilia or flagella, the centrioles reproduce themselves, and the "daughter" centrioles migrate to the cell membrane and become basal bodies, extending the A and B subtubules of each triplet to push the cell membrane out and form the rest of the cilium or flagellum.

CILIA AND FLAGELLA.

MICROTUBULES⋆
　DOUBLET⋆
　　SUBTUBULE A$_C$
　　SUBTUBULE B$_D$
　SINGLET$_E$
　SUBTUBULE C$_F$
SPOKE$_G$
SHEATH ELEMENT$_H$
NEXIN$_I$
DYNEIN$_J$
CELL MEMBRANE$_K$

CELLS WITH CILIA$_A$
CELLS WITH A FLAGELLUM$_B$

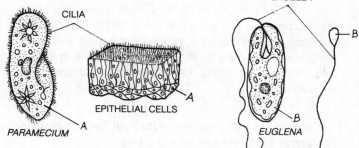

CILIA

A

PARAMECIUM

EPITHELIAL CELLS

A

FLAGELLA

B

B

EUGLENA

HUMAN SPERM CELL

MECHANISM OF BENDING⋆

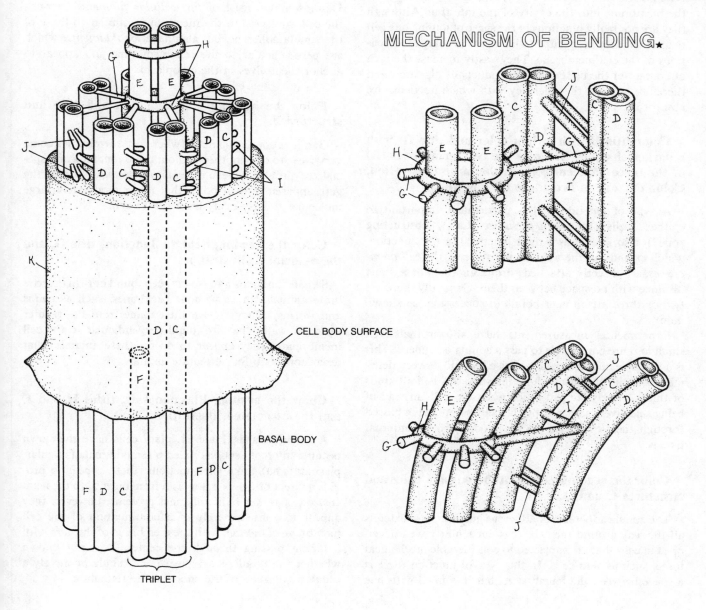

CELL BODY SURFACE

BASAL BODY

TRIPLET

Although a few cells move about entirely independently, most cells of a multicellular organism must have some way of remaining attached to other cells. The electron microscope has revealed a variety of complex structures for this purpose. Nearly all of them are found in epithelial cells, which line all the various cavities and vessels of animals.

Color title A and the intestinal epithelium cells in the upper right corner of the plate.

The epithelial cells lining the intestine show structures called microvilli, which are tiny, fingerlike projections of the membrane into the cavity of the intestine. Although they are not really junctions in the sense of joining one cell to another, they are related to junctions in being modifications of the cell membrane. They vastly increase the area of contact of the cell with the products of digestion and therefore increase the efficiency with which nutrients are absorbed.

Color title B with a light color and title B¹ with a darker shade of B. Color the cell membranes (B) of the three adjacent cells in the large illustration. Color the zonula occludens (B¹) as well.

A type of junction that is particularly important to epithelial cells is the *zonula occludens* (Latin, "obstructing zone"), also known as a tight junction. In this junction, which extends all the way around the cell, the *cell membranes* are directly attached to one another for a short distance with no space between them. Generally there are two or three attachment points in succession, as shown here.

Experimental measurements have shown that even small ions are barely able to pass a zonula occludens. This is a valuable characteristic because it could be very detrimental to the organism whose intestine this is if all kinds of things merely slipped by between the cells instead of being selectively screened by the cells before being passed through their cytoplasm to the blood or other internal tissues.

Color the heading Zonula Adherens and titles and structures C and D.

The zonula adherens, or adherens junction, also extends all the way around the cell. It is particularly well developed in cells that are subjected to considerable mechanical force, such as skin cells. In this type of junction there is a space between the membranes, but it is filled with fine filaments (called *intercellular filaments*, since they are between the cells), which appear to act as a sort of cement to hold the cells together. There are additional filaments within the cell *(intracellular filaments)*, which form a band around the inside of the cell.

Color the heading Desmosome and titles and structures E and E¹.

Desmosomes provide reinforcement of connections between cells where strength is needed, but not as much strength as a zonula adherens. Very often they occur as spot reinforcements in addition to a zonula adherens. Desmosomes consist of *intercellular filaments* between the cells and next to the membranes, similar to those of the zonula adherens, but also have *tonofilaments*, which are perpendicular to the cell membrane and appear to anchor themselves in the cytoplasm.

Color the heading Gap Junction and title and structure F.

Gap junctions are patches where cell membranes come very close to one another without quite touching and specialized protein molecules *(channel proteins)* bridge the gap, apparently with pores that allow moderately large molecules to pass easily from cell to cell.

Color the heading Septate Junction, title G, and the associated illustration.

Septate junctions have so far been found only in invertebrate animals (those without backbones, such as clams and worms). They receive their name from the regular *septa*, or walls, that are found perpendicular to the cell membranes. They appear to do for these animals what desmosomes do for vertebrate animals.

Color the heading Plasmodesma, titles H and I, and the associated illustration.

As mentioned in Plate 31, plant cells have their own peculiar interconnections called plasmodesmata (singular, plasmodesma). Like gap junctions, these appear to provide a direct bridge of cytoplasm from one cell to the next. Instead of rather bulky channel proteins, however, they appear to consist merely of a continuation of the cell membrane of one cell with the membrane of the next with a *tubule* passing through the center. It is not known whether the tubule is a typical microtubule or merely a tubular extension of the endoplasmic reticulum.

CELL JUNCTIONS.

INTESTINAL EPITHELIUM CELLS_A
CELL MEMBRANE_B
ZONULA OCCLUDENS_{B¹}
ZONULA ADHERENS★
INTERCELLULAR FILAMENTS_C
INTRACELLULAR FILAMENTS_D
DESMOSOME★
INTERCELLULAR FILAMENTS_E
TONOFILAMENTS_{E¹}
GAP JUNCTION★
CHANNEL PROTEINS_F

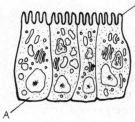

MICROVILLI

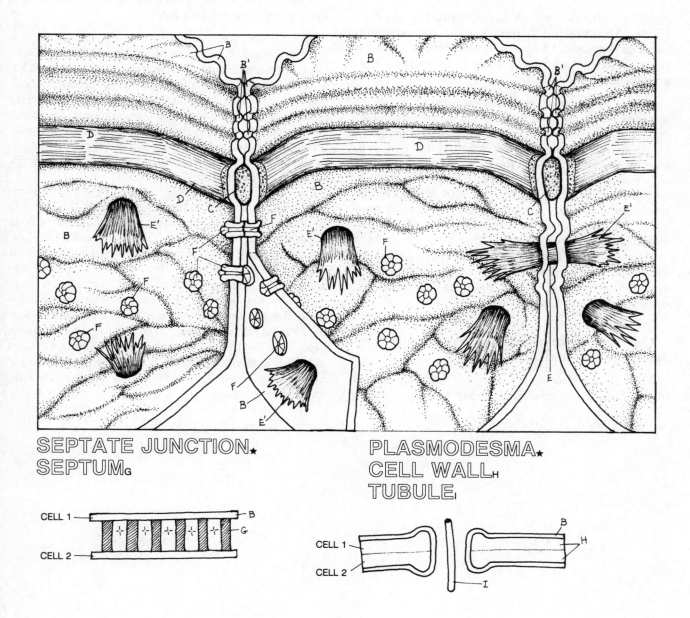

SEPTATE JUNCTION★
SEPTUM_G

PLASMODESMA★
CELL WALL_H
TUBULE_I

CELL 1
CELL 2
B
G

CELL 1
CELL 2
B
H
I

FLOW OF ENERGY AND MATTER IN THE BIOSPHERE

One of the requirements of life is a constant flow of energy. Life involves activity, and activity requires energy. If the supply of energy stops, life stops. A constant flow of matter is also necessary, since matter is intimately involved in trapping energy and transporting it from one place to another within the living organism or from one organism to another. The food we eat, for example, consists of matter organized as carbohydrates, proteins, and lipids. These molecules contain usable energy, but when the same atoms are combined as carbon dioxide and water, they contain virtually no usable energy.

Color titles A through I, including the headings Carbon Dioxide and Water, and the corresponding parts of the plate. Choose a light color for C. Leave the oxygen and carbon dioxide next to the rabbit uncolored for now.

All the life processes on earth obtain their energy directly or indirectly from the *sun*. *Plants* absorb *light energy* and convert it into chemical energy in the process of photosynthesis. Plants conduct photosynthesis by combining carbon dioxide from the air with water and minerals taken up from the soil to make *carbohydrates, proteins,* and *lipids.* In the daytime, when the plant is photosynthesizing, the *oxygen* of the water molecules is a waste product as far as the plant is concerned, so it releases that oxygen into the atmosphere. The identical process goes on in plants and algae that live in lakes, streams, and oceans, except that they are immersed in water and don't have to depend on soil for it. (At night, photosynthetic organisms use oxygen just as animals do.)

Color title J, the animal, the oxygen it consumes, and the carbon dioxide it releases.

Although *animals* cannot carry out photosynthesis to obtain energy from light directly, they obtain it indirectly by eating plants or eating animals that eat plants (or eating animals that eat animals that eat plants, etc.). To extract the energy from the food they eat, animals must combine the food molecules with oxygen. This process is called oxidation and results in the production of carbon dioxide and water, which are released into the atmosphere when the animal exhales (although some of the water may be excreted in liquid or semisolid form). Plants carry on oxidation also, both to grow and to maintain themselves during the hours of darkness.

Color titles K and K[1] and the arrows representing heat energy gained and lost.

Anyone who has ever been out in the sun knows that the sun radiates *heat* as well as light, and that heat keeps the earth warm enough for living things to survive. What is not so obvious is that even light energy is sooner or later converted to heat. No chemical process is 100 percent efficient, and the reactions of photosynthesis lose a little of the trapped light energy as heat. Much more heat is produced by the oxidation of the products of photosynthesis as a plant grows or as an animal converts them into energy for its own life processes.

Eventually the heat energy received by the earth is radiated away into outer space. If you find this hard to believe, take notice in the winter how much colder it is on a morning following a night of clear skies than it is following a night with a heavy overcast to reflect radiating heat back to the earth. Energy, then, flows through the biosphere—the thin layer of our planet's surface that supports life—and back out into space. Matter, on the other hand, flows in constant cycles, and no significant amount of matter is added to the earth or lost from it. The cyclic flow of carbon from plants to animals and back to plants again is commonly called the "carbon cycle." Many other kinds of matter also flow in cycles, such as water, nitrogen, oxygen, and sulfur.

FLOW OF ENERGY AND MATTER IN THE BIOSPHERE.

SUN_A

LIGHT ENERGY_B

PLANT_C

CARBON DIOXIDE (CO_2)⋆

CARBON ATOM_D

OXYGEN ATOM_E

WATER (H_2O)⋆

HYDROGEN ATOM_F

CARBOHYDRATE_G

PROTEIN_H

LIPID_I

ANIMAL_J

HEAT ENERGY GAINED_K

HEAT ENERGY LOST_{K'}

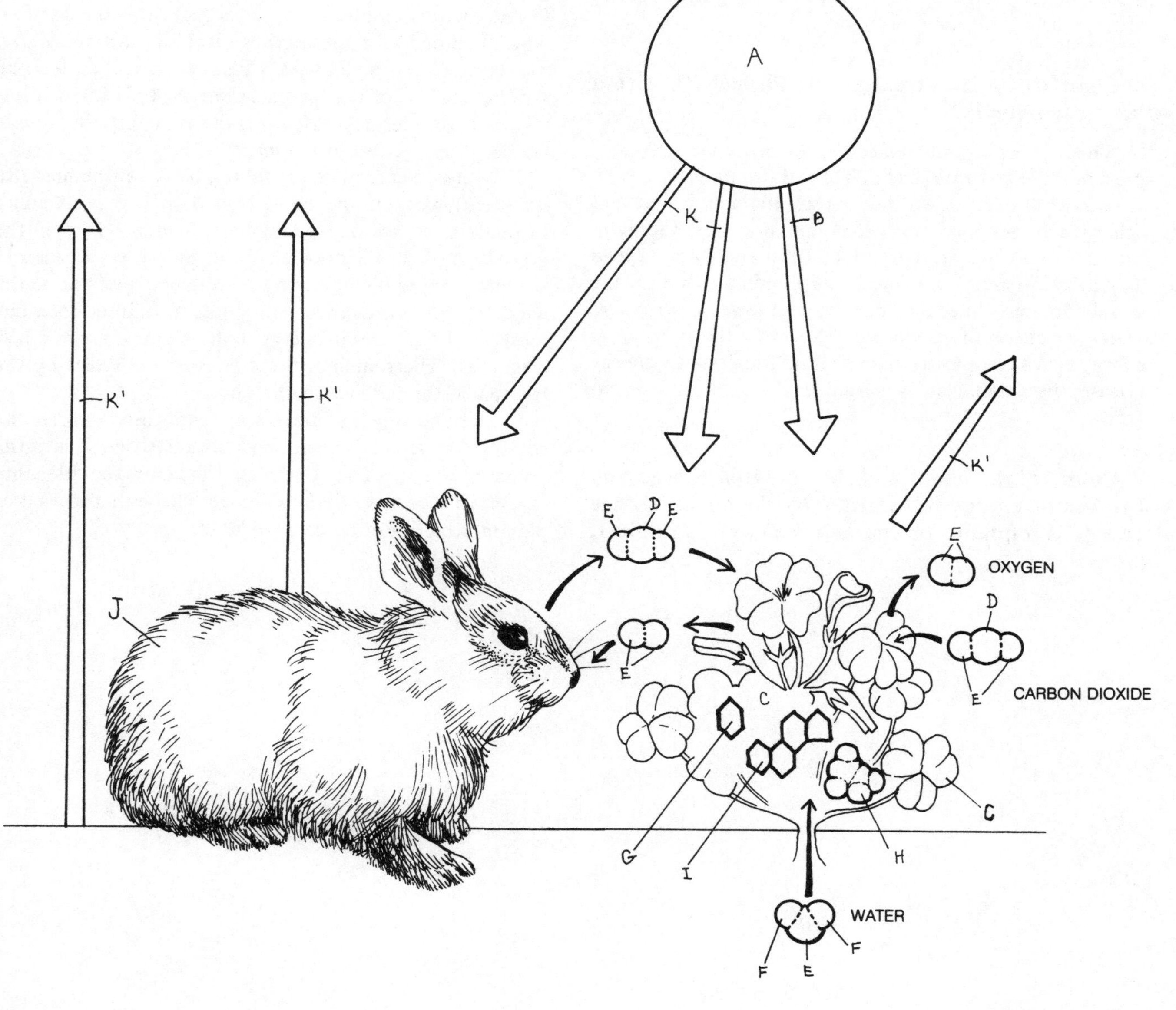

METABOLISM

Metabolism is defined as the sum of all the chemical reactions in a living organism. That is quite a large sum, since every activity we associate with life is powered by chemical reactions, be it movement, growth, or just awareness of a change in the environment. Even if you are not interested in chemistry, you are probably interested in food, and metabolism also includes what happens to food after you eat it.

Metabolism is customarily divided into two parts, catabolism and anabolism. Catabolism includes all the reactions that break large molecules down into smaller ones to trap the energy from them for life processes. Anabolism includes all the reactions that put smaller molecules together to build larger ones to allow a living organism to grow, repair itself, or reproduce.

Color titles and structures A through H. Use a light color for F.

When *simple organic molecules* are absorbed into *cells,* most of them go through the chemical reactions of *catabolism* to be broken down into carbon dioxide, water, and other small *inorganic molecules,* releasing their energy in the process. A moderate fraction of this energy is trapped as *chemical energy* and used to recombine some of the simple organic molecules into *complex organic molecules* in the reactions of *anabolism.* However, the transfer of energy is not completely efficient, and much of the energy released by catabolism is wasted in the form of *heat energy.*

Color titles I and J and the potential energy and the kinetic energy represented by the books. (Since energy is a quality or characteristic, not an object,

we really can't color it, so we color the books containing it.)

The chemical energy in an organic molecule is often referred to as potential energy, which is similar to the *potential energy* in a book being held out the window. The energy isn't doing anything yet, so it is merely potential. When the book falls or the chemical reactions of catabolism occur, the energy is called *kinetic energy* (energy of motion). As the book falls, some of the kinetic energy is transferred to the surrounding air molecules, speeding them up a little, which is another way of saying that the air molecules are heated a little. When the book hits the ground, the kinetic energy of the book as a whole is transferred to those molecules of the book and the ground, which collided with one another. (If this is hard to believe, drive a few large nails with a hammer and then feel the striking surface of the hammerhead. You will find it has warmed considerably.) Heat energy is simply the kinetic energy of molecules in motion.

The kinetic energy of the falling book can be made to do useful work if the book is tied to a rope running through a pulley to a somewhat lighter book on the ground. If the heavier book is dropped, the kinetic energy is transferred to the lighter book, which is raised up to the window. When the movement stops, the lighter book has most of the potential energy that the heavy book had originally. The remainder was converted to heat by the friction of the pulley and the air.

In a similar way, kinetic energy that is used to drive the reactions of anabolism becomes converted back into the potential energy of the complex organic molecules being made. That energy will be released whenever those complex molecules are broken down.

METABOLISM.

SIMPLE ORGANIC
   MOLECULES$_A$
CELL MEMBRANE$_B$
CATABOLISM$_C$
SIMPLE INORGANIC
   MOLECULES$_D$

ANABOLISM$_E$
COMPLEX ORGANIC
   MOLECULES$_F$
CHEMICAL ENERGY$_G$
HEAT ENERGY$_H$

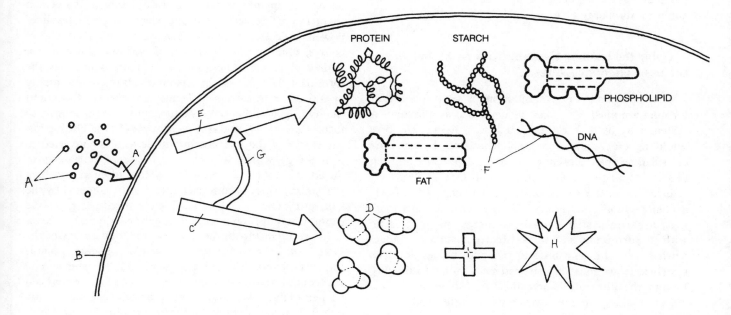

PROTEIN   STARCH

PHOSPHOLIPID

FAT

DNA

POTENTIAL ENERGY OF BOOK$_I$
KINETIC ENERGY OF BOOK$_J$

ENERGY CHANGES

Energy changes have been found to follow certain laws known as the laws of thermodynamics (*thermo,* "heat"; *dynamics,* "movement"). They were given that name because they were first discovered in studies of heat transfer, but they have since been found to be universal laws that apply to all energy changes.

Color the heading System, titles A, B, and C, and the associated structures.

To a scientist, any combination of parts or objects forming a unified whole can be described as a system. A system may be a living one, such as a living *cell,* or a nonliving one, such as a gasoline *engine.* Systems are classified as open, closed, or isolated, depending on their relation to all the other things around them, collectively referred to as the *surroundings.* An open system freely exchanges both matter and energy with its surroundings, a closed system exchanges energy but not matter with its surroundings, and an isolated system exchanges neither. Living organisms and gasoline engines are open systems, as are most of the systems familiar to us. (Although the illustration here does not show any actual objects making up the surroundings, water and air are important parts of the surroundings for cells and gasoline engines.)

Color title and quotation D, the headings Energy of System and Energy of Surroundings, and titles and structures E through F¹.

The *first law of thermodynamics* is most simply stated as "Energy is conserved." (It is sometimes called the law of conservation of energy.) This means that for any system, the total amount of energy remains unchanged, even though the energy may change form and place. In any chemical reaction (or any physical event, such as the falling of the book described in the preceding plate), the sum of the energy of the system and the energy of the surroundings before the reaction is equal to the sum of the energy of the system and the energy of the surroundings after the reaction.

Color title and quotation G; then color the remainder of the plate.

The *second law of thermodynamics* is most concisely stated as "Total entropy always increases." In the simplest terms, entropy means disorder or randomness. If a person living in a wilderness area wants electric light, a gasoline-powered electric *generator* can be hauled in, with, of course, some *gasoline* to run it. Gasoline is a mixture of hydrocarbon molecules, such as octane, which has the formula C_8H_{18}. When the gasoline is burned, the highly organized hydrocarbon molecules are broken down, and the hydrogen and carbon atoms combine with *oxygen* to form *water* (H_2O) and *carbon dioxide* (CO_2). The chemical energy of the hydrocarbon molecules is converted to heat energy, which causes the expansion of the gases produced. This is converted into mechanical energy to turn the engine, which is converted into electrical energy by the generator. The *wires* conduct the electrical energy to the cabin, where it is converted into light energy in the *lamp.*

In accordance with the first law, the total energy of the system plus the surroundings—gasoline, engine, generator, wires, lamp plus the atmosphere, soil, the cabin and its contents, and so on—is the same after the gasoline has been burned to light the lamp as it was before. The second law tells us that the final state of the system plus the surroundings must have greater entropy—that is, must be more disorderly—than the initial state. Not only are the hydrogen, carbon, and oxygen atoms more disorderly as carbon dioxide and water than they were as gasoline and oxygen gas in the air, but a great deal of additional disorder has been created by heating up a lot of the molecules in the engine, generator, wires, cabin, and the atmosphere, making them move faster and farther apart from one another.

Some time ago it was thought that the second law of thermodynamics did not apply apply to living things. After all, they maintain a high level of orderliness and they even increase orderliness by producing more living things. Today we know that the second law does apply to living things, as do all the laws of physics and chemistry, and living things increase their own orderliness only by increasing the disorderliness of their surroundings, just as the gasoline engine increased the disorderliness of the atoms put into its gas tank.

ENERGY CHANGES.

SYSTEM.★
  CELL_A
    ENGINE_B
SURROUNDINGS_C

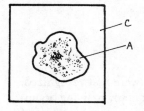

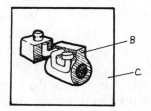

FIRST LAW OF THERMODYNAMICS_D

"ENERGY IS CONSERVED"

ENERGY OF SYSTEM.★   ENERGY OF SURROUNDINGS.★
BEFORE REACTION_E   BEFORE REACTION_F
AFTER REACTION_{E^1}   AFTER REACTION_{F^1}

SECOND LAW OF THERMODYNAMICS_G

"TOTAL ENTROPY ALWAYS INCREASES"

GASOLINE_H   GENERATOR_L
OXYGEN_I   WIRING_M
CARBON DIOXIDE_J   LAMP_N
WATER_K

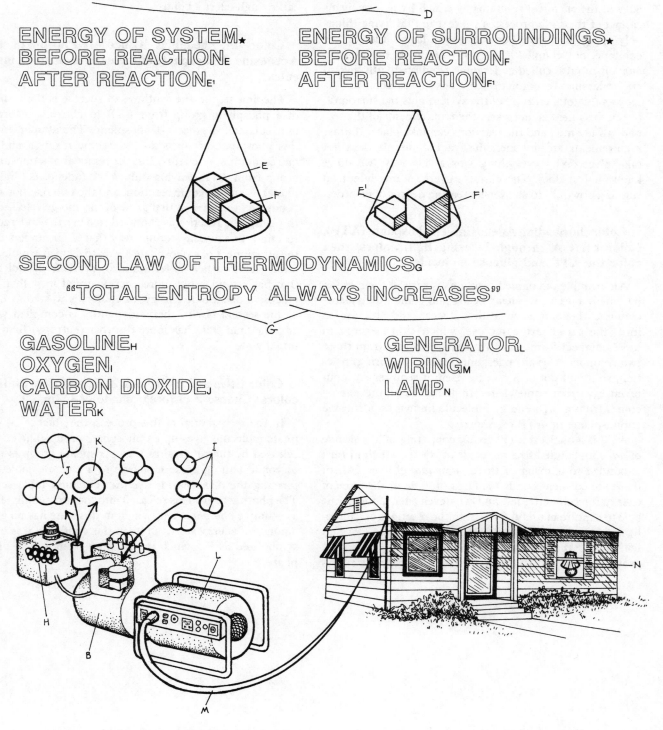

When the laws of thermodynamics were first recognized, many people thought they could not apply to living organisms. After all, anything that is alive is going to a state of greater order, not disorder, as it grows, learns, and produces more living organisms like itself from the disordered molecules in its surroundings. But with improved knowledge of the chemical reactions by which living organisms carry out their life processes we realize that living things only appear to "cheat" on the second law; actually, they combine, or "couple," the energy-requiring reactions of anabolism (also called endergonic reactions) with the energy-releasing (exergonic) reactions of catabolism. As long as the total entropy of the system plus the surrounding environment is increased, the requirements of the second law are met, and the reactions can take place. If those requirements are not met, the reaction simply does not take place. Why everything works this way, we don't know; it just does. The entire universe seems intent on running downhill to an ever-increasing degree of disorder.

Color the heading Adenosine Triphosphate (ATP). Color titles A through D using light colors; then color the ATP and glucose molecules.

An example of coupled reactions is seen in the method by which sugarcane, sugar beets, and some other plants combine the monosaccharides glucose and fructose to form the disaccharide sucrose, which stores energy in more compact form. The synthesis of sucrose from those two components is an endergonic (energy-requiring) reaction, so it will not occur all by itself; it requires an input of energy from somewhere. In living cells, the energy comes from a high-energy molecule known as adenosine triphosphate, or ATP for short.

ATP is a nucleotide (Plate 22) consisting of a molecule of *adenine* bonded to a molecule of *ribose*, which in turn is bonded to a string of three *phosphate groups*. (All of these are covalent bonds.) ATP is the principal source of energy for cell activities, and whatever energy the cell is able to capture in useful form from the reactions of catabolism is almost always captured in the form of ATP. You will notice that the bonds joining the two phosphate groups on the end to the rest of the molecule are repre-

sented by wavy lines rather than straight ones. This is to indicate that a larger than average amount of energy is associated with those bonds. It takes quite a bit of energy to attach each of these phosphate groups to the rest of the molecule, and quite a bit is released when each of these phosphate groups is removed. As a consequence, they are called high-energy bonds.

Color the headings Glucose-1-Phosphate and Adenosine Diphosphate and the corresponding molecules.

The first step in the synthesis of sucrose is the transfer of a phosphate group from ATP to *glucose* to form a molecule called glucose-1-phosphate. The attachment of this phosphate group is an endergonic reaction and requires an input of energy. But the removal of a phosphate group from ATP, leaving only ADP (adenosine diphosphate), is an exergonic reaction, and the two reactions are "coupled" together so that some of the energy released by the exergonic reaction is captured to drive the endergonic reaction. In some mysterious way that is not understood, the attachment of a phosphate group to a glucose molecule (and various other molecules also) produces a molecule that has considerably more energy stored in it than the glucose alone. Some energy, of course, is wasted as heat, so the second law of thermodynamics is complied with, and the final state has more disorder (entropy) than the initial state.

Color titles and molecules E, F, and C^1 with light colors. Choose a different shade of C for C^1.

In the second step of this process, the glucose-1-phosphate molecule gives up its phosphate group. The energy released by the removal of that phosphate group is used to connect the glucose molecule to a *fructose* molecule, forming the disaccharide *sucrose*, common table sugar. The phosphate group is released into the surrounding fluid as a simple *phosphate ion*, which in this state has no great amount of energy stored in it. In this condition it is often symbolized as P_i, which means simply "inorganic phosphate."

COUPLED REACTIONS.

ADENOSINE TRIPHOSPHATE (ATP)★
  ADENINE A
  RIBOSE B
  PHOSPHATE GROUP C
GLUCOSE D
GLUCOSE-1-PHOSPHATE ★
ADENOSINE DIPHOSPHATE (ADP) ★
FRUCTOSE E
SUCROSE F
PHOSPHATE ION C'

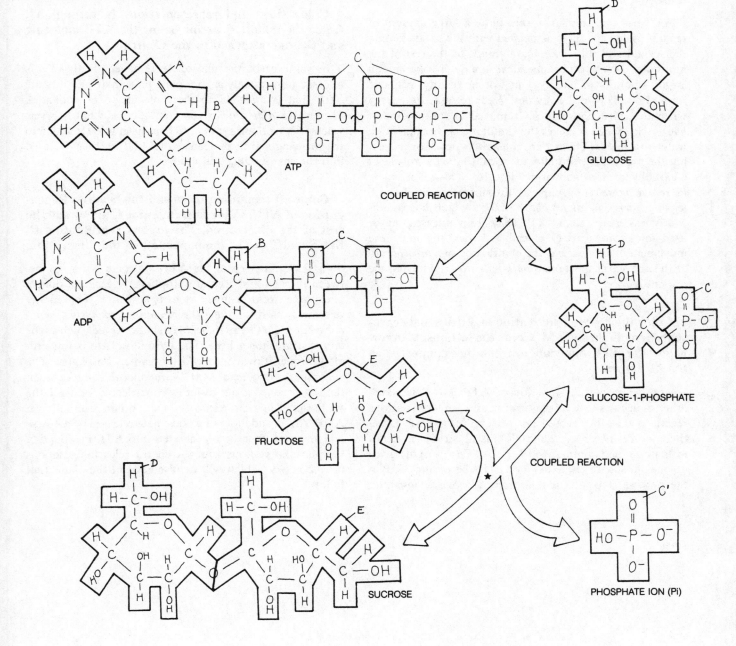

ATP AND ENERGY TRANSFER

Adenosine triphosphate (ATP) is commonly called the "energy currency" of the cell because it is used as a sort of intracellular money to "pay" for nearly every energy-requiring activity cells carry out.

Color the heading Adenosine Triphosphate (ATP) and titles A through C. Then color the heading Movement/Active Transport and titles D through J. Color the upper part of the illustration, leaving C for later.

ATP has the peculiarity that quite a large amount of readily usable energy is associated with the bonds joining the second and third *phosphate groups* to the rest of the molecule. Many other chemical bonds involve much larger amounts of energy, but not in readily available form. For many cell activities, ATP releases its energy when the third phosphate is removed in the process of *hydrolysis,* which involves the simultaneous breaking of a water molecule. (The water molecule is not shown here, but the process is essentially identical to the hydrolysis of a carbohydrate, as shown in Plate 16.) This releases one *inorganic phosphate* group into the surrounding cell fluid, leaving *adenosine diphosphate (ADP),* which has significantly less energy than ATP. The energy released can be used for a wide variety of cell activities, including the movement of a *flagellum,* the contraction of a *muscle cell,* or the active transport of a *small molecule* by a membrane protein *(carrier).*

Color the heading Activation and titles and representations K through M. Color the leftmost C arrow below the ATP molecule and the one connected to the M arrow.

Another important function of ATP is to "activate" other molecules so that they will readily undergo various reactions that they would not undergo otherwise. Often this involves *phosphorylation* (K), as illustrated in detail in Plate 46 and summarized here. ATP (top right) gives up one phosphate group to *glucose* (L), becoming ADP in the process (top left) and forming *glucose-1-phosphate.*

Most of the energy released by the removal of the phosphate from ATP is captured in the bond joining the phosphate group to glucose. This energy makes it possible to use the glucose to *synthesize sucrose* or various other glucose-containing molecules important to the cell. The synthesis requires energy, which comes from breaking the bond between glucose and the phosphate group. The phosphate is released into the surrounding cell fluid as a low-energy inorganic phosphate group (P_i).

Color titles and representations N through O. Color the middle C arrow below the ATP molecule and the one attached to the O arrow.

In some cases the phosphate removed from ATP is released immediately as inorganic phosphate and it is the *ADP* that is *attached* to the molecule to be activated, forming an *ADP-glucose complex.* This is what happens when cells *synthesize starch.* When the glucose is added to the growing starch molecule, the ADP is then released into the surrounding cell fluid.

Color all remaining titles and the heading Regeneration of ATP. Color the remaining C arrow and the rest of the illustration. The arrows representing titles S and T are at the upper left of the illustration.

In some other cases the ATP molecule is broken between the first and second phosphate, releasing a double phosphate group known as *pyrophosphate.* The energy released is used to *attach* the remaining *adenosine monophosphate (AMP)* to a molecule to be activated, such as the *amino acid* shown here. The pyrophosphate is immediately broken down into two inorganic phosphates. The activated amino acid *(AMP–amino acid complex)* can then be attached to a transfer RNA molecule for use in the synthesis of new protein or in some other reaction. The ADP, AMP, and inorganic phosphate released by all these reactions can then be regenerated into ATP by coupling the phosphorylations with the energy-releasing reactions of *catabolism,* which will be discussed in the plates that follow.

ATP AND ENERGY TRANSFER.

ADENOSINE TRIPHOSPHATE
   (ATP)★
   ADENOSINE$_A$
   PHOSPHATE GROUP$_{B^1+B^2+B^3}$
DIFFUSION$_C$
MOVEMENT/ACTIVE
   TRANSPORT★
HYDROLYSIS$_D$
ADENOSINE DIPHOSPHATE
   (ADP)$_{A+B^1+B^2}$
INORGANIC PHOSPHATE
   (Pi)$_E$
ENERGY$_F$
FLAGELLUM$_G$
MUSCLE CELL$_H$
CARRIER$_I$
SMALL MOLECULE$_J$
ACTIVATION★
PHOSPHORYLATION$_K$
GLUCOSE$_L$
GLUCOSE-1-PHOSPHATE$_{L+B^3}$
TO SUCROSE SYNTHESIS$_M$
ADP ATTACHMENT$_N$
ADP-GLUCOSE
   COMPLEX$_{A+B^1+B^2+L}$
TO STARCH SYNTHESIS$_O$
ADENOSINE
   MONOPHOSPHATE
   (AMP) ATTACHMENT$_P$
AMINO ACID$_Q$
PYROPHOSPHATE$_{B^2+B^3}$
AMP–AMINO ACID
   COMPLEX$_{A+B^1+Q}$
TO TRANSFER RNA$_R$
REGENERATION OF ATP★
CATABOLIC REACTIONS$_S$
REPHOSPHORYLATION$_T$

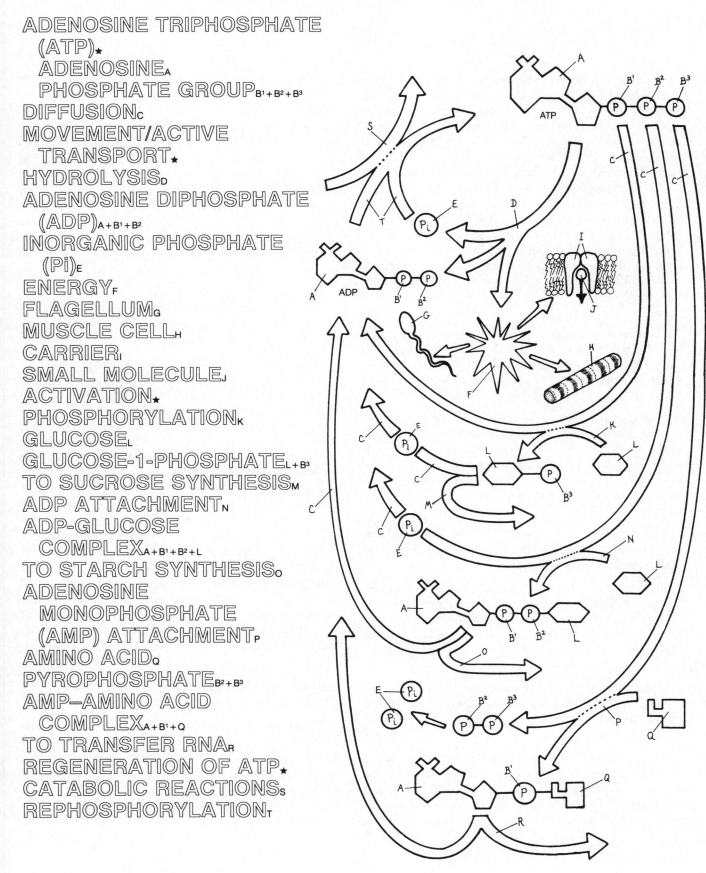

Most of the important energy changes occurring in living organisms involve types of chemical reactions known as oxidations and reductions. Originally, the term "oxidation" was confined to reactions where some substance combined with oxygen to form a chemical compound. "Reduction" was the term used for reactions in which the compound was broken down to remove the oxygen. As the science of chemistry progressed, it became clear that other kinds of chemical reactions were similar enough to be put in the same category as oxidation. Unfortunately, the early chemists did not invent a new term to cover the expanded meaning but continued to use the term "oxidation," even if oxygen was not involved. In this plate we will examine just what is involved in oxidation and reduction reactions.

Color the headings Combustion, Carbon Dioxide, and Water Vapor. Color titles and structures A through E. You may want to use the standard colors for the atoms: red for oxygen, black for carbon, and white or yellow for hydrogen.

One familiar example of oxidation is the process we call combustion. Combustion is merely *oxidation* that is rapid enough to release a lot of heat and light at once. Although *wood* contains many different chemical compounds, much of it is cellulose, and cellulose combines with *oxygen* to form carbon dioxide and water vapor.

Color the heading "Rusting," titles F and G, and the associated illustration, including the chemical equation.

If you hang a shiny, new horseshoe out in the weather, it gradually undergoes a slow oxidation to form rust *(iron oxide),* which is a compound of two atoms of iron and three atoms of oxygen, Fe_2O_3. The oxidation occurs slowly, so we don't notice the release of the heat energy.

Color the heading Electron Removal, titles H through L, and the associated illustration. Use a light color for H. Choose the color for K to contrast with C.

The reaction of metals like iron with acids was one of the kinds of reactions that led chemists to expand the definition of oxidation to include cases where no oxygen was involved. In rusting, iron transfers three electrons to oxygen but remains attached to the oxygen. In an *acid,*

iron also *transfers* three *electrons,* but it transfers them up to the *hydrogen ions* of the acid. The hydrogen ions, instead of combining with the iron to form a compound, accept the electrons to become *reduced* to hydrogen atoms, which then pair up to form hydrogen molecules and bubble out of the solution as hydrogen gas. (The upward-pointing arrow after H_2 is the chemist's way of indicating that the hydrogen is in the gas state.) Each iron atom then has three more protons than it has electrons to balance them, so it is then known as *ferric ion* (from the Latin name for iron, *ferrum*) and is symbolized as Fe^{+++}. The previously insoluble iron is now very soluble as ferric ion, so the iron seems to disappear.

Color the heading Hydrogen Removal and titles M, N, and O, using pale colors for M and N. Color the associated illustration.

In the cells of living organisms, oxidation usually involves the removal of electrons belonging to hydrogen atoms; in many cases the proton (hydrogen has only one) goes along with the electron so that the entire hydrogen atom is removed. This portion of the plate illustrates one of the oxidations that occur in mitochondria when energy is extracted from molecules in food. Here a *succinate ion* (the negative ion of succinic acid) is oxidized by removal of a hydrogen atom from each of the two carbon atoms in the center, resulting in the formation of a *fumarate ion* (the negative ion of fumaric acid), which has two hydrogen atoms fewer and a double bond (two pairs of shared electrons) between the two center carbon atoms where there was only a single bond before. The two removed electrons (along with the two removed protons) are picked up by an *electron acceptor molecule,* which thereby becomes reduced. You should note that the oxidation of one substance always requires that something else be reduced. (In combustion and rusting, oxygen is reduced.)

Color the rectangles enclosing the tables at the bottom of the plate.

The tables summarize the definitions of oxidation and reduction. Oxidations release energy and consist of combining with oxygen, losing one or more electrons, or losing one or more whole hydrogen atoms. Reductions take up energy and consist of separation from oxygen, gaining one or more electrons, or gaining one or more whole hydrogen atoms.

OXIDATION AND REDUCTION.

COMBUSTION.★
WOOD$_A$
OXYGEN$_B$
OXIDATION (FLAME)$_C$
CARBON DIOXIDE.★
  CARBON$_D$
WATER VAPOR.★
  HYDROGEN$_E$

RUSTING.★
IRON$_F$
IRON OXIDE$_G$

ELECTRON REMOVAL.★
ACID$_H$
ELECTRON TRANSFER$_I$
HYDROGEN ION$_J$
REDUCTION$_K$
FERRIC ION$_L$

HYDROGEN REMOVAL.★
SUCCINATE ION$_M$

FUMARATE ION$_N$
ELECTRON ACCEPTOR
  MOLECULE$_O$

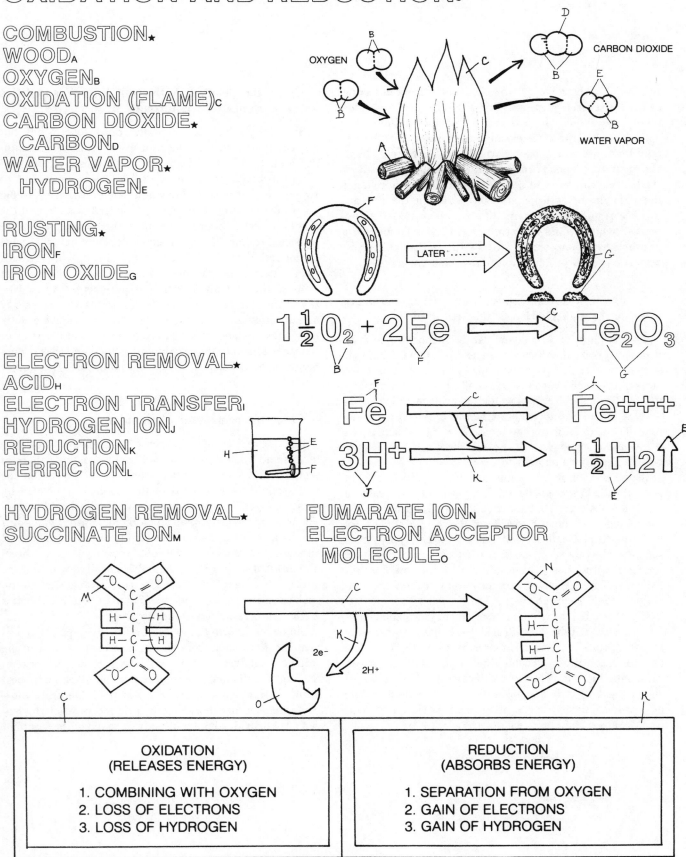

OXYGEN

CARBON DIOXIDE

WATER VAPOR

LATER ·······

$$1\tfrac{1}{2}O_2 + 2Fe \longrightarrow Fe_2O_3$$

$$Fe \qquad Fe^{+++}$$

$$3H^+ \qquad 1\tfrac{1}{2}H_2$$

$2e^-$

$2H^+$

OXIDATION (RELEASES ENERGY)	REDUCTION (ABSORBS ENERGY)
1. COMBINING WITH OXYGEN	1. SEPARATION FROM OXYGEN
2. LOSS OF ELECTRONS	2. GAIN OF ELECTRONS
3. LOSS OF HYDROGEN	3. GAIN OF HYDROGEN

In obtaining their energy from chemical reactions, living cells face several problems. For one, chemical reactions take place only if the atoms or molecules collide hard enough and in the right orientation to react. But the energy for those collisions is normally in the form of heat, and many reactions require so much energy to get them started that the cell would cook itself to death before it got hot enough. In addition, the molecules in the cell are capable of undergoing many different chemical reactions, most of which would be harmful to the cell. To make the desirable reactions occur without the undesirable ones, cells use special protein molecules called enzymes.

Color the heading Catalysis and titles and structures A through D. Use a light color for A.

The function of an enzyme can be demonstrated with *hydrogen peroxide,* which is used as a disinfectant and hair bleach. It is made up of two atoms of hydrogen and two atoms of oxygen (H_2O_2). It gradually breaks down all by itself to form a molecule of water and an atom of hydrogen—releasing a few bubbles of *oxygen* as in the leftmost test tube. The activation energy required to start this reaction is low enough that even at cool temperatures a few molecules collide hard enough to react.

If you drop in a few grains of a black inorganic compound called *manganese dioxide (MnO_2),* the rate of breakdown is greatly increased, and bubbles form rapidly. The manganese dioxide speeds up the reaction but is not itself used up. Chemists call this action catalysis, and a substance having such an effect is called a catalyst. Since the temperature hasn't been raised, the molecules haven't gained energy; somehow the catalyst has lowered the activation energy requirement.

If you add an equal quantity of an enzyme called *catalase* to a third tube of hydrogen peroxide, the peroxide breaks down so fast that it foams all over. Enzymes are usually much more efficient than inorganic catalysts in catalyzing the same reaction. Enzymes are also very specific, catalyzing only one, or at most two, reactions, while inorganic catalysts such as manganese dioxide may catalyze dozens of different reactions.

Color the heading Activation Energy and titles and representations E through H.

Suppose you needed a boulder rolled down a hill, but a small mound blocked the path. By brute force, the boulder could be rolled up the mound until it reached the point where it would continue by itself. This is equivalent to supplying *heat* to start a reaction. An alternative is to dig away at the mound, creating a downward slope so that the potential energy already present in the boulder will cause it to start rolling. This is the equivalent of what an *enzyme* (or any other catalyst) does. Instead of supplying the activation energy by the brute force of heat, the catalyst reduces the activation energy requirement *(energy barrier),* and the reaction can be started with the energy already present in the molecules. No one knows exactly how enzymes and other catalysts do this, but it is clear that they attach to the reacting molecules, and they probably steer the molecules into specific collisions of particular atoms.

Color the remainder of the plate.

Enzymes are easily "denatured" (inactivated) by heavy metals and other poisons as well as by drastic changes in temperature or pH. The two graphs at the bottom of the plate show how the speed of the reaction is affected by these changes. A typical *enzyme* causes a particular reaction to go much faster than an *inorganic catalyst* at moderate temperatures, but if the temperature is increased enough, the weak bonds that hold the enzyme in its three-dimensional shape (Plates 20 and 21) break down, and the enzyme no longer functions. The inorganic catalyst has no such sensitivity; it continues to function better and better as the temperature continues to rise.

Similarly, enzyme activity is affected by pH. The enzyme *pepsin,* which works in the acid environment of the stomach, catalyzes best at pH 2. *Trypsin,* which works in the slightly alkaline environment of the intestine, is best at pH 9. *Most enzymes* are best near pH 7 (neutral). *Inorganic catalysts* are unaffected by pH, except where hydrogen or hydroxide ions are directly involved in the reaction.

ENZYME PROPERTIES.

CATALYSIS.★
HYDROGEN PEROXIDE (H_2O_2)$_A$
OXYGEN (O_2)$_B$
MANGANESE DIOXIDE (MnO_2)$_C$
CATALASE$_D$

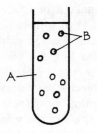

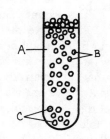

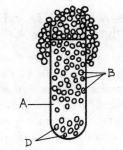

ACTIVATION ENERGY.★
ENERGY BARRIER$_E$
REACTANT MOLECULES$_F$

HEAT$_G$
ENZYME$_H$

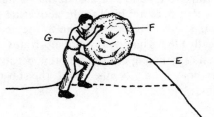

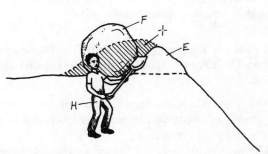

TEMPERATURE EFFECT.★
WITH INORGANIC
  CATALYST$_I$
WITH ENZYME$_J$

pH EFFECT.★
PEPSIN$_K$
TRYPSIN$_L$
MOST ENZYMES$_M$
INORGANIC CATALYSTS$_N$

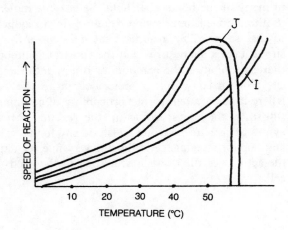

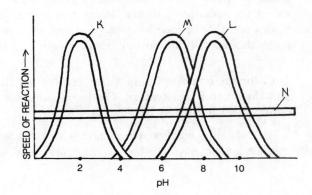

Although we can't actually see how an enzyme works, indirect evidence provides a fair degree of understanding and some interesting hypotheses.

Color titles A through D and the equation at the right. Use a light color for A.

This equation symbolizes the events we believe occur in an enzyme-catalyzed reaction. The *enzyme* binds (attaches) to the molecule or molecules (called the *substrate*) it is going to affect, forming an *enzyme-substrate complex*. Binding of enzyme to substrate is usually by means of the same kinds of relatively weak chemical bonds that hold proteins in their complex three-dimensional folds (Plates 20 and 21). Formation of the complex encourages a particular chemical reaction to take place, after which those weak bonds break, releasing the *product* of the reaction and returning the enzyme to its initial state, free to catalyze another reaction. (The small reverse arrows are reminders that the reaction can occur in the reverse direction, although normally at a much slower rate.)

Color the headings Lock-and-Key Theory and Induced-Fit Theory, title A^1 and the related illustrations.

This plate illustrates hypothetical enzymes catalyzing the joining of two different substrate molecules to form a single molecule (product).

The portion of the enzyme where the substrate binds (called the *active site*) must match the shape of the substrate rather closely. The original version of this idea was the lock-and-key theory, which stated that the active site of the enzyme has a shape that fits its proper substrate as closely as a lock must fit a key. However, modern research evidence suggests that the binding of the substrate induces a change in the shape of the enzyme so that it fits the substrate much more closely after it binds. This idea is called the induced-fit theory. It appears that most enzymes probably operate by induced fit, but there may be a few that actually do operate on a lock-and-key basis.

Color the heading Competitive Inhibition, title E, and the related illustration. Choose a color for E to contrast with B^1 and B^2.

Each chemical reaction within the cell must be carefully regulated. One way in which this occurs is known as competitive inhibition. In this process, the product of an enzyme so closely resembles the substrate of that enzyme or the substrate of an earlier enzyme in a sequential series that it *"competes"* for the active site and acts as an *inhibitor*. When that inhibitor binds to the active site, the enzyme cannot catalyze its usual reaction because it doesn't have the correct substrate. It also does not release the inhibitor right away, so its active site remains blocked for a while. Since the bonds holding the two together are relatively weak ones, they eventually break, and the enzyme is free to bind to a molecule of substrate or another molecule of inhibitor, depending on which one it runs into first. At any given time, the rate of the reaction depends on the relative amounts of substrate and inhibitor.

Color the heading Noncompetitive Inhibition, titles F and A^2, and the related illustration.

Sometimes the action of an enzyme is inhibited by a molecule that does not bind to the active site but binds to a separate site that is close enough to the active site that it physically blocks it and prevents substrate from binding. This is called noncompetitive inhibition because the *inhibitor* does not actually compete with the substrate for the active site.

Color the heading Allosteric Regulation, titles G, H, and A^3, and the related illustrations.

Enzymes are also controlled by allosteric regulation (Greek: *allo,* "other"; *stereos,* "solid"). This kind of regulation can be *negative* (inhibition), slowing down the rate of the reaction, or *positive* (enhancement), speeding the rate. The regulator molecule does not bind to the active site but to a different part of the enzyme molecule called the *allosteric site.* When it binds, it causes changes in many of the bonds that hold the enzyme molecule in its three-dimensional shape, resulting in a readjustment of that shape. That readjustment may cause the active site to take a shape to which the substrate cannot bind, thereby inhibiting the reaction, or it may move an active site from deep within a molecule out to the surface, where it has a much higher probability of colliding with substrate, thereby speeding up the reaction. Many enzymes appear to have one allosteric site for an inhibitor and one for an enhancer, allowing very fine regulation of the activity of the enzyme according to the needs of the cell.

ENYME ACTION.

ENZYME_A
SUBSTRATE_{B,B¹,B²}
ENZYME-SUBSTRATE
  COMPLEX_C
PRODUCT_D

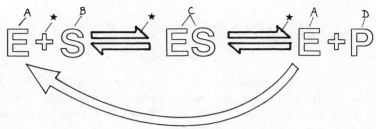

LOCK-AND-KEY THEORY. ★
ACTIVE SITE_{A¹}

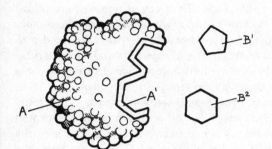

INDUCED-FIT THEORY. ★

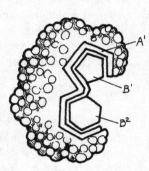

COMPETITIVE INHIBITION. ★
COMPETITIVE INHIBITOR_E

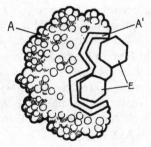

NONCOMPETITIVE INHIBITION. ★
NONCOMPETITIVE INHIBITOR_F
NONCOMPETITIVE INHIBITOR
  BINDING SITE_{A²}

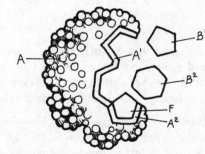

ALLOSTERIC REGULATION. ★
NEGATIVE REGULATOR_G
ALLOSTERIC SITE_{A³}

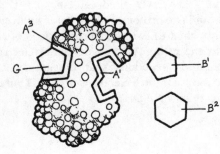

POSITIVE REGULATOR_H

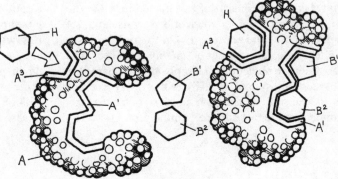

Although all enzymes are proteins, some are unable to act as catalysts unless they have a particular nonprotein group attached to them. Such a group is called a cofactor. It may consist of a single inorganic ion, such as zinc, magnesium, copper, or chloride ion, or of a small organic molecule of 200 or so atoms. (Remember, a protein has thousands of atoms.) Cofactors that are organic molecules are usually called coenzymes. ATP often functions as a cofactor, and many of the vitamins, which animals are unable to make for themselves and must obtain from their diets, are used in the cell as components of essential coenzymes.

Color titles and structures A through A¹ + C¹.

A *cofactor* is an essential part of the *active site*. Without the cofactor, the *substrate* either does not bind to the *enzyme* at all or, if it does bind, the reaction still does not proceed at any significant rate. Cofactors may function to hold the polypeptide chains in the correct shape to form the active site, or they may serve to attract the substrate to it. Some of them also function as electron acceptors or electron donors in oxidation and reduction reactions. The most common ones are held to the protein portion by weak bonds such as hydrogen and hydrophobic bonds (Plates 11, 20, and 21) and freely separate and reattach, carrying electrons to or from the active site.

Color the headings Nicotinamide Adenine Dinucleotide and Oxidized Form (NAD⁺). Color titles and structures D through H in the associated illustration. Use light colors for D through G.

One of the three coenzymes most widely used by cells is nicotinamide adenine dinucleotide (NAD). This one coenzyme is used by many different enzymes as an electron acceptor in reactions that oxidize (remove electrons from) molecules the cell uses as fuel (Plates 53 and 54). NAD receives its name from its structure: it is called a dinucleotide because it consists of two nucleotides. On the left you see nicotinamide nucleotide, consisting of *nicotinamide, ribose,* and *phosphate.* On the right you see adenine nucleotide, consisting of *adenine,* ribose, and phosphate. Adenine nucleotide is simply another name for adenosine monophosphate (AMP), which serves other functions in the cell in addition to that of coenzyme. Nicotinamide is made by adding an amino group ($-NH_2$)

to nicotinic acid, which is better known as the vitamin niacin. (These have no relation to nicotine, in spite of the similarity of the names.)

Color the heading Reduced Form (NADH + H⁺) and titles I through M. Color the associated illustration. Only part of the NAD molecule is shown.

In cellular *oxidations,* the *electrons* removed are almost always the electrons of *hydrogen atoms,* and the *protons* of those hydrogen atoms are almost always removed with them. They are also nearly always removed two at a time. The NAD molecule readily accepts two electrons but only one proton. One electron attaches to the nitrogen atom of the nicotinamide ring (small arrow in the ring), neutralizing its positive charge; the other electron and one proton remain together as a hydrogen atom and attach to the carbon atom at the top of the ring. (Other electrons in the ring shift around so that only two double bonds—pairs of shared electrons—remain instead of three.) The second proton is turned loose in the surrounding watery fluid, where we generally refer to it as a hydrogen ion (H^+). (Remember, a hydrogen ion is a single proton.)

The second widely used coenzyme is the closely related nicotinamide dinucleotide phosphate (NADP), which is not shown here because it has a structure identical to that of NAD except for an additional phosphate group at the point indicated on the drawing. NADP is used by cells as an electron donor for the reduction reactions of biosynthesis, the assembly of raw materials to make the specific molecules needed by the cell. NADP carries electrons in exactly the same way as NAD.

Color the headings Flavin Adenine Dinucleotide and Reduced Form (FADH₂), titles N and O, and the associated illustration.

The third widely used coenzyme is flavin adenine dinucleotide, which gets its *flavin* and *ribitol* portions from the vitamin riboflavin. When FAD is reduced (shown), it carries two electrons and two protons in the form of hydrogen atoms (arrows) attached to two of the three rings of the flavin portion. In the cell, FAD is used as an electron acceptor for oxidation reactions that do not release enough energy to reduce (attach electrons to) NAD but do release enough to reduce FAD.

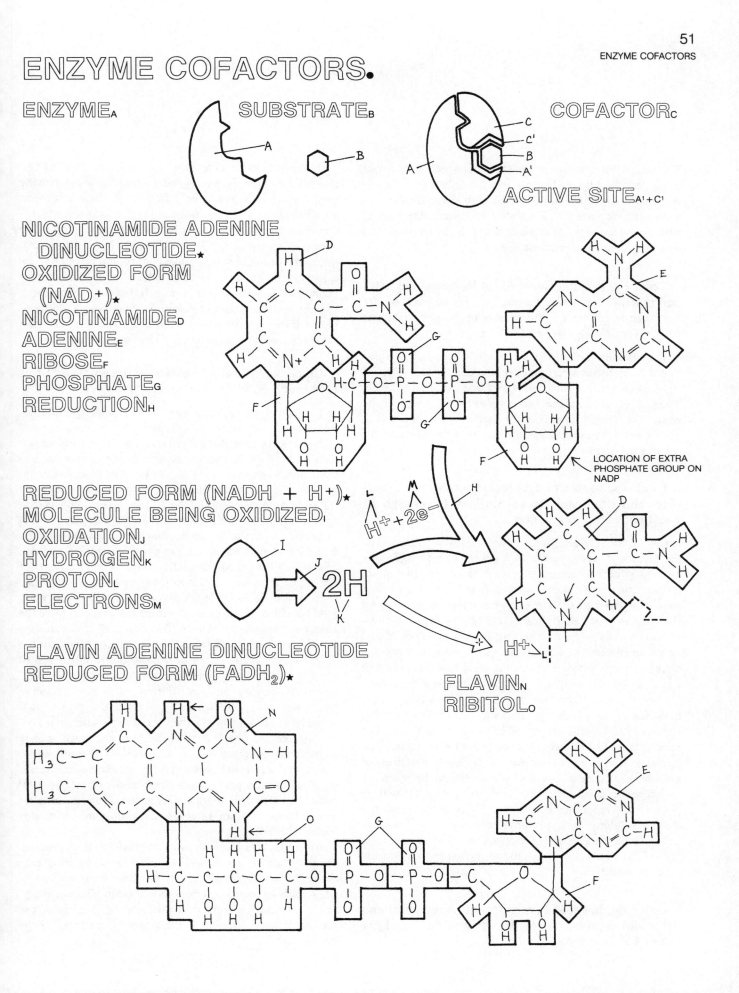

ENZYME COFACTORS.

ENZYME_A   SUBSTRATE_B   COFACTOR_C

ACTIVE SITE_{A1 + C1}

NICOTINAMIDE ADENINE
   DINUCLEOTIDE★
OXIDIZED FORM
   (NAD+)★
NICOTINAMIDE_D
ADENINE_E
RIBOSE_F
PHOSPHATE_G
REDUCTION_H

LOCATION OF EXTRA
PHOSPHATE GROUP ON
NADP

REDUCED FORM (NADH + H+)★
MOLECULE BEING OXIDIZED_I
OXIDATION_J
HYDROGEN_K
PROTON_L
ELECTRONS_M

$H^+ + 2e^-$

2H

FLAVIN_N
RIBITOL_O

FLAVIN ADENINE DINUCLEOTIDE
REDUCED FORM (FADH_2)★

This plate illustrates another trick by which cells avoid being destroyed by the heat exchanges of their own chemical reactions: the use of metabolic pathways. These pathways also increase a cell's ability to control the rate at which the reactions occur, since many of the enzymes can be inhibited or enhanced as needed.

Color the heading Generalized Pathway and titles and representations A through F. Since each of the enzymes is different, you may want to use different colors for B^1, B^2, B^3, and B^4.

A metabolic pathway is a series of chemical reactions that accomplishes a certain chemical change in a series of small steps, each of them catalyzed by a different enzyme. The product of each reaction serves as the *substrate* for the next, until the *final product* for that pathway is reached. That final product may in turn be the first substrate for some other pathway.

Color the heading Glycolysis (Aerobic or Anaerobic) and titles and representations G through K. Use light colors for G and K.

Glycolysis (Greek: *glycos,* "sugar"; *lysis,* "loosening" or "breaking") is the metabolic pathway by which *glucose,* having six carbon atoms, is broken down into two molecules of the three-carbon compound *pyruvate* (the ionized form of pyruvic acid). The energy released is captured by the formation of *ATP* from ADP and inorganic phosphate. Glycolysis may be aerobic (in the presence of air) or anaerobic (in the absence of air) because the process does not involve oxygen. As in any metabolic pathway, the reactions are all catalyzed by enzymes. These enzymes are located in the hyaloplasm of the cell. For each glucose molecule oxidized by the pathway, four *electrons* are removed (and transferred to NAD—not shown here, but recall Plate 51), and two molecules of ATP are made. The electrons that are removed in the oxidation reactions of glycolysis must be disposed of someplace or the reactions will come to a complete halt and no more ATP will be produced. Aerobic organisms do this by transferring the electrons (along with protons) to oxygen in a process called respiration. Anaerobic organisms transfer the electrons (also along with protons) to pyruvate in a process called fermentation.

Color the heading Fermentation (Anaerobic) and titles and representations L, M, and N. Use light colors for L and M.

Each pyruvate receives two electrons from NADH and two protons from the hyaloplasm, which combine to make two hydrogen atoms. The addition of these hydrogen atoms converts pyruvate into some other compound. Different compounds are made by different organisms. Two common examples are *lactate* (ionized lactic acid), produced by the bacteria that make yogurt (and by our muscle cells in strenuous exercise), and *ethyl alcohol,* produced by yeast. Yeast also produces *carbon dioxide* as a by-product, thereby causing bread to rise and giving beer and champagne their fizz. In all fermentations, the final electron acceptor is some organic (carbon-containing) molecule.

Color the heading Respiration (Aerobic) and titles O through R. Color the rest of the plate.

The glycolytic pathway and the reactions of fermentation are carried out in the cytoplasm of the cell. In most organisms, the resulting pyruvate enters the mitochondrion, where it is further oxidized in the process of respiration, which removes many more electrons than fermentation does, giving more energy than fermentation does from the same amount of food. In respiration, the final electron acceptor is an inorganic molecule—always oxygen, except in certain bacteria that use other things. Respiration is usually by a pathway known as the *Krebs cycle* (see Plate 55), which occurs in the matrix (inner compartment) of the mitochondrion. The electrons removed in the reactions of the Krebs cycle are all carried by the coenzymes NAD and FAD to a group of molecules that comprise the *electron transport system (ETS).* (In aerobic organisms, the electrons removed in glycolysis are also passed to the ETS.) These electron transport molecules are believed to be embedded in the inner membrane of the mitochondrion. As the electrons pass from molecule to molecule in this system, the potential energy of those electrons is used to make ATP, which can then serve as an energy source for the cell. At the end of the system, the electrons are transferred to *oxygen* (along with protons) to make *water.*

Originally it was thought that for one glucose molecule, the electron transport system would make exactly 34 ATP molecules. Modern evidence indicates that the number probably varies from 16 to 34, depending on conditions within the cell.

Although this plate shows the breakdown of glucose, other carbohydrates, as well as proteins and lipids, can be broken down for their energy by these same pathways. Most carbohydrates can be converted into glucose; everything else can be converted into one or another of the intermediate compounds in the glycolytic pathway or the Krebs cycle.

METABOLIC PATHWAYS.

GENERALIZED PATHWAY★
INITIAL SUBSTRATE$_A$
ENZYMES$_{B^1, B^2, B^3, B^4}$

INTERMEDIATE SUBSTRATE/
PRODUCT$_{C, D, E}$
FINAL PRODUCT$_F$

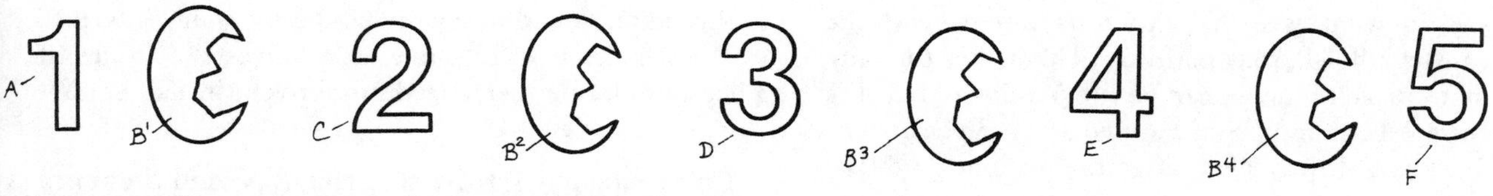

A 1 C$_{B^1}$ 2 C$_C$ C$_{B^2}$ 3 C$_D$ C$_{B^3}$ 4 C$_E$ C$_{B^4}$ 5 $_F$

GLYCOLYSIS (ANAEROBIC OR
AEROBIC)★
GLUCOSE$_G$
GLYCOLYTIC
PATHWAY$_H$
ATP$_I$

ELECTRONS$_J$
PYRUVATE$_K$

CH$_2$OH
G

I
2 ATP

H

J
4e$^-$

FERMENTATION
(ANAEROBIC)★
LACTATE$_L$
ETHYL ALCOHOL$_M$
CO$_{2N}$

RESPIRATION
(AEROBIC)★
KREBS CYCLE$_O$
ETS$_P$
O$_{2Q}$
H$_2$O$_R$

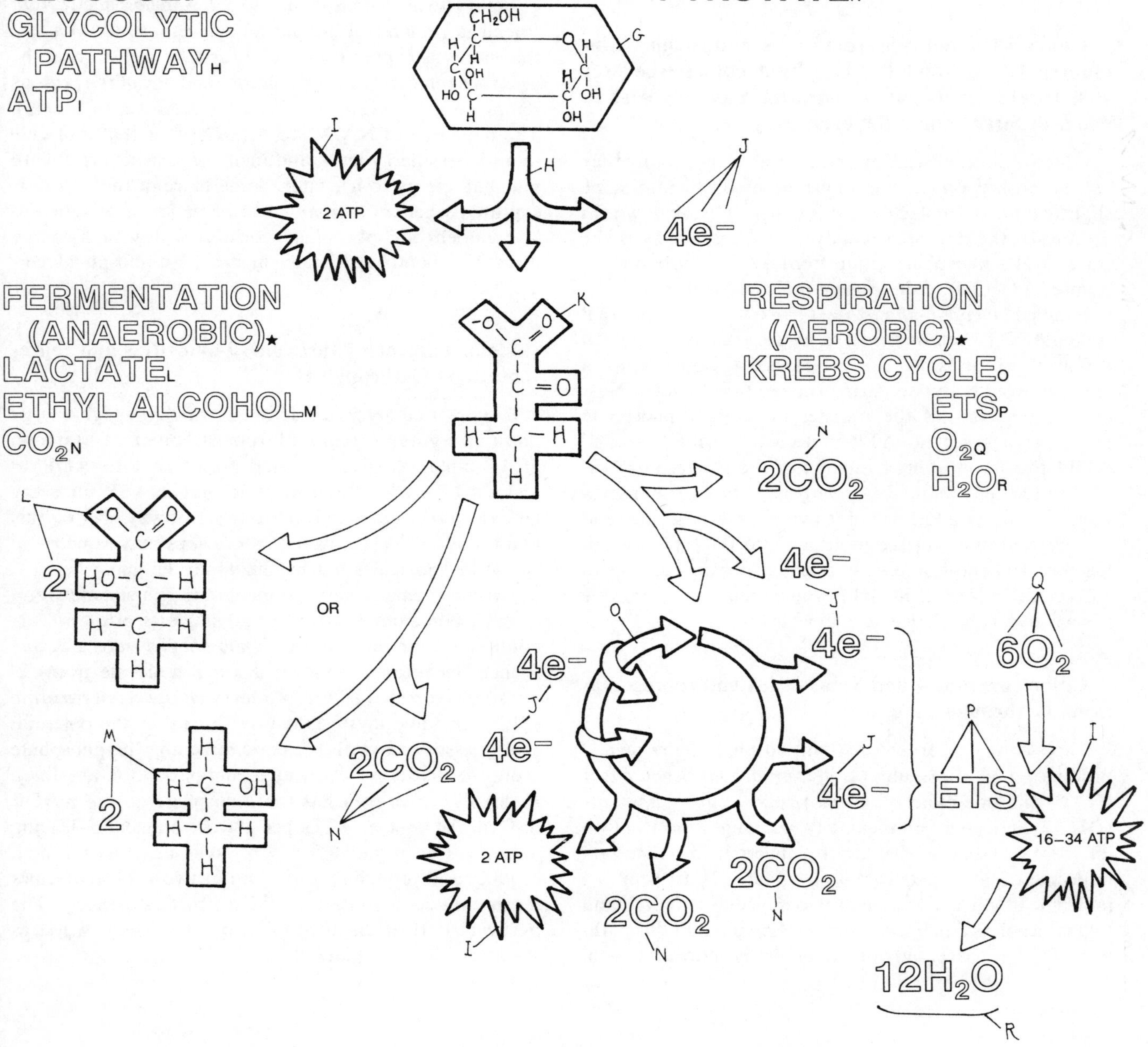

K

N
2CO$_2$

L
2 HO-C-H
H-C-H
H

OR

4e$^-$
J
4e$^-$
J

Q
6O$_2$

M
2 H-C-OH
H-C-H
H

N
2CO$_2$

4e$^-$
J
4e$^-$
J

O

4e$^-$
J
4e$^-$
J

P
ETS

I
16–34 ATP

2 ATP
I

2CO$_2$
N

2CO$_2$
N

12H$_2$O
R

This plate shows the details of the chemical reactions of glycolysis in what is called (after its discoverers) the Embden-Meyerhof-Parnas pathway. This is not the only pathway by which glucose can be broken down, but it is by far the most common. Some amino acids can be broken down by this pathway also; after their amino (—NH₂) groups are removed, they are converted into one or another of the intermediate compounds in this pathway, which are then oxidized in exactly the same way as if they came originally from glucose.

Color titles and representations A through G, including E¹, E², and E³. Use light colors wherever structural formulae are illustrated. You may wish to use a different color for each enzyme.

Glucose molecules are extremely stable, and before they can be broken down, they must be modified into some slightly different molecule that is easier to break down. In glycolysis, the first step (catalyzed by *enzyme 1*) is the transfer of a phosphate group from *ATP* to carbon atom number 6 (arrow) of the *glucose* molecule to form a molecule called *glucose-6-phosphate*. (The remnant of the ATP is then *ADP.*) Then *enzyme 2* catalyzes a reaction in which a few chemical bonds are rearranged ("isomerized," a chemist would say) to form *fructose-6-phosphate*. Next *enzyme 3* catalyzes the transfer of another phosphate group from a second ATP molecule (leaving a second ADP) to carbon atom 1 (arrow) of the fructose-6-phosphate to form *fructose-1,6-diphosphate*. For some inscrutable reason, the addition of the two phosphate groups and the rearrangement of the molecule into the form we call fructose somehow makes it possible for an enzyme to cleave the molecule in half, while there is no enzyme known that can do that to plain glucose.

Color enzymes 4 and 5 and titles and representations H through J.

The *cleavage* of fructose-1,6-diphosphate by *enzyme 4* produces one molecule of *glyceraldehyde-3-phosphate* (GAP) and one molecule of *dihydroxyacetone phosphate* (DHAP). *Enzyme 5* immediately rearranges the atoms in dihydroxyacetone phosphate to convert it into a second molecule of glyceraldehyde-3-phosphate. Thus from one molecule of glucose we obtain two molecules of GAP and two molecules of each compound following it in this pathway. To save space, only one structure is shown for each,

but a large 2 is written in front of each as a reminder. Up to this point, no usable energy has been obtained. Instead, two molecules of ATP have been "invested" to get the glucose molecule rearranged and broken in half.

Color enzyme 6, titles K through N, and the representations for L, M, and N.

The first reaction in which energy is actually released is the one in which *enzyme 6* catalyzes the oxidation of glyceraldehyde-3-phosphate to *1,3-diphosphoglycerate* (the negative ion of 1,3-diphosphoglyceric acid). Much of the energy is captured in useful form by transferring the electrons removed in the oxidation (and one of the protons removed) to the coenzyme *NAD⁺* (recall from Plate 51). The reduction of NAD⁺ to *NADH* raises that molecule to a higher energy level, and many organisms are able to put that energy to use later. Some more of the energy is captured when an *inorganic phosphate ion* is attached to glyceraldehyde-3-phosphate, forming a new high-energy phosphate bond (wavy line) in the 1,3-diphosphoglycerate.

Color enzymes 7 through 10 and titles and representations O through S.

Enzyme 7 catalyzes a reaction in which the high-energy phosphate group is removed from carbon atom 1 (arrow) of 1,3-diphosphoglycerate and transferred to ADP to make ATP—a total of two molecules of ATP for every glucose molecule that started the pathway. As of this point, the cell has recovered its previous "investment" of two ATP molecules but has made no net gain.

Enzyme 8 causes the movement of the phosphate group from carbon atom 3 (arrow) of *3-phosphoglycerate* to carbon atom 2 (arrow) to form *2-phosphoglycerate*. *Enzyme 9* then dehydrates (removes a *water* molecule from) 2-phosphoglycerate (2-PGA) to form *phosphoenol pyruvate* (PEP). In some mysterious way this raises the potential energy associated with the bond attaching the phosphate group so that it is now a high-energy bond (wavy line). *Enzyme 10* then transfers that phosphate group to ADP to form ATP—two ATPs per glucose molecule—leaving *pyruvate* (the negative ion of pyruvic acid) as the final product and providing a net gain of two ATP molecules for every glucose molecule oxidized by this pathway. The pyruvate is then disposed of in one of several ways, as shown in the next plate.

GLYCOLYSIS.

GLUCOSE$_A$
ENZYME$_{E^1...E^{10}}$
ATP$_B$
ADP$_C$
GLUCOSE-6-PHOSPHATE$_D$
FRUCTOSE-6-PHOSPHATE$_F$
FRUCTOSE-1,6-DIPHOSPHATE$_G$
CLEAVAGE$_H$
DIHYDROXYACETONE
  PHOSPHATE$_I$
GLYCERALDEHYDE-3-
  PHOSPHATE$_J$
INORGANIC PHOSPHATE$_K$
NAD$^+$$_L$
NADH + H$^+$$_M$
1,3-DIPHOSPHOGLYCERATE$_N$
3-PHOSPHOGLYCERATE$_O$
2-PHOSPHOGLYCERATE$_P$
WATER$_Q$
PHOSPHOENOL PYRUVATE$_R$
PYRUVATE$_S$

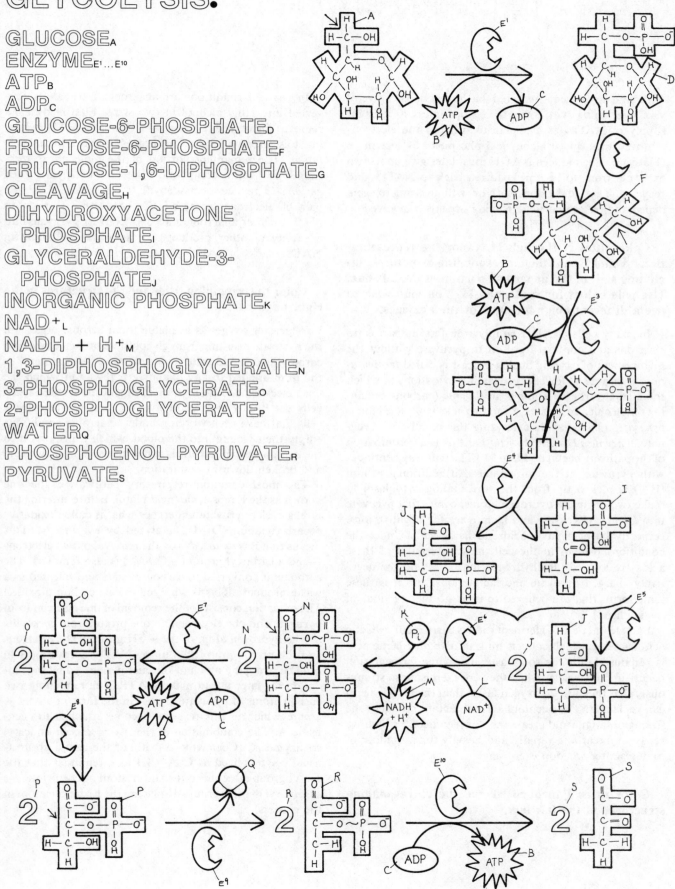

THE FATE OF PYRUVATE

Although no oxygen is required for the reactions of glycolysis, everything will come to a stop unless a constant supply of NAD$^+$ is available to pick up the electrons removed from glyceraldehyde-3-phosphate by enzyme 6 (Plate 53). The reduced NADH must later give up its two extra electrons to become oxidized back to NAD$^+$ and recycled. This can be done with or without using oxygen, depending on the particular living organism involved.

Color titles A through H. Color the representations A and B and the corresponding structures, including E^{11}, in the upper left portion of the left box. Use pale colors for A, D, and H. You may wish to use a different color for each of the enzymes.

In many cells, NADH simply transfers its two extra electrons and one extra proton to pyruvate without the involvement of oxygen. Such a process is called *fermentation,* a term used for energy-yielding oxidations in which the final electron acceptor is an organic (carbon-containing) molecule. In this case the final electron acceptor is *pyruvate.* (The enzyme catalyzing this reaction is designated *enzyme 11* to emphasize that this is a continuation of the pathway begun on Plate 53.) The pathway continues with pyruvate, at the top center. An additional proton (H$^+$) is picked up from the surrounding cytoplasm to make two complete hydrogen atoms, converting pyruvate into *lactate.* (Pyruvate and lactate are the negative ions, respectively, of pyruvic acid and lactic acid. Under the conditions present in the cell, most molecules of these acids are ionized into hydrogen ion, H$^+$, and the negatively charged pyruvate and lactate ions, and it is these ionic forms that are believed to undergo the reactions of metabolism.)

Lactate (lactic acid) fermentation is carried out by lactic acid bacteria, which convert milk into yogurt; lactic acid is responsible for the sour taste. The same fermentation occurs in our muscles when they are exercising so strenuously that they use up oxygen faster than the bloodstream can replace it. The accumulation of lactate produces the feeling of fatigue and creates an "oxygen debt," which we repay by breathing rapidly and heavily for a number of minutes after we stop exercising.

Color titles I through M and the corresponding structures in the left box.

In yeast, if conditions are anaerobic, pyruvate is converted into ethyl alcohol in two steps. First *enzyme 12* removes the carboxyl group (—COO$^-$, plate 17), releasing it as *carbon dioxide.* This process is called *decarboxylation.* The remaining portion of the molecule is rearranged to form a compound called *acetaldehyde.* Then *enzyme 13* reduces acetaldehyde to *ethyl alcohol* using a pair of electrons from *NADH,* converting it back to *NAD$^+$.* Other anaerobic microbes convert pyruvate into a variety of other products, all as a way of oxidizing NADH back to NAD$^+$.

Color the remaining titles and the structures in the right box.

Whenever oxygen is available to an aerobic organism, the pyruvate resulting from glycolysis is broken down into carbon dioxide and water by *respiration.* "Respiration" is the term applied to any oxidative pathway in which the final electron acceptor is an inorganic molecule. Yeast cells also live by respiration if oxygen is available. Muscle cells that have been working under anaerobic conditions empty their lactate into the blood, which transports it to the liver, where the lactate is converted back into pyruvate and broken down by respiration.

The most common respiratory pathway is the one known as the Krebs cycle (next plate). Before entering the Krebs cycle, pyruvate undergoes what is called oxidative decarboxylation (D+I), catalyzed by *enzyme 14.* This means that it is *oxidized*—by the removal of two electrons —and a carboxyl group (—COO$^-$) is also removed. The carboxyl is converted to carbon dioxide and released as a waste product. (This is why you exhale carbon dioxide.) The oxidation consists of the removal of one electron from pyruvate and one electron and one proton from *coenzyme A* (the hydrogen atom of the —SH group.) The electrons and the proton are picked up by NAD$^+$, reducing it to NADH, trapping much of the energy released by the oxidation for later use in making ATP. The remaining two carbon atoms of pyruvate are left in the form of an *acetyl group,* which in this reaction become attached to coenzyme A. The combination of the two is known as acetyl coenzyme A. (Coenzyme A without the acetyl group is usually symbolized as CoA—SH as a reminder that the acetyl group attaches to the sulfur atom of coenzyme A's sulfhydryl (—SH) group, displacing the hydrogen atom in the process).

THE FATE OF PYRUVATE.

PYRUVATE$_A$
FERMENTATION
(ANAEROBIC PATHWAYS)$_B$
ENZYME$_{E^{11}, E^{12}, E^{13}, E^{14}}$
REDUCTION$_C$
OXIDATION$_D$
NADH + H^+_F
NAD^+_G
LACTATE$_H$
DECARBOXYLATION$_I$
CO$_{2\,J}$
HYDROGEN ION$_K$
ACETALDEHYDE$_L$
ETHYL ALCOHOL$_M$

RESPIRATION
(AEROBIC PATHWAYS)$_N$
ACETYL COENZYME A$_\star$
COENZYME A$_O$
ACETYL GROUP$_P$

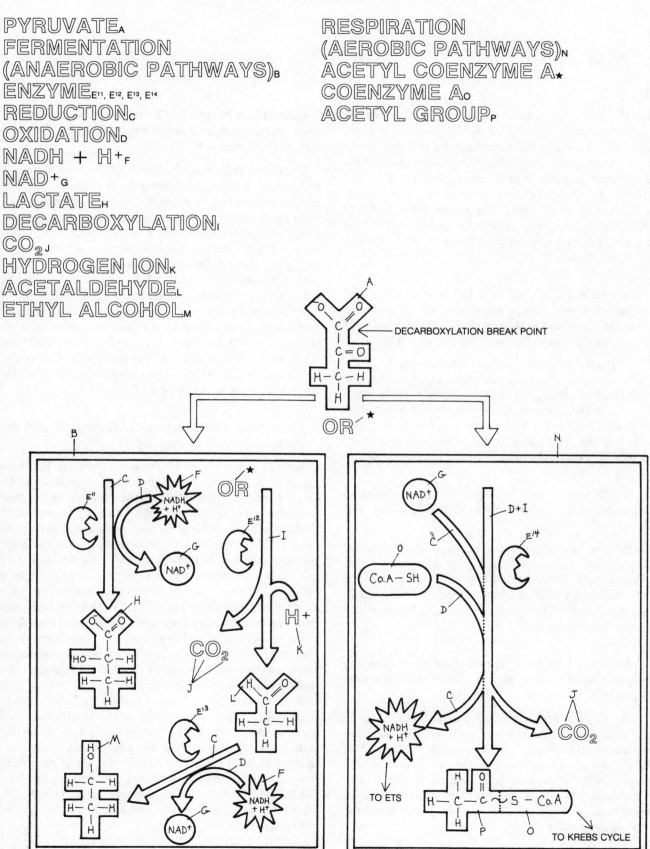

DECARBOXYLATION BREAK POINT

OR $\star$

TO ETS

TO KREBS CYCLE

The principal metabolic pathway for respiration is named the Krebs cycle, after its discoverer, Sir Hans Krebs. It is called a cycle because the starting compound is returned after the series of reactions is completed. Two other names commonly used are citric acid cycle and tricarboxylic acid (TCA) cycle, because citric acid is the first compound formed and it has three carboxyl ($-COO^-$) groups. (Today we know that most of it is in the form of citrate, the negative ion, but old names are hard to get rid of.)

Color titles A, B[1], C, E[15], D[1], and F and their corresponding structures, using pale colors to avoid obscuring the structural formulae. You may wish to use a different color for each of the enzymes.

To start the Krebs cycle, *enzyme 15* catalyzes the transfer of the 2-carbon acetyl group from *acetyl coenzyme A* (recall the preceding plate) to the 4-carbon *oxaloacetate ion* to form the 6-carbon *citrate ion*. A *water molecule* (D[1]) is also broken down to supply two hydrogen atoms and an oxygen atom. *Coenzyme A* is released to go back and pick up an acetyl group from the next pyruvate.

Color titles G and H and their corresponding structures, as well as structures E[16], E[17], G, D[2], and D[3].

In the next reaction, *enzyme 16* catalyzes the removal of two hydrogen atoms and an oxygen atom from citrate to form a water molecule and the product, *cis-aconitate*. In the reaction after that, a water molecule is broken down again to provide two hydrogens and an oxygen to be added again to form *isocitrate*.

Color titles I through M and their corresponding structures at the lower right. Also color structures E[18] and E[19].

In the next reaction, *enzyme 18* catalyzes the oxidation of isocitrate to *oxalosuccinate*, and the two electrons removed in the process reduce another molecule of *NAD+*, capturing some more energy, which is later used to make ATP in the electron transport system (ETS). Then *enzyme 19* catalyzes the decarboxylation of the 6-carbon oxalosuccinate to the 5-carbon α-*ketoglutarate*, releasing a molecule of *carbon dioxide*.

Color titles N through T and their corresponding structures. Color also structures I, J, L, B[2], E[20], and E[21] at the lower left.

Under the influence of *enzyme 20*, α-ketoglutarate is simultaneously oxidized and decarboxylated (releasing another carbon dioxide molecule) to form a 4-carbon succinyl group, which is immediately attached to coenzyme A to form *succinyl-coenzyme A*. Once again, the electrons removed in the oxidation are transferred to NAD+, capturing more energy for later synthesis of ATP in the electron transport system (ETS.) Next the succinyl group is converted to *succinate*, releasing the coenzyme A molecule. Enough energy is released by this step to form a new high-energy bond by attaching an inorganic phosphate ion (P_i) to a molecule of guanidine diphosphate (GDP)—a nucleotide very similar to ADP—converting it to guanidine triphosphate (GTP). The phosphate is immediately transferred to *ADP* to form *ATP*.

Color titles and structures U through X and the rest of the plate.

Succinate is oxidized to *fumarate*, but the electrons removed do not have enough energy to reduce NAD. Instead they are transferred to the more easily reduced coenzyme *FAD* to form *FADH₂*. The energy captured in $FADH_2$ will also be used later to make ATP, but not as much ATP as from NADH. Next another water molecule supplies hydrogen and oxygen atoms to convert fumarate into *malate*. Malate is then oxidized, with NAD+ picking up the electrons again, to re-form the oxaloacetate with which the cycle started. The net result of all this, including the transition reactions from pyruvate to acetyl coenzyme A shown in Plate 54 and the reactions of the Krebs cycle, is that pyruvate and two water molecules are consumed to phosphorylate one molecule of GDP to GTP, to produce three molecules of carbon dioxide, and to reduce four molecules of NAD+ to NADH and one molecule of FAD to $FADH_2$. Since one glucose molecule produces two pyruvates, these numbers must be doubled to obtain the yield from one glucose molecule. Except for the one reaction in which GTP is produced, all of the energy gain from this pathway comes when the the electrons captured by NAD+ and FAD are passed through the ETS (Plate 56), where that energy is used to make numerous additional molecules of ATP.

THE KREBS (TCA) CYCLE.

ACETYL COENZYME A_A
COENZYME A_{B^1, B^2}
ENZYME_{E^{15}...E^{24}}
OXALOACETATE_C
WATER_{D^1...D^4}
CITRATE_F
CIS-ACONITATE_G
ISOCITRATE_H
NAD+_I
NADH + H+_J

OXALOSUCCINATE_K
CO_{2L}
α-KETOGLUTARATE_M
SUCCINYL-CoA_N
GDP_O   P_{iP}   GTP_Q

ADP_R   ATP_S
SUCCINATE_T
FAD_U   FADH_{2V}
FUMARATE_W
MALATE_X

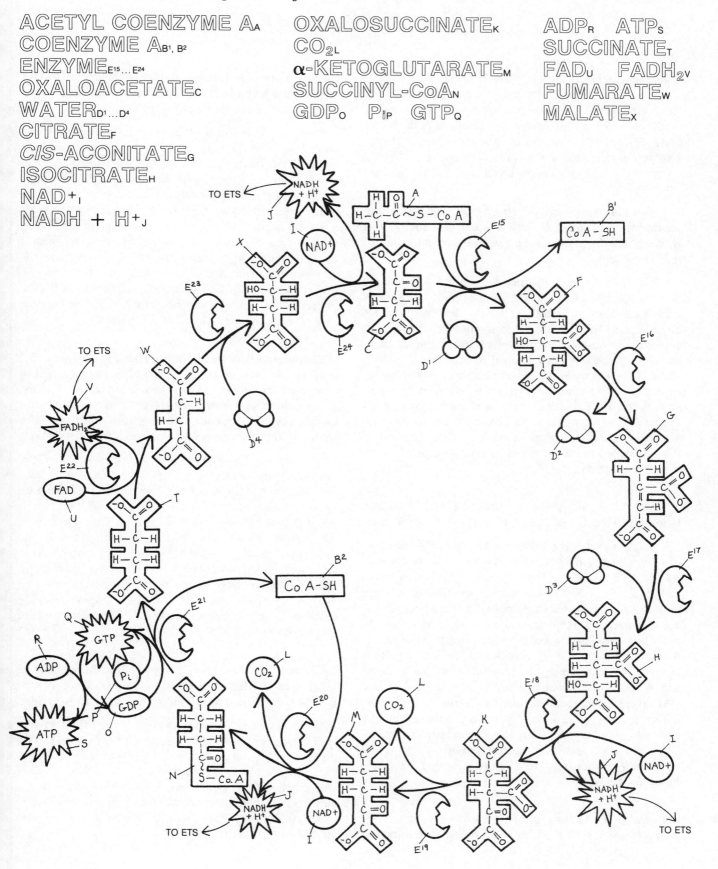

The reactions of glycolysis (Plate 53) occur in the cytoplasm, and the resulting pyruvate must diffuse into the inner compartment, or matrix, of a mitochondrion to reach the enzymes that catalyze the reactions of the Krebs cycle (Plate 55). It is only in the mitochondrion that oxygen becomes involved and a really large amount of useful energy can be obtained for the cell.

Color the heading Mitochondrion and titles and structures A through D. Use a light color for D. The tiny stalked F_1 particles on the inner membrane are left uncolored.

Most of the energy obtained from glycolysis and the Krebs cycle is captured in the form of high-energy electrons picked up by NAD and FAD. This energy is not directly useful for any significant number of the cell's activities. It must first be converted into the much more useful form of ATP by a process called oxidative phosphorylation. The electrons are transferred through a sequence of transport molecules known as the electron transport system (ETS). These molecules work something like a bucket brigade for electrons, passing them along from molecule to molecule until an enzyme at the very end of the line transfers the electrons—along with protons—to oxygen to form water.

Color the heading Chemiosmotic Hypothesis and titles and structures E and F, using light colors.

The lower part of the plate shows a portion of one crista of a mitochondrion, illustrating the chemiosmotic hypothesis for ATP formation. This hypothesis was first put forth in 1961 by Peter Mitchell, who received the Nobel Prize for it in 1978. Since that time, more and more evidence has accumulated to support it, and it is widely accepted. As described in the two preceding plates, much of the energy from the oxidations of glycolysis and the Krebs cycle is captured by transferring the removed electrons to *NAD* and *FAD;* numerous molecules of reduced NAD and FAD therefore begin to accumulate in the matrix of the mitochondrion. (The Krebs cycle occurs there, and the electrons from glycolysis in the hyaloplasm are transferred in.) NAD and FAD are recycled by giving up their electrons (becoming *oxidized*). The electrons are transferred to the ETS.

Color titles G through J, the proton pathway (I) at the lower left of the plate and the proton entering

that path, and the electron pathway (H) from NADH only through the first three transport molecules (G). Then color those three transport molecules.

It has long been known that some of the electron transport molecules accept electrons and protons when they are reduced, but others accept only electrons. According to the chemiosmotic hypothesis, the first *transport molecule* accepts electrons and *protons* on the matrix side of the membrane and passes both on to the second transport molecule on the other side of the membrane. The third transport molecule accepts only electrons, so the protons become "orphaned," so to speak, but on the opposite side of the membrane from where they started. All this has the effect of pumping protons across the membrane, out of the matrix of the mitochondrion into the intermembrane space (also known as the outer compartment).

Complete the coloring of the electron pathway (H) that is already started. Color also the remaining transport molecules (G) through which it passes. Then color the second proton pathway through the membrane, the protons entering that path, and the paths of the electrons and protons from $FADH_2$.

After leaving the protons on the opposite side of the membrane, the electrons apparently pass from one transport molecule to the next until they reach the matrix side again and once more reduce a transport molecule, which accepts protons along with the electrons. Then they repeat the process, transferring four more protons across the membrane, making a total of six protons pumped across for every NAD molecule. The electrons carried by $FADH_2$ do not have as much energy as the ones on NADH, so they do not reduce the first transport molecule in the series and hence pump only four electrons across the membrane instead of six.

Color the remainder of the plate. The phospholipid bilayer can be left uncolored, or you may color over it with color B to orient you to the upper illustration.

When the electrons reach the end of the transport system, they are transferred to *oxygen* to form *water*. The protons accumulating in the intermembrane space flow back into the matrix through passages in stalked F_1 *particles* attached to the inner membrane. Although the mechanism is not yet understood, that flow of protons *phosphorylates ADP* to *ATP* and provides the energy source for virtually all the activities of life.

ELECTRON TRANSPORT IN THE MITOCHONDRION.

MITOCHONDRION★
OUTER MEMBRANE A
INNER MEMBRANE B
INTERMEMBRANE SPACE C
MATRIX D

CHEMIOSMOTIC HYPOTHESIS★
REDUCED NAD/FAD E
OXIDIZED NAD/FAD F
TRANSPORT MOLECULE G
ELECTRON PATHWAY H
PROTON PATHWAY I
PROTON J
OXYGEN K
OXYGEN PATHWAY L
WATER M
F₁ PARTICLE N
ADP + Pᵢ O
PHOSPHORYLATION P
ATP Q
MISCELLANEOUS PROTEIN R

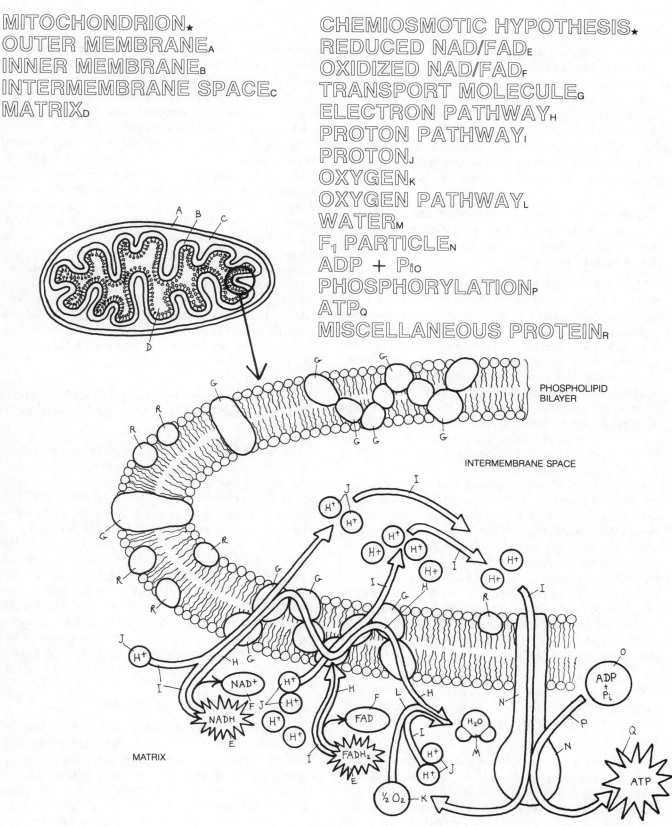

PHOSPHOLIPID
BILAYER

INTERMEMBRANE SPACE

MATRIX

Whereas a mitochondrion releases energy from molecules, a chloroplast stores energy in molecules by capturing the energy of light and using it to assemble carbon dioxide and water into carbohydrate and other energy-rich compounds. This process is called photosynthesis (Greek: *photo*, "light"; *syn*, "together"; *thesis*, "putting" or "placing"). Without photosynthesis, we would soon run out of food. We would also run out of oxygen, and the only life forms left on earth would be a few bacteria.

Color the heading Net Reaction and titles and structures A through E. Use two different shades of the same color for the oxygen in carbon dioxide and the oxygen in water.

The equation given here shows only the net change that results from photosynthesis. Six molecules of carbon dioxide and six molecules of water are used to produce one molecule of glucose, and six molecules of oxygen are released as a waste product. (Other carbohydrates can also be made, as well as amino acids, lipids, or anything else the plant needs.) The equation is deceptively simple and originally had scientists believing that photosynthesis was a simple process. We now know that it consists of a large number of complex chemical reactions and that all of the *oxygen* released in photosynthesis comes from *water* and all of the *oxygen* in carbohydrates comes from *carbon dioxide*.

The reactions of photosynthesis occur in two phases, the photochemical, or light-dependent, reactions, which must have light energy to drive them, and the thermochemical, or light-independent, reactions (see next plate), which use the high-energy molecules produced by the photochemical reactions but do not directly require light themselves. (At night, plants have to live by respiration, just as we do.) The photochemical reactions are believed to be carried out by molecules embedded in the thylakoid membranes (recall Plate 38), while the thermochemical reactions appear to occur in the stroma, the portion of the chloroplast outside the thylakoids.

Color the heading Photosystem I and titles F through L. Color the E arrows and structures F through L, but leave structures F¹, H¹, I¹, and J¹ for later.

The thylakoid membrane seems to have two groups of molecules, called photosystem I and photosystem II, each with different functions. Although both work simultaneously, we will have to discuss them one at a time. The first and most important event in either photosystem is the capture of *light energy* by a group of molecules acting as a sort of light-capturing antenna. These molecules are all brightly colored, so they are commonly referred to as *pigments*. They include two kinds of chlorophyll (three in some algae), which are responsible for the green color of plants, as well as several carotenes (yellow-orange) and xanthophylls (yellow). The absorbed light energy is passed along to a specialized molecule of chlorophyll a, which in photosystem I is called *pigment 700* (from the wavelength in nanometers of light it absorbs most efficiently). The energy it receives from the other pigments is transferred to a single electron, which is raised to such a high energy level that it breaks free of the pigment 700 molecule and passes (along the *electron pathway*) through a series of electron *transport molecules* similar to those in the mitochondrion.

When two photons (units of light energy) are absorbed, two electrons are passed through these transport molecules and reduce a molecule of *NADP* (a close analogue of NAD; recall Plate 51) to *NADPH*, joined by one *proton (hydrogen ion)*. Since this is a reduction and it is driven by light energy, it is called *photoreduction*.

Color the heading Photosystem II and structures F¹, H¹, I¹, and J¹. Color titles and structures M through Q.

The light-capturing pigments of photosystem II channel the energy absorbed to a slightly different molecule of chlorophyll a, called *pigment 680*, which also transfers the energy to an electron and causes it to break free. That *electron* is passed through a different series of transport molecules specific to photosystem II, and it replaces the electron lost by pigment 700. (The illustration shows two electrons doing this, since two were required for the reduction of NADP, but they apparently go through one after the other.) The loss of electrons leaves the pigment 680 molecule with such a strong attraction for replacement electrons that it removes electrons from *water*, releasing protons (hydrogen ions) into the thylakoid interior and also releasing an *oxygen atom* (half of an oxygen molecule). This process is called *photolysis*.

Color the remainder of the plate.

According to the chemiosmotic hypothesis, each electron carries a proton from the stroma into the interior of the thylakoid (*proton pathway*) in the same manner as protons are transported in the mitochondrion (Plate 56). The flow of protons out of the thylakoid, through stalked *CF₁ particles, phosphorylates ADP* to *ATP* to provide energy for the thermochemical reactions that follow.

PHOTOCHEMICAL REACTIONS.

NET REACTION★
CARBONᴀ
OXYGEN IN CARBON DIOXIDEʙ

HYDROGENᴄ
OXYGEN IN WATERᴅ
LIGHT ENERGYᴇ

PHOTOSYSTEM I★
LIGHT-CAPTURING PIGMENTSꜰ,ꜰ¹
PIGMENT 700ɢ
TRANSPORT MOLECULEʜ,ʜ¹
ELECTRON PATHWAYɪ,ɪ¹
PROTON (H+)ᴊ,ᴊ¹
PHOTOREDUCTIONᴋ
NADP+/NADPHʟ

PHOTOSYSTEM II★
PIGMENT 680ᴍ
PHOTOLYSISɴ
WATERᴏ
ELECTRONᴘ
OXYGENꞯ

PROTON PATHWAYʀ
PHOTOPHOSPHORYLATIONꜱ
CF₁ PARTICLEᴛ
CHEMIOSMOTIC FLOWᵤ
ADP + Pᵢ/ATPᵥ

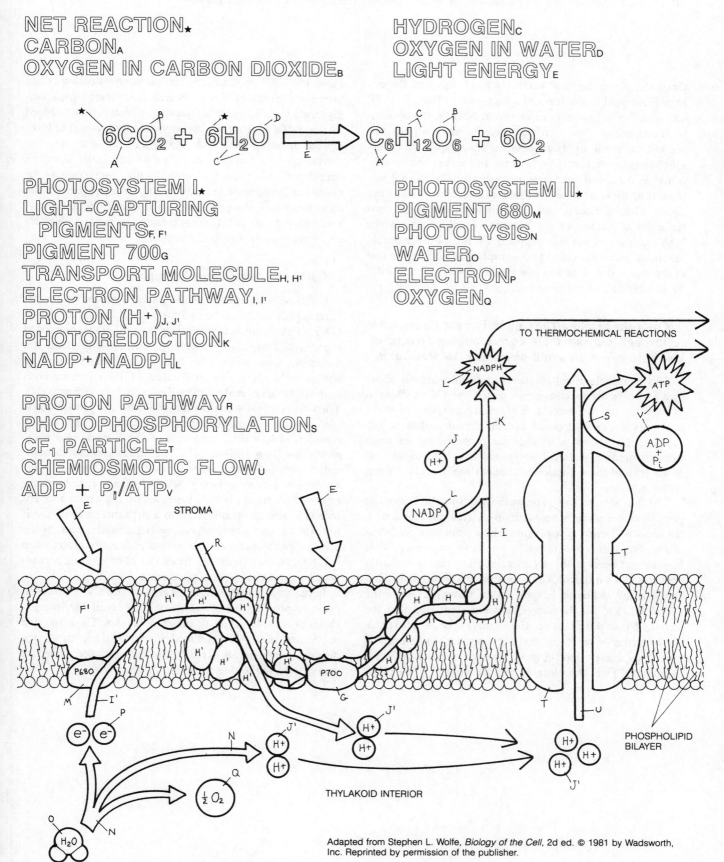

Adapted from Stephen L. Wolfe, *Biology of the Cell*, 2d ed. © 1981 by Wadsworth, Inc. Reprinted by permission of the publisher.

Once the photochemical reactions have captured the energy from light in the form of a high-energy bond in ATP and a pair of high-energy electrons in NADP, the remaining reactions of photosynthesis have no need of light and are known as the light-independent reactions or the thermochemical reactions. Sometimes the term "dark reactions" is used, but that term is misleading because it implies that these are the reactions plants carry out in the dark. When darkness first falls, these reactions do continue for a minute or so, but they quickly use up the high-energy compounds supplied by the photochemical reactions and soon come to a complete stop. For the rest of the night, the plant will live the same way you and I do: by glycolysis and respiration.

Color the heading The Calvin-Benson Cycle, titles A through G, and their corresponding structures. Use pale colors to avoid obscuring the structures.

The thermochemical reactions of photosynthesis occur in a cycle, which is commonly called the Calvin-Benson cycle after its discoverers. The entire purpose of these reactions is the fixation of carbon dioxide—that is, the capture of free carbon dioxide gas from the air (or from the surrounding water, if we're watching photosynthesis in water) and the reduction of that carbon dioxide to form carbohydrate.

The first step in this process is shown in the lower left portion of the plate, where *carbon dioxide* combines with the unusual 5-carbon sugar phosphate *ribulose-1,5-biphosphate.* This forms a 6-carbon intermediate compound that has not yet been positively identified because it instantly splits into two molecules of *3-phosphoglycerate.* This reaction is catalyzed by an enzyme on the outer surface of the thylakoid, and all the subsequent reactions occur in the stroma. From this point on, the formation of carbohydrate is by a series of reactions that are essentially a reversal of glycolysis, except that the enzymes are somewhat different and NADPH is the electron carrier instead of NADH.

First, the *ATP* produced by the photochemical reactions provides the high-energy phosphate to convert 3-phosphoglycerate to *1,3-diphosphoglycerate,* which is then reduced by the *NADPH* from the photochemical reactions to form *glyceraldehyde-3-phosphate* (GAP). GAP can then be converted into fructose and glucose and more complex carbohydrates or used to manufacture amino acids for proteins, lipids, nucleic acids, or any other organic molecules needed by the cell. Much of the GAP formed, however, must be recycled to regenerate the ribulose biphosphate used up in the first reaction.

Color the remainder of the plate.

Ribulose biphosphate (RuBP) is regenerated by several rearrangements of carbon atoms among five molecules of GAP. The compounds involved are all phosphorylated monosaccharides (single sugars), but most of them are unfamiliar ones, so only the numbers of carbon atoms are shown in the plate. Two molecules of GAP (three carbons each) combine to make a *phosphohexose* (six carbons) (H). Then one carbon atom is removed from the phosphohexose, leaving a *phosphopentose* (five carbons) (I^1), and is transferred to a third molecule of GAP, making a *phosphotetrose* (four carbons) (J). The phosphotetrose is combined with a fourth molecule of GAP to form a *phosphoheptose* (seven carbons) (K), and then two carbons are removed from the phosphoheptose, leaving a phosphopentose (I^2), and are transferred to a fifth molecule of GAP to form another phosphopentose (I^3). Finally, each of the phosphopentoses receives a second phosphate group from ATP molecules (received from the photochemical reactions) to form ribulose biphosphate.

It requires three complete turns of the Calvin-Benson cycle to provide one molecule of GAP available for synthesis of carbohydrates or other molecules. The other five molecules of GAP produced in those three turns are needed to regenerate RuBP. (Six turns of the cycle are required to make one molecule of glucose.)

THERMOCHEMICAL REACTIONS.

THE CALVIN-BENSON CYCLE.★
RIBULOSE-1,5-BIPHOSPHATE_A
CO₂ B
3-PHOSPHOGLYCERATE_C
ATP/ADP_D
1,3-DIPHOSPHOGLYCERATE_E
NADPH/NADP+_F
GLYCERALDEHYDE-3-
   PHOSPHATE_G
RuBP REGENERATION
   REACTIONS.★
PHOSPHOHEXOSE_H
PHOSPHOPENTOSE_{I1, I2, I3}
PHOSPHOTETROSE_J
PHOSPHOHEPTOSE_K

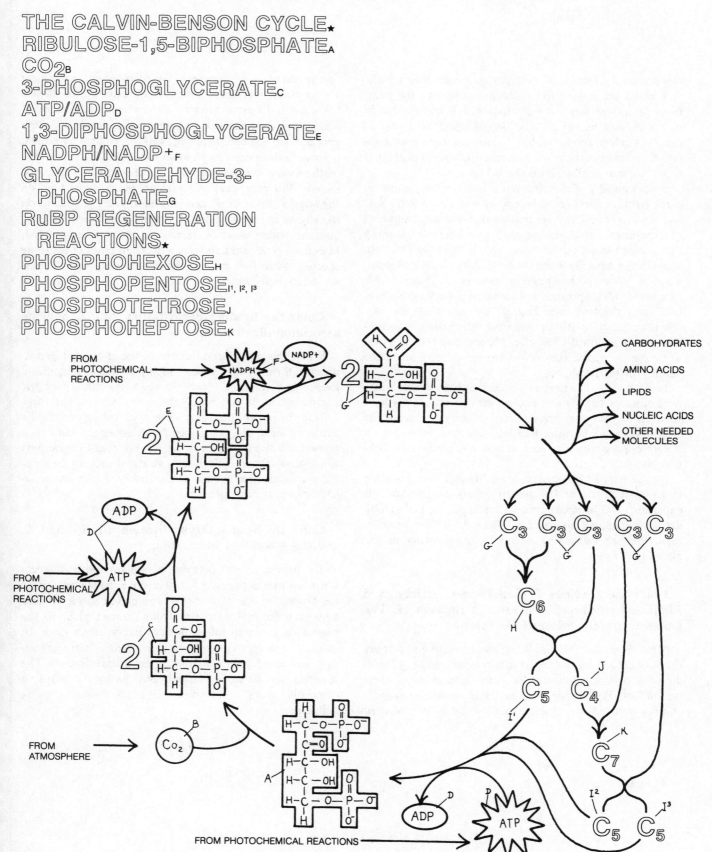

INTRODUCTION TO GENETICS

Genetics is the branch of biology that studies how hereditary traits are passed from one generation to the next. Even in ancient times it was already well known that in plants as well as in animals, the offspring of a pair of parents tends to show a highly variable mixture or a blending of the traits of the parents, frequently a great deal more from one parent than from the other.

By the middle of the nineteenth century the process of plant hybridization (crossbreeding) was in common use, and a number of scientists had carried out large numbers of experiments that also pointed to a variable blending that seemed to follow no rules. But in 1865 the first reliable laws of heredity were announced by Gregor Mendel, a monk in an Augustinian monastery in Austria. Although Mendel reported his findings at a meeting of scientists and published those findings the following year in a scientific journal with an extensive international circulation, it was not until 1900, after Mendel had been dead for a number of years, that other biologists recognized the validity of his results.

In beginning the study of genetics, it is well worth taking a close look at Mendel's actual experiments. They have such an elegant simplicity that they provide one of the easiest ways to be introduced to genetics as well as a superb opportunity for a look at how a scientific investigation should be designed.

At the time of his experiments, Mendel was a teacher in an agricultural and technical college and he did his experiments in his spare time with pea plants. Let us look first at the general approach to all kinds of plant hybridization before examining Mendel's experiments in the plates that follow.

Color the headings Typical Flower, Stamen, and Pistil and titles and structures A through H. The pollen is greatly enlarged for coloring.

Most people are familiar with *petals,* which give flowers much of their color; hay fever sufferers are certainly familiar with *pollen,* which is the immediate cause of their discomfort. Pollen contains the male reproductive cells and is produced in a structure called the *anther,* located on the end of a long *filament.* The filament and anther are known collectively as a stamen. In the center of the flower is the pistil, the expanded base of which is called the *ovary.* The ovary contains one or more *ovules,* each of which produces an egg cell. If the egg cell is fertilized by pollen, it grows into an embryo plant within a seed. The extension of the ovary is called the *style,* and its tip is called the *stigma.* For reproduction to occur, pollen must land on the stigma. In some plants, it is enough for it to land on the stigma of the same flower; other plants are self-sterile, and the pollen must be carried to the stigma by wind, insects, or other animals from a different plant of the same species. (Most flowers have the open arrangement of petals shown here, which encourages such cross-pollination.)

Color the heading Fertilization, title I, and the associated illustration.

When the pollen grain contacts the stigma, it germinates to form a *pollen tube,* which then grows longer and longer and migrates down through the tissues of the stigma and style and enters the ovary. One of the three cells in the pollen tube then unites with the egg cell in the ovule to form the zygote, or fertilized egg, which then grows and divides to form the embryo plant of the next generation. (Other cells of the ovule and pollen tube form the rest of the seed to protect the embryo plant and nourish it when it first sprouts.)

Color the heading Hybridization, titles J and K, and the associated illustration.

To form a hybrid between two plants with different traits, all that is necessary is to remove the stamens from the flowers of one plant before the pollen is mature and later transfer mature pollen from another plant to the stigma of the stamenless flower. Every effort must be made, of course, to prevent wind or insects from transferring unwanted pollen to the experimental flowers. The resulting seeds can be planted, and the hybrid offspring from this "cross" can be observed in the following growing season.

INTRODUCTION TO GENETICS.

TYPICAL FLOWER.★
PETALₐ
STAMEN.★
FILAMENTᴮ
ANTHERᴄ
POLLENᴅ

PISTIL.★
OVARYᴇ
STYLEꜰ
STIGMAɢ
OVULEʜ

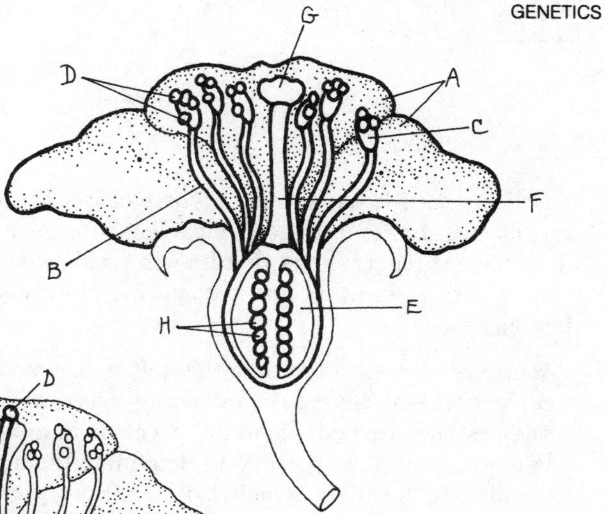

FERTILIZATION.★
POLLEN TUBEᵢ

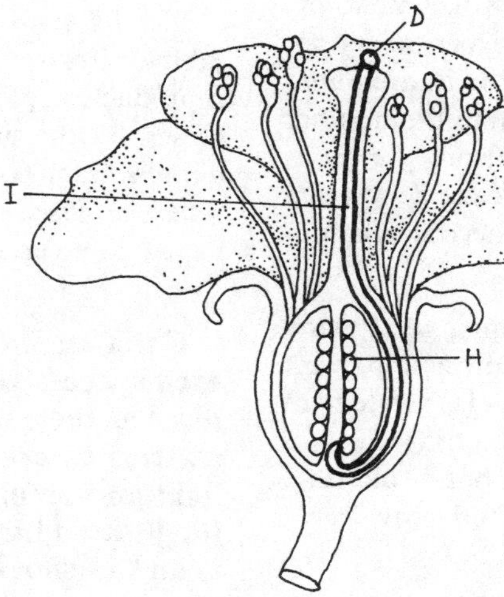

HYBRIDIZATION.★
FORCEPSⱼ
BRUSHᴋ

FLOWER 1

FLOWER 2

MATURE
FLOWER 1

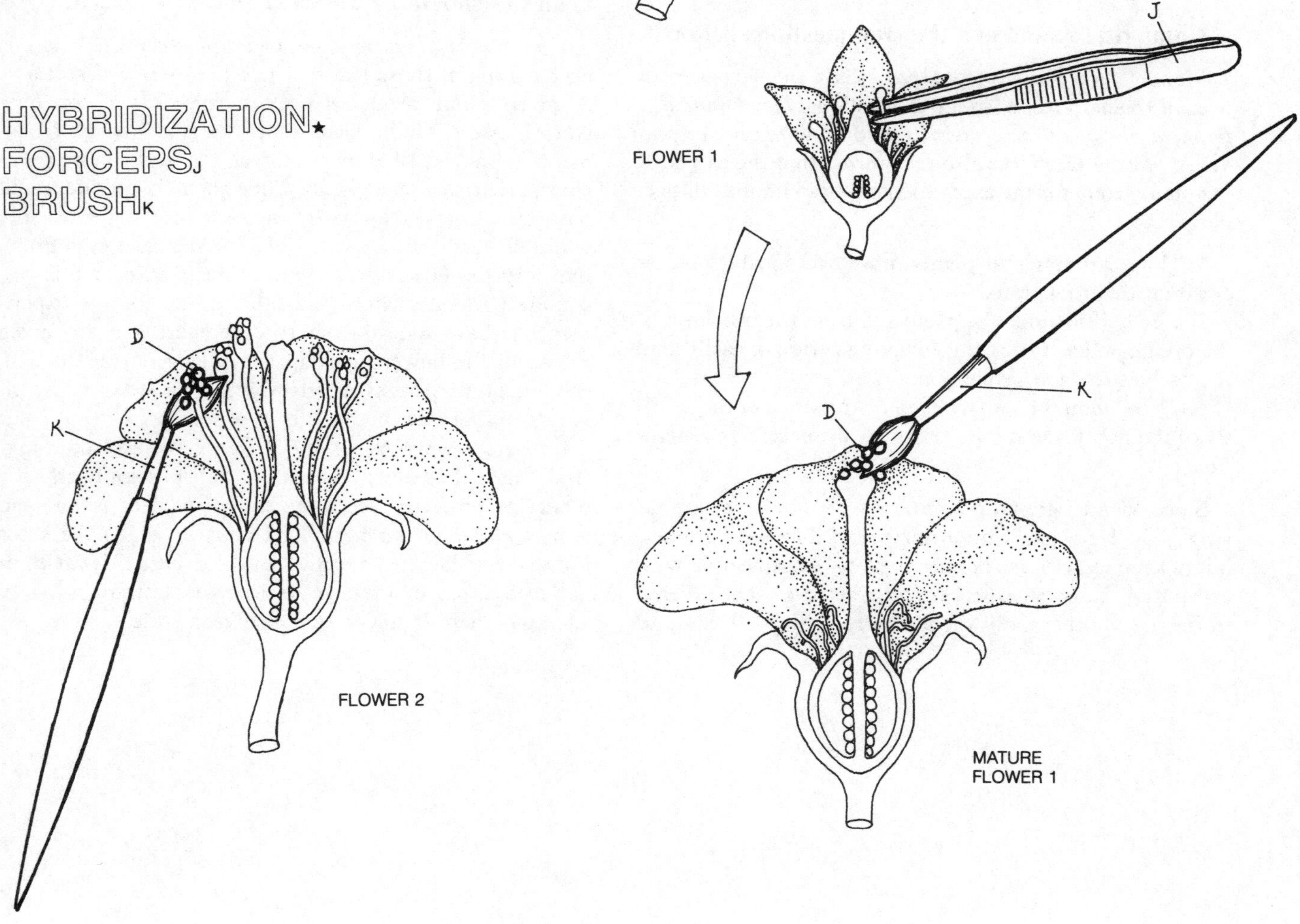

60
MENDEL'S PEAS

It is clear that Mendel's success where others had failed was not just the result of good luck. The introduction to his published report shows that he was familiar with the work of other scientists and recognized what mistakes they had made:

> Whoever surveys the work in this field will come to the conviction that among the numerous experiments not one has been carried out to an extent or in a manner that would make it possible to determine the number of different forms in which hybrid offspring appear, permit classification of these forms in each generation with certainty, and ascertain their numerical relationships.

In this paragraph we see Mendel performing what many great scientists have emphasized is the most important step in a scientific investigation: making a clear statement of the experimental question. He didn't phrase it as a question, with a question mark at the end, but he did state exactly what answers his experiment should provide.

Color title A and and the two questions below it.

Mendel correctly recognized that the experiment needed to answer two essential questions: *How many different kinds of offspring result?* and *How many are produced of each kind?* He also recognized that the choice of the plant group for the experiments was an important one. He wrote:

1. [The experimental plants must necessarily] possess constant differing traits.
2. The hybrids must be protected from the influence of all foreign pollen during the flowering period or easily lend themselves to such protection.
3. There should be no marked disturbance in the fertility of the hybrids and their offspring in successive generations.

Since Mendel grew up on a farm, he was very familiar with peas. He began with a number of different varieties, but in keeping with his requirement of "classification with certainty," he rejected many characteristics that differed from one another only on a "more or less" basis and

settled on seven pairs of traits that always showed up as either one thing or the other.

Color the heading Pea Blossom and titles and structures B, C, and D.

Mendel knew that peas had a flower structure that inhibited cross-pollination. As the illustration shows, the reproductive parts of the pea flower are completely enclosed by the petals. Before the flower even opens, the *anthers* burst and dust pollen all over the *stigma.* Thus unless an insect (or a scientist) interferes, self-pollination is virtually certain.

Color the heading Pea Plant Characteristics and each successive heading and pair of titles and structures as they are mentioned in the text. Choose contrasting colors for each pair wherever possible. Use light purple for E and white or some very light color for F. K and P should be different shades of yellow; L and O should be different shades of green.

Pea flowers are always either *purple* or *white;* there is no blending of those traits, which are clearly hereditary. A given plant produces only purple flowers or only white flowers. (It happens that this trait can be recognized in the seed before it is even planted. If the semitransparent seed coat is gray, the plant it produces will have only purple flowers. If the seed coat is white, only white flowers will be produced. For the sake of simplicity, only the flower color will be dealt with in this discussion.) Mendel chose the other traits for his experiments to have the same sharp differences. On any given plant, all the flowers are *axial* (growing out of the axil, the junction where a leaf grows out of a stem) or *terminal* (growing out of the end of a stem). In height, pea plants are always either *tall* (over 2 meters) or *dwarf* (less than 1 meter). Medium-tall or semidwarf pea plants never occur. Seeds are either *green* or *yellow* and either *smooth* or *angular.* The pods in which the seeds grow are either *green* or *yellow* and either *inflated* or *deflated.* Each of these traits is distinct from its opposite, and there is never an in-between state.

MENDEL'S PEAS.

THE QUESTIONS_A
   HOW MANY KINDS? ————————— A
   HOW MANY OF EACH? ——————

PEA BLOSSOM★
OVULE_B
STIGMA_C
ANTHER_D

PEA PLANT CHARACTERISTICS★

FLOWER★	PURPLE_E	WHITE_F
	AXIAL_G	TERMINAL_H
HEIGHT★	TALL_I	DWARF_J
SEED★	YELLOW_K	GREEN_L
	ROUND_M	ANGULAR_N
POD★	GREEN_O	YELLOW_P
	INFLATED_Q	WRINKLED_R

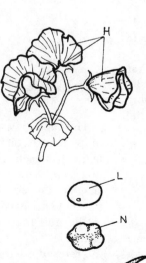

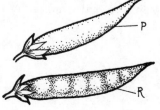

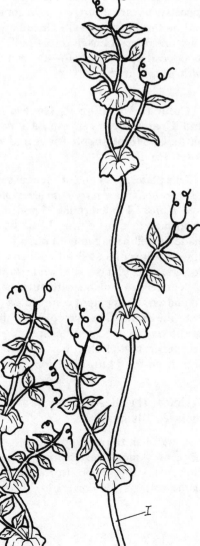

Another key to Mendel's success was his decision to begin his experiments with plants that were pure-breeding, that is, would produce nothing but the same trait generation after generation. Other investigators had merely cross-pollinated thousands of plants without paying attention to this. Mendel clearly recognized in advance what his experiments ultimately proved, that some hereditary traits that do not show at all in one generation of plants might be carried in invisible form and passed on to show up in the next generation.

To be sure that all his batches of seeds were really pure-breeding, Mendel planted them in different parts of the garden and allowed them to self-pollinate for two successive years. (Sure enough, some did not breed true and had to be discarded.) Once he was sure he had only pure-breeding strains, he cross-pollinated each strain with its opposite (see Plate 60)—tall with dwarf, white flower with purple, and so on.

Color the heading P_1 Generation and titles A, B, and C and their representations at the top of the plate. Use light purple for A and white or a very pale color for B.

This plate shows one of Mendel's experiments in which he *cross-pollinated purple-flowered plants* with *white-flowered plants.* To determine if pollen or ovule were more important in transmitting heredity, all his crosses were made by pollinating in both directions: plants with purple flowers were pollinated with pollen from plants with white flowers, and plants with white flowers were pollinated with pollen from plants with purple flowers. He found that no matter which parent contributed the pollen, the results were the same. (The symbol P_1 in the plate is commonly used today to indicate the parental generation and F_1 the first generation of offspring—called the first filial generation, from the Latin word for "son" or "daughter").

Color the heading F_1 Generation and the associated illustration.

A peculiar thing happened in the first generation of offspring. Among the offspring, only one trait of each pair showed up. The opposite trait just vanished. Although plants with purple flowers had been crossed with plants with white flowers, all the F_1 plants had only purple flowers. Even this, however, was nothing new. Other investigators had done similar experiments, including some with pea plants, and had obtained similar results. But Mendel had the genius to select a species of plant that produced fertile hybrids, so he was able to allow the F_1 plants to self-pollinate and produce an F_2 (second filial) generation, which he could also observe.

Color the heading F_2 Generation, title D, and the associated illustration.

When the hybrid plants of the F_1 generation were allowed to *self-pollinate,* they produced an F_2 generation with the proportion of flower colors shown here. Of 929 plants, 705 had purple flowers and 224 had white flowers. Because of his mathematical approach, Mendel recognized what no one else had recognized in similar cases, that this was a ratio of almost exactly 3 to 1. His results with the other pairs of traits he studied were almost identical, as shown in this table:

P_1 Cross	F_1 Generation	F_2 Generation
Axial flowers × terminal	All axial	651 axial, 207 terminal
Tall × dwarf	All tall	787 tall, 277 dwarf
Yellow seed × green	All yellow	6022 yellow, 2001 green
Round seed × angular	All round	5474 round, 1850 angular
Green pod × yellow	All green	428 green, 152 yellow
Inflated pod × deflated	All inflated	882 inflated, 229 deflated

The table shows the actual numbers of plants and seeds that Mendel counted. (You can see he was a thorough fellow.) In each case, one trait vanished in the F_1 generation and then reappeared in the F_2, though in much smaller numbers. Mendel applied the term "dominant" to the trait that persisted in the F_1 and the term "recessive" to the trait that receded from sight in the F_1 but reappeared in the F_2. For each pair of traits, Mendel calculated the ratio of dominants to recessives in the F_2 generation and found all of them to be very close to 3:1.

FLOWER COLOR HYBRIDS.

P₁ GENERATION.★
PURPLE FLOWER ₐ
WHITE FLOWER ᵦ
CROSS-POLLINATION ᴄ

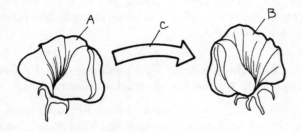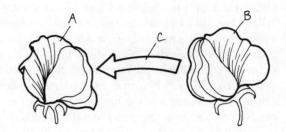

F₁ GENERATION.★

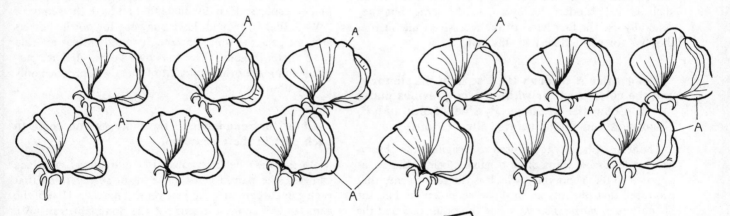

F₂ GENERATION.★
SELF-POLLINATION ᴅ

To explain the results of his experiments, Mendel proposed certain hypotheses. First, hereditary traits must be passed from one generation to the next as discrete units, rather than by some variable sort of blending. (Mendel called those units elements, but to avoid confusing them with chemical elements they are known today as genes.) Second, each individual plant must have a pair of these genes for each characteristic, one of them received from each parent. Third, when an individual has two conflicting genes, one dominates the other. (There are exceptions to this, but not among the traits Mendel studied.)

To illustrate how the experimental results could be produced, Mendel symbolized dominant genes with capital letters, such as "A," and recessive genes with small letters, such as "a." Modern biologists do the same, but they generally use the first letter of the recessive trait, such as the "w" for white flower in this plate.

Color titles A through B[1]. Use light purple for A and the color used for white on the previous plates for B. Color the headings P_1 Phenotypes and P_1 Genotypes and the associated illustrations.

The term "phenotype" is used to designate the observable or detectable traits of an organism being studied, as opposed to its "genotype," which is the set of genes that produced that phenotype. In this case, one plant had the phenotype of *purple flowers* while the other one had the phenotype of *white flowers*. (Be aware, however, that phenotype is really the result of both the genotype and the environment. Flower color is not easily influenced by the environment, but many traits are.)

Since Mendel made certain that the plants he started with were pure-breeding for their respective traits, each of the P_1 plants must have had two *genes* of the same kind. Today we would say they are "homozygous" (Greek: *homo,* "same"; *zygos,* "yoked" or "joined"). One is homozygous dominant (WW), while the other is homozygous recessive (ww).

Color the heading P_1 Gametes, titles C and D, and the associated illustration.

This section illustrates the fourth hypothesis Mendel made to explain his results: in the formation of pollen and ovules (today known collectively as *gametes*), the genes of each pair *"segregate"* into different gametes so that any gamete has only one gene of each pair. This fourth hypothesis is known as "Mendel's first law" or the "law of segregation." We see that because of segregation, each gamete of the homozygous purple-flowered plant has only one gene for flower color (purple), and each gamete of the homozygous white-flowered plant has only one gene for flower color (white).

Color the headings F_1 Phenotype and F_1 Genotype, title E, and the associated illustrations.

When two gametes fuse to form a new individual, the genes will be in pairs once again and will remain so as cell division occurs again and again to produce the mature plant. Here we see that *cross-pollination* of these particular plants results in F_1 individuals that all have the genotype "Ww"; that is, they all have one gene for purple flowers and one gene for white flowers. Such individuals are said to be "heterozygous" (Greek: *hetero,* "other"). Since purple is dominant over white, all of them will produce only purple flowers.

Color the remainder of the plate as you come to each part in the discussion.

When the heterozygous F_1 individuals produce gametes, the paired genes once again segregate so that each gamete contains only one gene of the pair. Half of the gametes will contain a gene for the dominant trait (W), and half will contain a gene for the recessive trait (w). When the F_1 plants are allowed to self-pollinate, four combinations are possible. The most error-free way of keeping track of these combinations is by means of a "Punnett square," named after its inventor, Reginald Punnett, an eminent British geneticist of the early 1900s. The gametes of one parent are listed along the top margin and the gametes of the other along the side margin. Then each possible combination of one gamete from each parent is written in the square where the appropriate column and row intersect. These combinations of genes are the genotypes of the F_2 generation. Since there are two different ways of getting one gene of each kind, half of the F_2 plants are heterozygous. One-fourth of them are homozygous dominant, and one-fourth are homozygous recessive. If you toss two coins 100 times and keep track of the resulting combinations of heads and tails, you will get approximately the same 1:2:1 ratio. With peas, however, purple is dominant over white, so the *phenotypic ratio* is three purple to one white.

GENE SYMBOLS.

PURPLE FLOWER_A
PURPLE GENE (DOMINANT)_{A¹}
WHITE FLOWER_B
WHITE GENE (RECESSIVE)_{B¹}

SEGREGATION_C
GAMETE_D
CROSS-POLLINATION_E

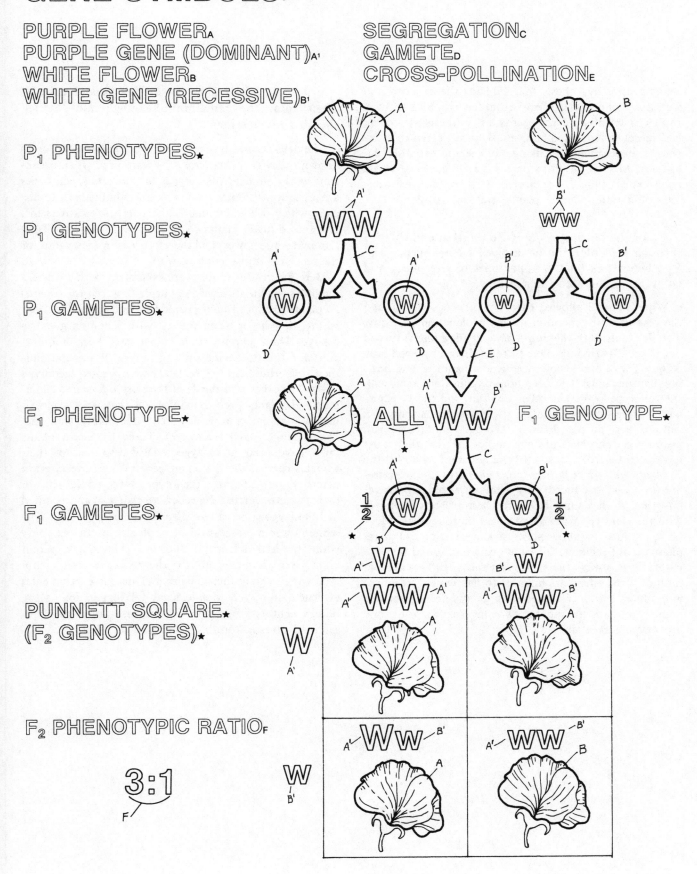

P₁ PHENOTYPES.★

P₁ GENOTYPES.★

P₁ GAMETES.★

F₁ PHENOTYPE.★

ALL Ww   F₁ GENOTYPE.★

F₁ GAMETES.★

PUNNETT SQUARE.★
(F₂ GENOTYPES).★

F₂ PHENOTYPIC RATIO_F

3:1

MENDEL'S FURTHER TESTS

Making up a hypothesis that explains the experimental facts does not necessarily prove that the hypothesis is true. In fact, a scientist never regards a hypothesis as proven; it is merely "supported" and must be subjected to every possible experimental challenge. In keeping with this philosophy, Mendel did not stop when he thought he had a good explanation. He subjected it to two further tests: self-pollination of the F_2 plants and test crosses.

Color the heading F_2 Self-Pollination and titles A through C. Color the headings F_2 Phenotypes and F_3 Phenotypes on the upper part of the plate, and color the associated illustrations.

When Mendel allowed the F_2 plants to *self-pollinate,* the resulting F_3 generation fit his explanation perfectly. The F_2 plants with the *purple flowers* (dominant) turned out to be of two kinds. One-third of them produced only offspring with purple flowers (the dominant trait), indicating that they must have been homozygous dominant, with the genotype WW. The other two-thirds of the F_2 plants with purple flowers (dominant) produced offspring in a ratio of three purple (dominant) to one white (recessive), just as their parents had done, indicating that they must have been heterozygous, with the genotype Ww. All the F_2 plants with *white flowers* (recessive) proved to be pure-breeding and produced only white-flowered (recessive) offspring, so they must have been homozygous recessive, with the genotype ww. With typical thoroughness, Mendel continued to allow successive generations of these plants to self-pollinate for eight years and found that the pattern was always the same. Obviously, the phenotypic ratio of three purple (dominant) to one white (recessive) was almost certainly caused by a genotypic ratio of one homozygous dominant to two heterozygous to one homozygous recessive.

Color the heading F_2 Test Crosses and title D. Then color the remaining headings and the associated illustrations.

Mendel's second test was a type of experimental cross known today as a *"test cross,"* a term used when organisms with dominant phenotypes but uncertain genotypes are crossed with homozygous recessive individuals to disclose which of the dominants also carry a recessive gene. He crossed the F_2 plants (WW, Ww, and ww) with plants that had white flowers and therefore were homozygous for the recessive trait (genotype ww).

Here again the results were consistent with Mendel's hypotheses. One-third of the purple-flowered (dominant) F_2 plants produced only purple-flowered (dominant) offspring, in spite of being crossed with a homozygous recessive. These parents, then, must have been homozygous for the dominant trait (genotype WW) and thus their offspring had to be all heterozygous (genotype Ww). The other two-thirds of the purple-flowered (dominant) F_2 plants produced offspring that were approximately one-half purple-flowered (dominant) and one-half white-flowered (recessive). These F_2 parent plants must have been heterozygous (Ww), so half of their gametes carried the dominant gene (W) and resulted in heterozygous offspring (genotype Ww) while half of their gametes carried the recessive gene (w) and resulted in homozygous recessive offspring (genotype ww). The white-flowered (recessive) F_2 plants produced only white-flowered offspring, showing that those parent plants were homozygous for the recessive trait (genotype ww), as were their offspring. Thus the experimental evidence supported Mendel's law of segregation. More than a century of additional experimentation by thousands of investigators has continued to support it.

MENDEL'S FURTHER TESTS.

F₂ SELF-POLLINATION.★
PURPLE FLOWER**ₐ**/GENE**ₐ₁**
WHITE FLOWER**ᵦ**/GENE**ᵦ₁**
SELF-POLLINATION**c**

F₂ PHENOTYPES.★

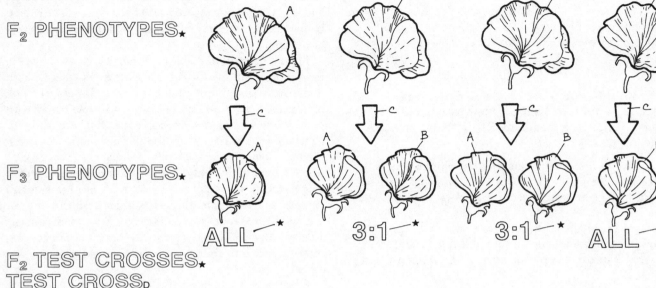

F₂ TEST CROSSES.★
TEST CROSS**ᴅ**

F₂ PHENOTYPES.★

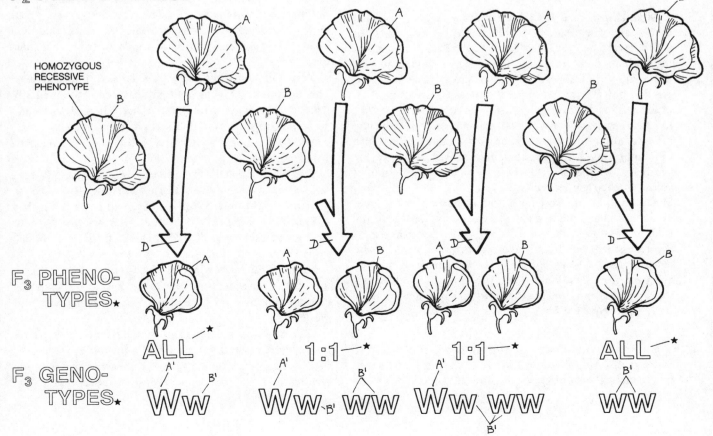

TWO-CHARACTER CROSS

Having satisfied himself with the test of his first law, Mendel went on to see if one characteristic influenced the way another one was inherited. He crossed plants that were pure-breeding for two different dominant traits with plants that were pure-breeding for the two opposite recessive traits.

Color titles A, B, C, D, and E (leaving A^1, B^1, C^1, and D^1 for later) and all their associated illustrations. Color the headings P$_1$ Phenotype and F$_1$ Phenotype.

As one of his experiments, Mendel *crossed* plants that were pure-breeding for *tallness* and *purple flowers* with plants that were pure-breeding for *dwarfness* and *white flowers*.

All of the plants in the first generation of offspring (F$_1$) showed both dominant traits, tall plant and purple flowers. None of the plants showed any recessive trait.

Color titles A^1, B^1, C^1, D^1, F, G, and H. Color the headings P$_1$ Genotype, P$_1$ Gametes, and F$_1$ Genotype (Dihybrid) and the associated illustrations.

Mendel reasoned that these results would be explained by his law of segregation, as illustrated by the gene symbols shown. The P$_1$ plants were pure-breeding, so the dominant plants must have been homozygous dominant with the genotype DDWW (P$_1$ genotype). (Mendel used AABB, but we are following the more common modern practice of using the first letters of the recessive traits, "d" for dwarf and "w" for white flower.) According to the law of segregation, the *genes* in each pair should separate into different *gametes*. Also, each gamete must have a gene for the height of the plant as well as a gene for the flower color, so each gamete must have one gene of each pair. Since the P$_1$ plants were both homozygous, every gamete of the dominant plant must therefore have one gene for each of the dominant traits (DW), while every gamete of the recessive plant must have one gene for each of the recessive traits (dw). When *cross-pollination* combined one gamete of each kind to form each new F$_1$ individual, the F$_1$ plants must all have been heterozygous for both traits (genotype DdWw).

Color the remainder of the plate as each part is discussed. Use pale colors for I, J, K, and L so as not to obscure the genotypes within the boxes.

Mendel then allowed the F$_1$ plants to self-pollinate. Since each of the F$_1$ plants is a hybrid for two characters and is in effect being crossed with itself, you will often find biology books referring to this self-pollination as a "dihybrid cross."

The F$_2$ generation resulting from this cross showed a ratio of nine tall plants with purple flowers (that is, both dominant traits) to three tall with white flowers (the first dominant and the second recessive) to three dwarf with purple flowers (the first recessive and the second dominant) to one dwarf with white flowers (both recessive). Mendel pointed out that this peculiar ratio of 9:3:3:1 would come about only if each of the two characters, height and flower color, were assorted into the gametes independently of each other. He summarized this in what is now called Mendel's second law, the law of independent assortment, which can be stated: "Genes for different characters assort into the gametes independently of one another."

When the F$_1$ individuals formed gametes, each of them must have formed four different kinds of gametes: DW, Dw, dW, and dw. With two parents each producing four different kinds of gametes, there are 16 possible pollination combinations. These are best worked out with a Punnett square as in Plate 63, since that will show every combination of one gamete from each parent.

You will note that every box labeled I has at least one dominant D gene and at least one dominant W gene. Nine of the 16 combinations fall into this category, and regardless of the second gene these plants receive for these characters, they will all be *tall* and have *purple flowers*.

Every box labeled J has at least one dominant D gene and two recessive w genes. Regardless of what the second gene for height may be, all of these will be *tall*, but they will all have *white flowers* because they are homozygous for the recessive trait. Three of the 16 possible combinations fall into this category.

Similarly, every box labeled K has two genes for dwarfness but at least one gene for purple flower, so these three combinations will produce plants that are *dwarf* with *purple flowers*.

Last, one combination (L) is homozygous recessive for both characters and will be *dwarf* with *white flowers*. All of this supports Mendel's view that genes for different characters assort independently of one another.

TWO-CHARACTER CROSS.

TALL PLANT$_A$/GENE$_{A^1}$
DWARF PLANT$_B$/GENE$_{B^1}$
PURPLE FLOWER$_C$/GENE$_{C^1}$

WHITE FLOWER$_D$/GENE$_{D^1}$
CROSS$_E$
SEGREGATION$_F$
GAMETE$_G$
CROSS-POLLINATION$_H$

P$_1$ PHENOTYPE.★

P$_1$ GENOTYPE.★

DDWW   X $_E$ ddww

P$_1$ GAMETES.★   DW   DW   dw   dw

F$_1$ PHENOTYPE.★

DdWw   F$_1$ GENOTYPE (DIHYBRID)★

★ALL

F$_1$ GAMETES.★ $\frac{1}{4}$ DW $\frac{1}{4}$ Dw $\frac{1}{4}$ dW $\frac{1}{4}$ dw

PUNNETT SQUARE.★
(F$_2$ GENOTYPES)★
F$_2$ PHENOTYPES.★
TALL PURPLE$_I$
TALL WHITE$_J$
DWARF PURPLE$_K$
DWARF WHITE$_L$

PHENOTYPIC RATIO★

9 : 3 : 3 : 1
  $_I$ $_J$ $_K$ $_L$

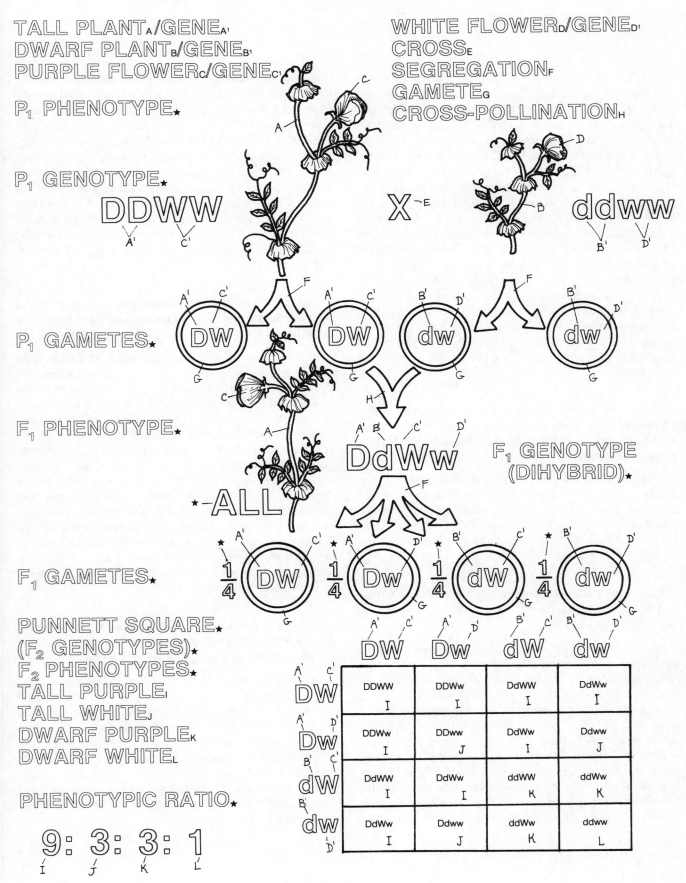

	DW	Dw	dW	dw
DW	DDWW I	DDWw I	DdWW I	DdWw I
Dw	DDWw I	DDww J	DdWw I	Ddww J
dW	DdWW I	DdWw I	ddWW K	ddWw K
dw	DdWw I	Ddww J	ddWw K	ddww L

Mendel's work went unappreciated for 35 years, by which time Mendel had died. Other biologists just weren't convinced that numbers had anything to do with biology. But during that 35-year period (1865–1900), great improvements were made in microscope lenses, and improved techniques for cell study (Plates 27 and 29) were developed. These advances led to the discovery of chromatin (Plate 30), so named because of the deep stain it takes from the appropriate dyes (Greek: *chromos*, "color"). In cell division the chromatin was seen to coil up into compact bundles called chromosomes as cells prepare to divide. In 1882 Walther Fleming worked out the details of the most common type of cell division, which is called mitosis (Greek: *mitos*, "thread") because of the threadlike appearance of the chromatin.

Although cell division is a continuous process, it is customary to divide it into four phases to make discussion easier. The period between divisions, when the cell is growing or just carrying out its life functions, is called interphase.

Color the heading Interphase at the upper right. Then color the headings Cell and Nucleus and titles A, D, E, and F. Color the cell beside the "Interphase" heading.

During interphase, the nucleus is surrounded by the nuclear envelope, and the *chromatin* is in the form of numerous loose threads. In cells of animals and primitive plants, a pair of *centrioles* is located in the cytoplasm just outside the nucleus. This plate illustrates division of such a centriole-containing cell.

Color the heading Prophase, all remaining titles, and the two prophase cells, one above and one below the Prophase heading.

Mitosis begins with prophase, in which the chromatin coils up and condenses into compact structures called *chromosomes.* Two prophase cells are shown, one early and one late, to emphasize the continuous nature of the process. As the chromosomes become shorter and thicker, it becomes clear that each one is made up of two subunits, which we call *chromatids.* If centrioles are present, a second pair of centrioles is synthesized during interphase, and a starlike array of microtubules called an *aster* forms around the centrioles. As prophase proceeds, the two pairs of centrioles move toward opposite sides of the cell, each with its own aster. During this migration, numerous addi-

tional microtubules are assembled between the centrioles to form the *spindle apparatus,* which is so named because of its similarity to the spindle of a spinning wheel. Many microfilaments of actin (one of the muscle proteins) become associated with the microtubules in the spindle. Any nucleoli present gradually become smaller and disappear. Eventually the asters arrive at opposite ends of the cell, and the spindle apparatus extends along one side of the nucleus. The nuclear envelope then disintegrates, marking the generally accepted end of prophase. (Cells without centrioles do not form asters but do form a very similar spindle, which is known as an anastral spindle, Latin: *an,* "without.")

Color the heading Metaphase and the structures in the two metaphase cells, one above and one below the heading.

In the portion of mitosis designated as metaphase, the spindle apparatus moves into the area of the nucleus. As it does so, the chromosomes move to the center of the spindle. Each chromosome attaches by a specialized portion *(centromere)* to a different bundle of spindle microtubules. In animals and seed plants, virtually all the cells have their chromosomes in homologous ("same-proportioned") pairs. This is referred to as the "diploid" condition. As you will see in the next plate, gametes are "haploid" (Greek: *haplos,* "single"), having only one chromosome of each pair. (Haploid cells can also divide by mitosis in some organisms but not in animals.) The number, size, and shape of the chromosomes is constant for any given species. Humans have 23 pairs of chromosomes; other species have chromosome numbers ranging from one pair to more than 200 pairs.

Color the headings Anaphase and Telophase and the corresponding cell structures. Then color the two daughter cells in the upper left corner.

In anaphase, the chromatids of each chromosome separate to form two daughter chromosomes. The spindle tubules pull these daughter chromosomes to opposite sides of the cell so that each end of the daughter cells has an identical set. In telophase, the spindle dissolves, the chromosomes uncoil and become a diffuse chromatin network again, and the nucleolus and nuclear envelope reappear. In most cells, the cytoplasm is then divided to form two separate cells, each surrounded by its own membrane, as illustrated at the upper left.

MITOSIS.

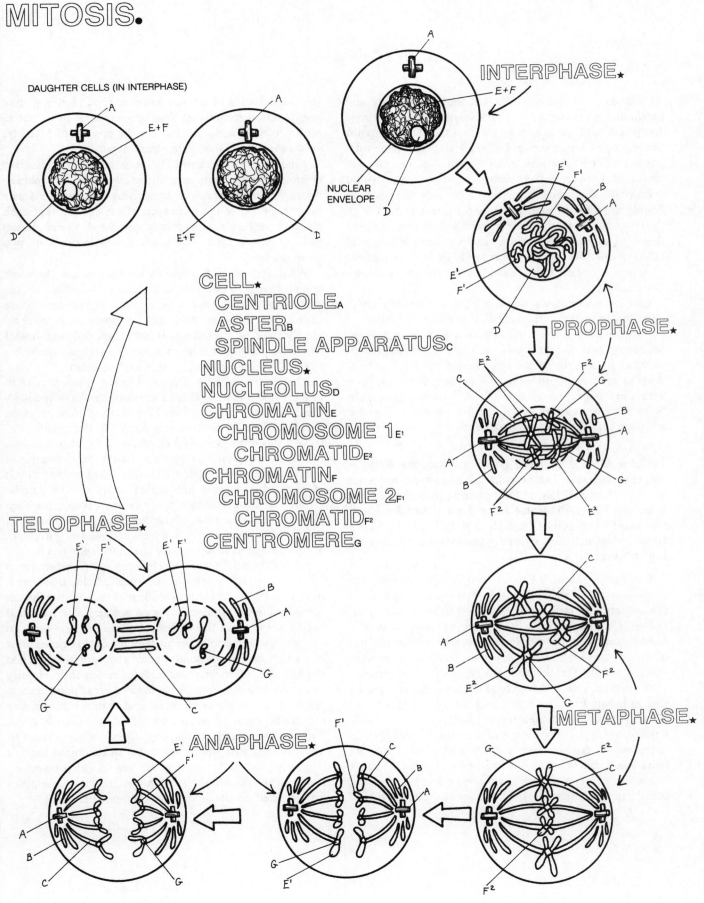

INTERPHASE★

DAUGHTER CELLS (IN INTERPHASE)

NUCLEAR ENVELOPE

PROPHASE★

CELL★
CENTRIOLE A
ASTER B
SPINDLE APPARATUS C
NUCLEUS★
NUCLEOLUS D
CHROMATIN E
CHROMOSOME 1 E¹
CHROMATID E²
CHROMATIN F
CHROMOSOME 2 F¹
CHROMATID F²
CENTROMERE G

TELOPHASE★

METAPHASE★

ANAPHASE★

As studies of sexual reproduction showed that egg and sperm nuclei fuse together in fertilization to form a single composite nucleus, it became obvious that at some point there must be a mechanism by which the cell reduces the number of chromosomes by half when such gametes are produced. Otherwise the number of chromosomes would double with each generation, and cells would soon have to double in size with each generation to have room for all the chromosomes. In 1890 Hertwig and Boveri discovered that eggs and sperm in animals are produced by a special kind of cell division that accomplishes that reduction. This type of cell division is called meiosis (Greek: *meioun*, "to diminish").

Like mitosis, meiosis begins with the chromosomes in homologous pairs, with each chromosome consisting of two chromatids. But instead of one division as in mitosis, producing two daughter cells with the same number of chromosomes as the parent cell, meiosis involves two divisions in succession, resulting in four daughter cells, each with only one chromosome of each homologous pair. Since there are two divisions, the four phases of division are gone through twice in succession.

Color titles A through B². Note that, for simplicity, the centrioles, aster, and spindle apparatus need not be colored. Begin at the upper right and color the headings Early Prophase I and Late Prophase I and the associated cells. Read the related text and continue in this manner, coloring the headings and reading the text.

Prophase I is much like the prophase of mitosis: the *chromatin* coils up to form *chromosomes,* a spindle is formed, and the nucleoli and nuclear envelope disappear. However, unlike mitosis, where the chromosomes of a homologous pair are randomly scattered within the nucleus, in meiosis the chromosomes undergo a process called synapsis (Greek: "union"), in which each chromosome pairs up with its homologue (the one that is the same size and shape) to form a compact bundle called a tetrad (Greek: "four," because each consists of four chromatids). During synapsis, one *chromatid* of each homologous chromosome crosses over a chromatid of the opposite homologous chromosome at one or more points. Such a crossover point is called a chiasma (Greek: "crossing point"; plural, chiasmata). The other two chromatids of

that tetrad usually do not cross over. Chiasmata may persist into metaphase I. This crossing over turns out to have profound consequences for the process of heredity, as we shall see a few plates from now.

Anaphase I is the same as in mitosis except that the two chromatids making up each chromosome do not separate; instead, the chromosomes comprising a homologous pair become pulled to opposite ends of the cell, so that each daughter cell ends up with only one chromosome of each pair, although each chromosome still consists of two chromatids.

What follows next depends on the species. In some species there is a complete telophase I, including the formation of nuclear envelopes, and an intermediate phase called interkinesis, in which the chromosomes uncoil completely as if in interphase. In other species, anaphase I proceeds directly into metaphase II. All gradations between these two extremes are found in nature.

The second meiotic division is even more similar to mitosis than the first. If the chromosomes have uncoiled, they coil back up in prophase II, and any nuclear envelope that was formed disintegrates again. If the spindle dissolved, it is re-formed. In metaphase II, the chromosomes attach to the spindle tubules and line up in the center of the cell. Anaphase II follows, in which the two chromatids of each chromosome are pulled apart by the spindle tubules to form daughter chromosomes, which are then pulled to opposite ends of the cell. Last, in telophase II, the spindle dissolves, the nucleoli and nuclear envelopes are formed, and four separate daughter cells result.

In 1900 Mendel's work was independently rediscovered by three separate scientists and received the attention it deserved. In 1902 a medical student named Walter Sutton compared this new information on genetics with what had been discovered in the meantime about chromosomes in cell division and published his hypothesis that genes must be carried on chromosomes, since they exactly matched the behavior of chromosomes. Both retain their identity from generation to generation, both occur in pairs in most cells and are single in gametes, and both are restored to the double number when two cells fuse in fertilization. Sutton also pointed out that the number of genes must be many times greater than the number of chromosomes and correctly predicted that Mendel's law of independent assortment would not apply to characteristics whose genes were carried on the same chromosome pair.

MEIOSIS.

CHROMATIN$_A$
  CHROMOSOME 1$_{A^1}$
   CHROMATID$_{A^2}$
CHROMATIN$_B$
  CHROMOSOME 2$_{B^1}$
   CHROMATID$_{B^2}$

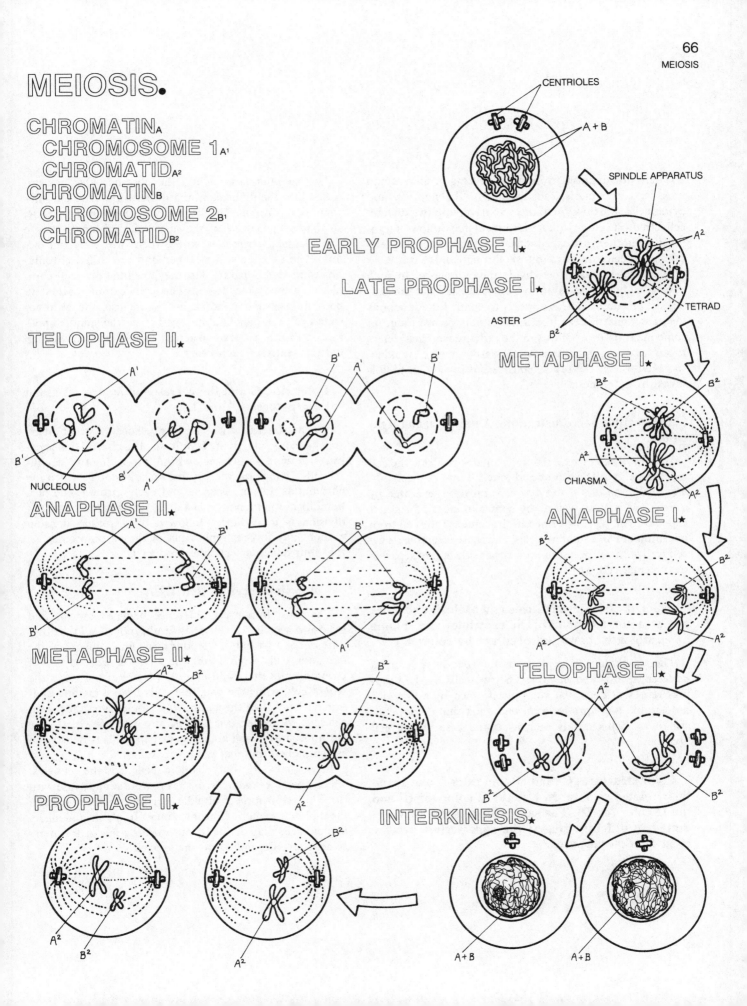

CENTRIOLES

A + B

EARLY PROPHASE I.★

SPINDLE APPARATUS

A^2

LATE PROPHASE I.★

ASTER

B^2

TETRAD

METAPHASE I.★

B^2 B^2

A^2

CHIASMA

A^2

ANAPHASE I.★

B^2 B^2

A^2 A^2

TELOPHASE I.★

A^2

B^2 B^2

INTERKINESIS★

A + B   A + B

TELOPHASE II.★

A^1

B^1

B^1 A^1 B^1

NUCLEOLUS B^1 A^1

ANAPHASE II.★

A^1

B^1

B^1

A^1

METAPHASE II.★

A^2 B^2

B^2

A^2

PROPHASE II.★

A^2

B^2

B^2

A^2

SUMMARY OF
MITOSIS AND MEIOSIS

Biologists distinguish two separate events in mitosis and meiosis: karyokinesis, the division of the nucleus, and cytokinesis, the division of the cytoplasm into two distinct cells. In most cells, cytokinesis immediately follows karyokinesis, but there are exceptions. Some cells, such as skeletal muscle cells, undergo mitosis through all the processes of karyokinesis but never divide the cytoplasm. In such cases it is difficult to say where one cell ends and the next begins, but each nucleus appears to control the functions in the cytoplasm near it. Similar situations are found in some plant tissues. Mitosis in the strict sense, then, refers to karyokinesis and does not necessarily involve cytokinesis. Meiosis, on the other hand, seems always to be followed by cytokinesis.

Color titles and illustrations A and B, using light colors.

In *animal cell cytokinesis,* microfilaments draw the cell membrane in at the center and pinch it into two cells. In *plant cell cytokinesis,* the new cells must be separated by a new cell wall as well as by a membrane. A structure called the cell plate is formed in the center of the cell from microtubules and small vesicles, and these assemble a new cell wall with cell membrane on either side of it to separate the original cell into two.

Color the headings Mitosis and Meiosis and titles and structures C and D. The centrioles and spindle apparatus are shown but need not be colored.

The remainder of the plate summarizes the differences between mitosis and meiosis. The two cells by the headings represent cells that were just formed by a previous mitosis and therefore have *chromosomes* that do not consist of two chromatids each as is the case during cell division.

Color structures C^1 and D^1 and titles E and F and their related structures. Use the C color for C^1 and the D color for D^1. The same chromosome is represented in different stages; the superscripts are for identification purposes.

During *interphase,* while the chromosomes are uncoiled into the diffuse chromatin network, each chromosome is *replicated.* That is, an exact copy of it is made, so that by prophase of the next cell division there are two identical structures where there was only one before. The two are still attached and are called chromatids until they separate, but they are in effect each complete chromosomes-to-be. This replication occurs in both mitosis and meiosis, but in meiosis the chromosomes of each pair line up together on the spindle apparatus to form a tetrad, whereas in mitosis they line up independently of each other.

Color title G and the corresponding arrows along with structures C^2 and D^2.

When the actual *division* of the cell occurs, other differences become apparent. In mitosis, the chromatids separate and there is only one division, producing two daughter cells. In meiosis, there is first one division in which the chromatids do not separate, but each chromosome of a homologous pair ends up in a different daughter cell. That division is immediately followed by a second division without any further replication of the chromosomes, so that four daughter cells are produced.

Color structures C^3 and D^3.

The most important difference between mitosis and meiosis is seen in the cells that result. Indeed, all the other differences exist only to produce this final difference: the daughter cells from a mitotic division have a set of chromosomes that is an exact duplicate of the set in the parent cell, while the daughter cells from a meiotic division have only half the number of chromosomes the parent cell had. Moreover, it is not just any half, but exactly one chromosome of each homologous pair. This is referred to as the haploid number of chromosomes.

In animals, the only haploid cells are the gametes, which must meet gametes of the opposite sex and fuse with them in fertilization or else die. Fertilization reestablishes the diploid number of chromosomes. In plants and algae, haploid cells may divide by mitosis for a few or for many generations, depending on the species.

SUMMARY OF MITOSIS AND MEIOSIS.

ANIMAL CELL CYTOKINESIS~A~

PLANT CELL CYTOKINESIS~B~

MITOSIS★   MEIOSIS★

CHROMOSOME 1~C,\ C^1,\ C^2,\ C^3~

CHROMOSOME 2~D,\ D^1,\ D^2,\ D^3~

INTERPHASE AND REPLICATION~E~

TETRAD FORMATION~F~

DIVISION~G~

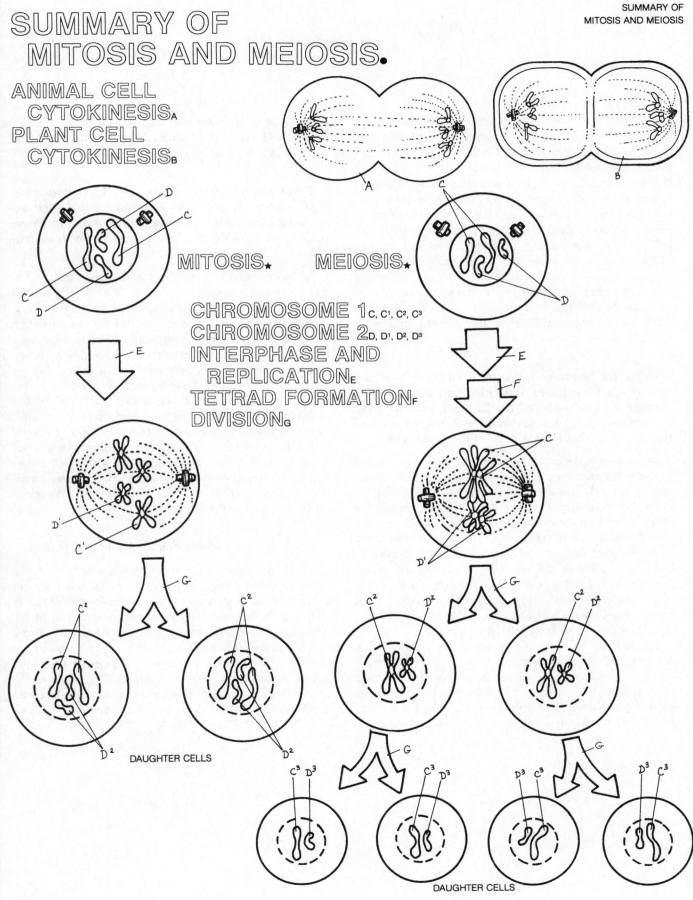

DAUGHTER CELLS

DAUGHTER CELLS

LINKAGE IN FRUIT FLIES

In 1910 Thomas Hunt Morgan began a project of studying the genetics of a tiny fruit fly with the scientific name of *Drosophila melanogaster,* which means "black-bellied dew lover." *Drosophila* feeds on the wild yeasts that grow in the dew that collects on ripe fruit, and it is an even better organism for genetic experiments than the pea plant. It is only a few millimeters long, and the 100 or more offspring resulting from a single mating can easily be raised in a half-pint bottle containing mashed bananas or a mixture of cornmeal and molasses with a little yeast sprinkled on top. A new generation is produced about every two weeks, so large numbers of offspring in several successive generations can be obtained in a very short time. *Drosophila* is still a favorite organism for experiments in student laboratory classes.

Color the headings *Drosophila melanogaster* and Wild-Type Characters and titles and structures A through E. Use dark red for C, black to color over the stripes (D), and a light tan for E, since these are the colors normally found in wild *Drosophila*.

Drosophila males are slightly smaller than *females* and have a more prominent black area on the tip of the abdomen. The small symbols you see next to the two flies are the biologist's symbols for male and female. Although the ideas from which these symbols originated would today be regarded as sexist stereotypes, the symbols are useful and a knowledge of their origins makes them easier to remember. The circle with the arrow point sticking out represents the shield and spear of Mars, the god of war, and is used to indicate a male. The circle with the cross at the bottom represents the hand mirror of Venus, the goddess of love, and is used to indicate a female.

The wisdom of Morgan's choice of *Drosophila* as his experimental organism became evident over the subsequent decade as one important discovery after another came out of his laboratory. Not only was *Drosophila* the right organism, but Morgan himself was a highly competent scientist and attracted some outstandingly capable students to study with him and do research under his direction. His laboratory, with hundreds of bottles of fruit flies, quickly became known as "the fly room."

Color the headings Linkage, P₁ Phenotype, and P₁ Genotype. Color titles F through J and their associated structures. Use the palest possible colors for G and I, since flies' wings are really transparent.

As Morgan and his coworkers studied their fruit flies to find more and more observable traits and determine how they were inherited, they found that Mendel's law of segregation held true with essentially 100 percent reliability. The law of independent assortment, however, failed more often than not. Many characteristics tended to be inherited together. This tendency came to be called linkage. Two traits in *Drosophila* that turned out to be linked were *black body* and *dumpy* (short) *wings,* both of which are recessive to the "wild-type" *tan body* and *long wings.* After obtaining pure-breeding (homozygous) strains of flies with tan bodies and long wings and separate pure-breeding (homozygous) strains of flies with black bodies and short wings, Morgan *crossed* the two strains. As expected, the F₁ offspring all showed the two dominant traits, tan body and long wings.

Color title K and the remainder of the plate.

When the hybrid flies were *interbred,* nearly all of the offspring showed only the combinations of traits present in the original P₁ parents. Nearly three-fourths of the F₂ offspring showed both dominant traits, and most of the remaining one-fourth showed both recessive traits. But a very few of the offspring showed the new combinations not present in the original P₁ parents: black body with long wings and tan body with short wings. The dashes in the genotypes indicate that the second gene of that pair could be either dominant or recessive, since we couldn't tell one from the other. Thus, B_ could be BB or Bb.

LINKAGE IN FRUIT FLIES.

DROSOPHILA MELANOGASTER★
MALE_A FEMALE_B
WILD-TYPE
  CHARACTERS★
EYES_C
STRIPES_D
BODY_E

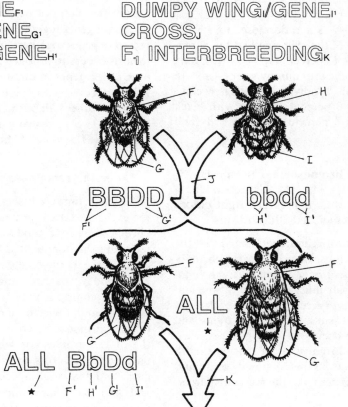

LINKAGE★
TAN BODY_F/GENE_F'   DUMPY WING/GENE_I'
LONG WING_G/GENE_G'   CROSS_J
BLACK BODY_H/GENE_H'   F_1 INTERBREEDING_K

P_1 PHENOTYPE★

P_1 GENOTYPE★   BBDD   bbdd

F_1 PHENOTYPE★

F_1 GENOTYPE★   ALL BbDd   ALL

F_2 PHENOTYPE★

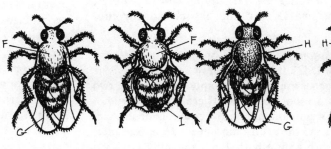

F_2 GENOTYPE★   B_D_   B_dd   bbD_   bbdd

CHROMOSOME CROSSOVERS

Morgan's observation of linkage in fruit flies was no great surprise, since Walter Sutton had predicted it and some cases had already been observed but not explained. Morgan wanted to investigate it thoroughly and find an explanation for it, so he did a test cross of the dihybrid F_1 progeny (Plate 68) with flies that were homozygous for both recessive traits, black body and dumpy wings. If complete linkage occurred, only the two parental combinations of traits (tan body with long wings and black body with dumpy wings) should appear in the offspring, in a ratio of 1:1. If independent assortment occurred, all four possible combinations of these traits should occur in the offspring in a ratio of 1:1:1:1.

What actually occurred was a ratio closer to 5:5:1:1. Approximately 84 percent of the offspring had the parental combinations of traits—42 percent tan body with long wings, 42 percent black body with dumpy wings—and 16 percent were "recombinants," individuals with new combinations of those genes—8 percent of them with black body and long wings and 8 percent with tan body and dumpy wings.

Color the headings Chromosome/Chromatid, P_1 Genotype/Chromosomes, and P_1 Gametes/Chromosomes; titles A, B, and C, using very light colors for A and B; and the associated illustrations.

To explain these results, Morgan hypothesized that the genes for these characters must be on the same chromosome. In the *homozygous dominant parent,* each chromosome of one homologous pair must have one B gene and one D gene. In the *homozygous recessive parent,* each chromosome of one homologous pair must have one b gene and one d gene. When *meiosis* occurs to form gametes, each gamete receives only one chromosome of that pair, both dominant genes in gametes from the dominant parent and both recessive genes in gametes from the recessive parent.

Color the heading F_1 Genotype/Chromosomes, title and structure D, and the associated chromosomes.

When *fertilization* occurs, the F_1 offspring not only have the genotype BbDd, but they have it in a special way: the two dominant genes are on one chromosome, and the two recessive genes are on the other. There are no chromosomes with one dominant and one recessive. (For simplicity, we omit all the other chromosomes.) In appearance, all of the F_1 offspring show both dominant traits, tan body and long wings.

Color titles and structures C¹ and E and the heading F_1 Gametes/Chromosomes. Color the chromosomes above the word OR and to the left of the E arrow. Color the C arrow at the left and the chromosomes below it.

Morgan reasoned that there must be an exchange of genes between chromosomes to produce the recombinants and that the chiasmata, or crossovers, seen in the *tetrads* of prophase I of meiosis must be points where the chromosomes actually break and trade parts. In some cases, the exchange either does not occur (as illustrated on the left side of the plate) or occurs in a way that keeps the two dominant genes together on one chromosome and the two recessive genes together on the other chromosome. When meiosis occurs to produce the F_1 gametes *without crossing over* of the genes in question, half of the gametes receive a chromosome with the two dominant genes, B and D, while the other half of the gametes receive a chromosome with the two recessive genes, b and d. (These gametes comprise *84 percent* of the total.)

Color the remainder of the plate.

At other times, *crossing over* occurs in such a way as to exchange these genes so that some recombinant chromosomes are produced, some gametes (8 percent) receiving the dominant B with the recessive d, and others (8 percent) the recessive b with the dominant D. We know today that crossing over occurs at some point (even at multiple points) in virtually every tetrad in every meiotic division, and in any tetrad one chromatid of one chromosome crosses over with one chromatid of the other chromosome while the other two chromatids usually do not cross over. Thus when crossing over occurs, half the resulting chromosomes are recombinants and half are nonrecombinants. When crossed with a homozygous recessive in a test cross, the phenotypes (appearances) of the resulting F_2 generation (not shown) are determined by whether or not the gametes from this F_1 generation carry chromosomes that are recombinants for these genes. The percentages of the phenotypes thus indicate the percentages present in the F_1 gametes. Different combinations of characteristics show different percentages of crossover, and A. H. Sturtevant, a student of Morgan's, developed a system of mapping chromosomes based on the hypothesis that genes farther apart on chromosomes should cross over more often than genes closer together. That method is still in use today.

CHROMOSOME CROSSOVERS.

CHROMOSOME/CHROMATID★
   PATERNAL (HOMOZYGOUS
      DOMINANT)A
   MATERNAL (HOMOZYGOUS
      RECESSIVE)B
MEIOSISC
FERTILIZATIOND

TETRAD FORMATIONC'
   NON–CROSSING OVERE
   CROSSING OVERF
RECOMBINANT CHROMOSOMES★
PATERNAL PORTIONA'
MATERNAL PORTIONB'

P₁ GENOTYPE/
CHROMOSOMES★

P₁ GAMETES/
CHROMOSOMES★

F₁ GENOTYPE/
CHROMOSOMES★

F₁ GAMETES/
CHROMOSOMES★

PERCENTAGE OF
TOTAL GAMETESH

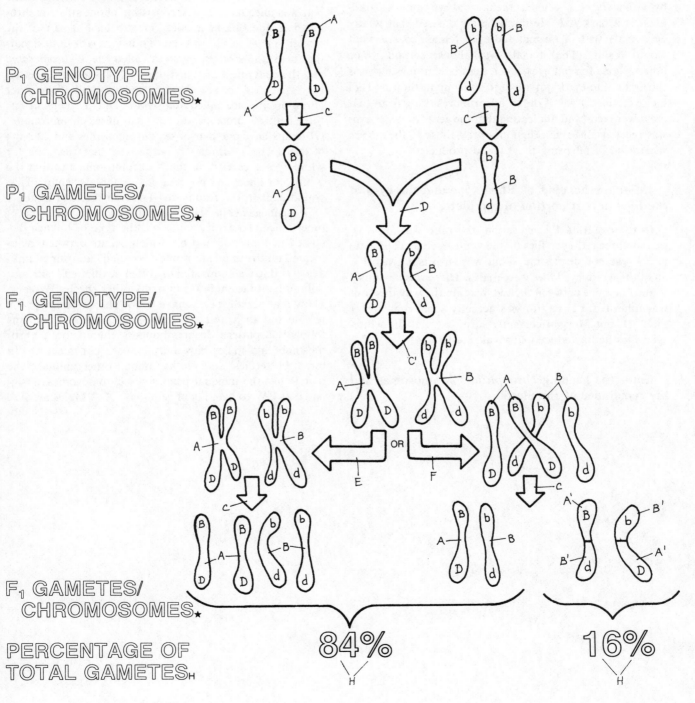

84%H

16%H

Color the headings P₁ Phenotype and F₁ Phenotype, titles and structures A through D, and the associated illustrations.

Morgan began his study of *Drosophila* by looking for various hereditary traits to study. He soon found one male fly with *white eyes,* whereas the normal eye color of *Drosophila* is a dark *red.* He immediately crossed that white-eyed male with a normal red-eyed female to see what would result. (That female was almost certain to be homozygous for red eyes, since the male in question was the first white-eyed individual to show up in thousands of flies examined.) All of the F₁ offspring resulting from that *cross* were red-eyed, indicating that the gene for white eyes was recessive. Morgan then *interbred* those F₁ flies to see what kind of offspring they would produce.

Color the heading F₂ Phenotype and the eyes of the flies in that portion of the plate.

In the resulting F₂ generation, the ratio was approximately three red-eyed flies to one white-eyed fly. The ratio of 3:1 was not surprising. What was surprising was that all of the white-eyed flies were males. There was obviously some difference between the two sexes in the way this trait was inherited. (The ratio was actually somewhat higher than 3:1, but Morgan correctly surmised that the white-eyed flies had a reduced survival rate.)

Color the heading *Drosophila* Chromosomes and the remainder of the plate.

Morgan hypothesized that this difference in the way the two sexes inherited the trait for white eyes could be explained if there were some difference in the chromosomes between the two sexes. He found that just a few years earlier someone had studied the chromosomes of *Drosophila* and found that there were four pairs of them, but one pair was unequal in males, consisting of one straight chromosome and one bent one. Females have a pair of the straight ones but no bent one. To have a way of designating them, the straight one was called the *X chromosome* and the bent one was called the *Y chromosome.* Together they are called the sex chromosomes, and all of the other chromosomes are called *autosomes.* Thus *Drosophila* has three pairs of autosomes and one pair of sex chromosomes. (Humans have one pair of sex chromosomes and 22 pairs of autosomes.) Morgan hypothesized that the gene for white eyes is carried on the X chromosome and that the Y chromosome has no gene at all for eye color. That would explain the results, as shown in the next plate.

Morgan gave the name "sex linkage" to this type of inheritance. To avoid confusion, the type of linkage discussed in Plates 68 and 69, which occurs between genes located on the same autosomes, is called "autosomal linkage." Today we know of many other sex-linked traits, not only in fruit flies but also in many other species. We also know that not all species have two X chromosomes in the female and an X and a Y in the male, although humans follow that pattern. In grasshoppers, there is no Y chromosome, and males have only one sex chromosome. In birds, butterflies, and moths, among other animals, the female has the unequal pair of sex chromosomes, designated as ZW to avoid confusion with XY. Males are ZZ.

SEX LINKAGE I.

WHITE EYE_A
RED EYE_B
CROSS_C
INTERBREEDING_D

P_1 PHENOTYPE.★

F_1 PHENOTYPE.★

ALL.★

F_2 PHENOTYPE.★

2♀
★

:1 ♂
★

:1 ♂
★

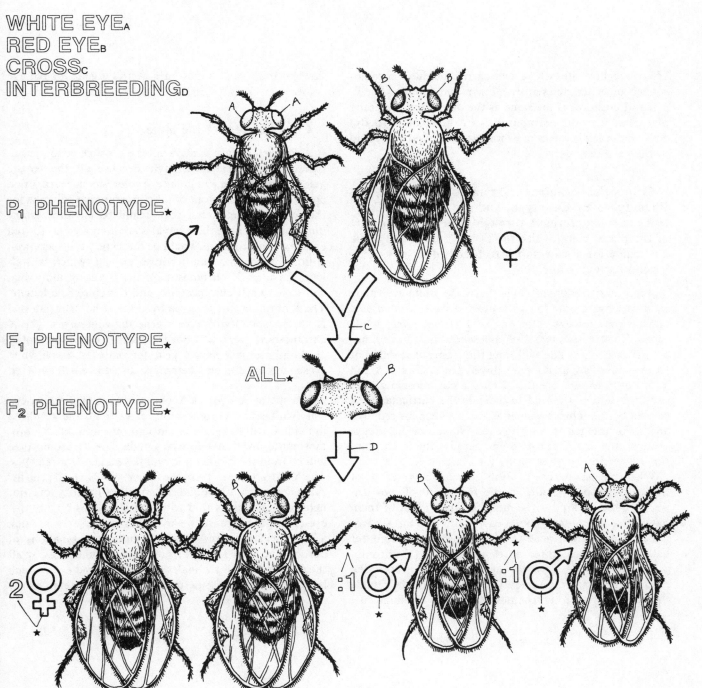

DROSOPHILA
   CHROMOSOMES.★
AUTOSOME_E
SEX CHROMOSOMES.★
   X CHROMOSOME_F
   Y CHROMOSOME_G

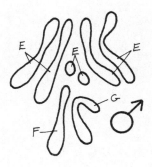

SEX LINKAGE II

This plate illustrates why a gene carried on the X chromosome is inherited differently in males and females. The P_1 generation shown is the same as the one in the preceding plate, but here the chromosomes in the somatic (body) cells and in the gametes are shown and colored according to the gene they carry.

Color the headings Chromosome/Symbol, P_1 Phenotype, P_1 Genotype, and P_1 Gametes. Color titles and structures A through G in the upper part of the plate, using white for D, red for E, and some dull color for F to emphasize that it does not carry a gene for eye color.

According to Morgan's hypothesis, the white-eyed male fly must have a gene for *white eyes* on its *X chromosome* and no gene for eye color on the *Y chromosome,* while each of the female's two *X chromosomes* must have genes for *red eye color.* You will note that somewhat different letter symbols are used for sex-linked genes. Capital X and Y are used to indicate the X and Y chromosomes, and subscript letters are used to indicate the particular gene carried by that chromosome: W for the gene for red eyes and w for the gene for white eyes. No subscript letter is written after the Y chromosome because the Y chromosome does not carry any gene for eye color.

When meiosis occurs to form gametes, *segregation* of the chromosomes results in the formation of two distinctly different kinds of gametes in males: half of them have three *autosomes* (one of each pair) plus an X chromosome carrying a gene for white eyes, and the other half of them have three autosomes plus a Y chromosome carrying no gene for eye color. Since both sex chromosomes are the same in females, all of the female gametes have three autosomes plus an X chromosome.

Each of those X chromosomes carries a gene for red eye color.

Color the rest of the plate.

A Punnett square to show all the possible genotypes in this case becomes a rectangle because all the female gametes are the same as far as eye color is concerned. (You could make a square out of it by making two columns for the female gametes, but that would be an unnecessary duplication because these females are homozygous for red eyes and both columns would be the same.) If a sperm cell with an X chromosome fertilizes an egg (which will always have an X chromosome), the resulting individual will have two X chromosomes and therefore be a female. The X chromosome received from the father (the paternal X chromosome) will have a gene for white eyes. The X chromosome received from the mother (the maternal X chromosome) will have a gene for red eyes. Since white eyes is recessive, such a heterozygous female will have red eyes.

If a sperm cell with a Y chromosome fertilizes an egg cell (with an X chromosome, of course), the resulting individual will have one X chromosome and one Y chromosome and will therefore be a male. The X chromosome will be from the mother and will have a gene for red eyes. The Y chromosome, from the father, will have no gene for eye color. Such an individual is neither homozygous nor heterozygous but is said to be "hemizygous." Only one eye-color gene is present, and that is for red eyes, so such a male will have red eyes. No other combinations of chromosomes are possible from this particular cross, so all the females and all the males should have red eyes, which is exactly how the experiment turned out (refer to the middle of Plate 70).

SEX LINKAGE II.

WHITE EYE$_A$
RED EYE$_B$
CHROMOSOME/SYMBOL★
AUTOSOME$_C$
WHITE-BEARING X$_D$
RED-BEARING X$_E$

Y CHROMOSOME$_F$
SEGREGATION$_G$

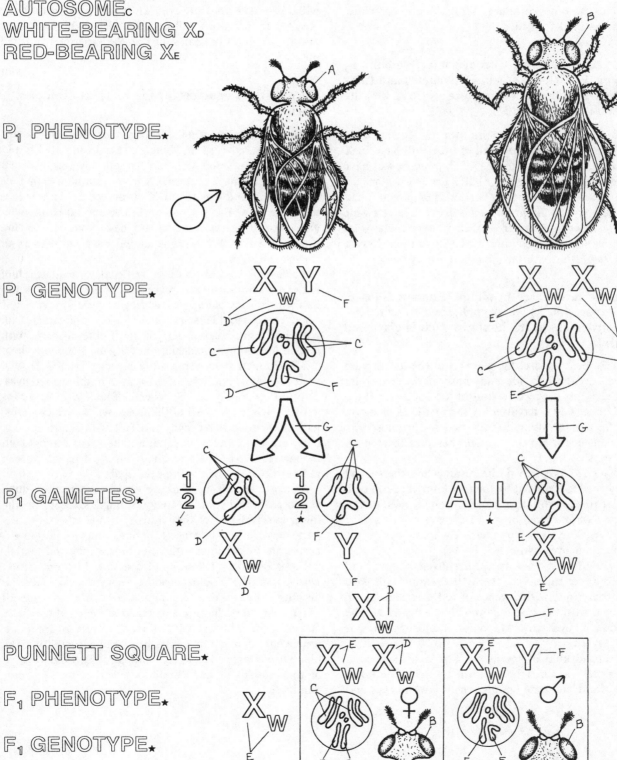

P$_1$ PHENOTYPE★

P$_1$ GENOTYPE★

P$_1$ GAMETES★

PUNNETT SQUARE★

F$_1$ PHENOTYPE★

F$_1$ GENOTYPE★

SEX LINKAGE III

This plate continues where Plate 71 left off in illustrating how the chromosomes sort out in Morgan's cross of white-eyed and red-eyed fruit flies.

Color the headings Chromosome/Symbol, F₁ Phenotype, F₁ Genotype, and F₁ Gametes and titles and structures B through G in those sections. Use the same colors as in Plate 71.

Continuing Morgan's reasoning, the F₁ females must produce two kinds of gametes: all of them will have an X chromosome, but half of those *X chromosomes* will have a gene for *red eyes* and the other half a gene for *white eyes*. The F₁ males will also produce two kinds of gametes: half of them will have an X chromosome that will have a gene for red eyes (since it came from their mothers, who were homozygous for red eyes), and half of them will have a *Y chromosome* that will have no gene for eye color.

Color the next three headings (Punnett Square, F₂ Phenotype, and F₂ Genotype). Then color title A and the structures in the boxes as each is discussed in the text.

In this Punnett square, the gametes of the females are symbolized on the left side and those of the males are symbolized across the top. Starting at the upper left, if the first kind of egg cell is fertilized by the first kind of sperm cell, the fly that results will have two X chromosomes, each with a gene for red eyes. It will therefore be a female with red eyes.

The lower left box shows the result when the second kind of egg cell is fertilized by the first kind of sperm cell: the fly that results will have two X chromosomes, one of them with a gene for red eyes and the other with a gene for white eyes. This fly will be a female with red eyes, since red is dominant over white.

If the second kind of sperm cell fertilizes the first kind of egg cell (upper right box), the fly that results will have one X chromosome with a gene for red eyes and one Y chromosome with no gene for eye color. This fly will be a male with red eyes, since the only gene for eye color is the gene for red.

If the second kind of sperm cell fertilizes the second kind of egg cell (lower right box), the fly that results will have one X chromosome with a gene for white eyes and one Y chromosome with no gene for eye color. This fly will be a male with white eyes. Thus only one-fourth of the possible fertilizations will produce flies with white eyes, and they will all be males, just as the experiment turned out.

Color the remainder of the plate as each part is discussed.

To test his hypothesis, Morgan *crossed* white-eyed males with red-eyed F₁ females. According to his hypothesis, these females should be heterozygous; each female should have a maternal X chromosome with a gene for red eyes and a paternal X chromosome with a gene for white eyes. Each white-eyed male should be hemizygous; his X chromosome should have a gene for white eyes and his Y chromosome should have no gene at all for eye color.

Each male should produce two kinds of gametes, half with an X chromosome carrying a gene for white eyes and half with a Y chromosome carrying no gene for eye color. Each female should also produce two kinds of gametes, all with an X chromosome, half of the X chromosomes with a gene for red eyes and half with a gene for white eyes.

The Punnett square shows the expected results. Half of the F₂ females should be heterozygous and have red eyes, since red is dominant; the other half should be homozygous for white eyes. All of the males will be hemizygous, half of them with red eyes and half with white eyes. If these are all produced in equal numbers, the ratio should be one red-eyed female to one white-eyed female to one red-eyed male to one white-eyed male.

When the resulting offspring were counted, the results supported the hypothesis very solidly. Except for the lower survival rate of the white-eyed flies, the ratio was quite close to 1:1:1:1. Since Morgan's time, numerous experiments by other investigators have supported his hypothesis beyond all shadow of a doubt. In humans and *Drosophila,* the Y chromosome seems to carry almost nothing but genes determining maleness, and virtually all other characters that are sex-linked are carried on the X chromosome. Three such "X-linked" traits that are found somewhat often in humans are hemophilia (inability of the blood to clot properly), muscular dystrophy (deterioration of the muscles in late childhood), and red-green color blindness.

SEX LINKAGE III.

WHITE EYE_A
RED EYE_B
CHROMOSOME/
SYMBOL★
AUTOSOME_C

F₁ PHENOTYPE★
F₁ GENOTYPE★

F₁ GAMETES★

PUNNETT SQUARE★

F₂ PHENOTYPE★

F₂ GENOTYPE★

TEST CROSS★

TEST CROSS_H

GAMETES★

PUNNETT SQUARE★

F₂ PHENOTYPES★

F₂ GENOTYPES★

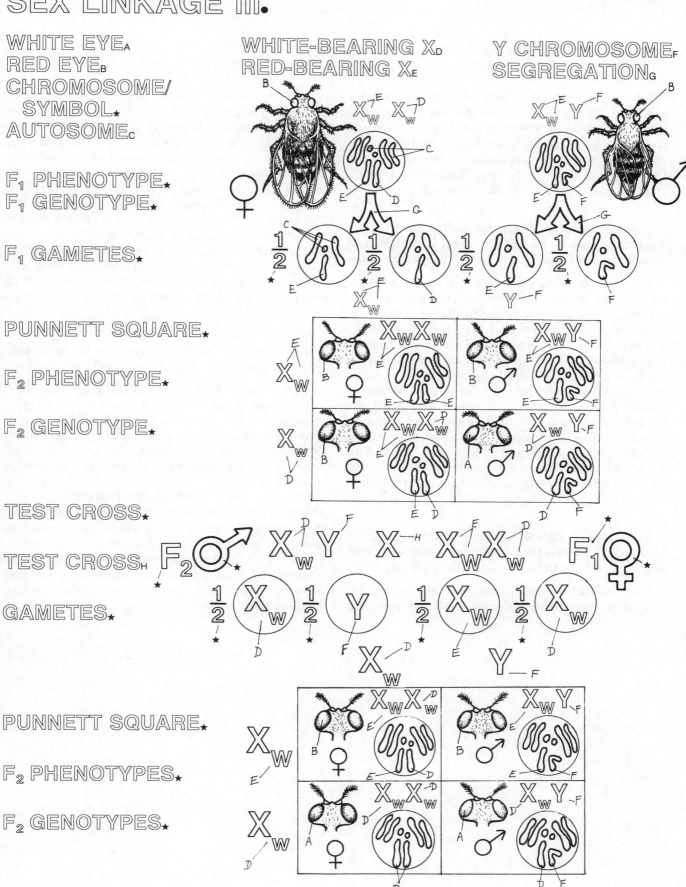

BLENDING INHERITANCE I

Not too long after Mendel's work was rediscovered, many cases were found in which dominance was lacking. (You will sometimes see these referred to as "incomplete dominance" or "codominance," but there is really no dominance at all, so "lack of dominance" is a better term, and "blending inheritance" is perhaps clearest of all.) An example of this is found in the color of snapdragon flowers.

Color the headings Snapdragons, P_1 Phenotype, F_1 Phenotype, and F_2 Phenotype, along with titles and structures A through E in the upper half of the plate.

If a snapdragon with *red flowers* is *crossed* with one having *white flowers,* neither color dominates the other, and the F_1 offspring all have flowers that are *pink.* If the F_1 plants are allowed to *self-pollinate,* their offspring (F_2) will be in a ratio of one red-flowered to two pink-flowered to one white-flowered. This result is explained only if the red-flowered plants are homozygous for red, the white-flowered plants are homozygous for white, and neither trait dominates the other. The pink flowers, then, are heterozygous, with one gene for red flowers and one gene for white flowers. If each gene produces its effect regardless of the other, the heterozygous plants will have both red and white pigments, and the result is the same as what we would get from mixing red and white paint: the flowers are pink.

Color the headings Genes, P_1 Genotypes, P_1 Gametes, and F_1 Genotypes and titles and structures A^1, B^1, and F in the associated illustrations.

The lower half of the plate shows how the *genes* sort out to produce the experimental results obtained above. To emphasize that there is no dominance, biologists usually symbolize the genotypes using a format similar to that used for sex linkage. Instead of capital and small letters, a capital letter is chosen for the general character being studied—C for color in this case—and different subscript letters for the different genes: R for red and W for white. Both subscript letters are capitals to remind you that there is no dominance. (Of course, you must also remember that this is not sex-linked, but since there is no such thing as a "C chromosome," that shouldn't be too hard.)

The two P_1 parents are both homozygous, and therefore each produces only one kind of gamete. After cross-pollination, the resulting F_1 plants are all heterozygous, with one gene for red flower and one for white, so their flowers are pink.

Color the heading F_1 Gametes and the remainder of the plate.

Each of the F_1 plants produces two kinds of gametes, one carrying the gene for red flower color and the other carrying the gene for white flower color. The Punnett square shows the offspring that result in the F_2 generation. The ratio predicted is the same 1:2:1 ratio actually obtained in the experimental cross, supporting the hypothesis of blending.

Today we know of many other characteristics in various animals and plants that show this pattern. In the breed of dogs known as Dalmatians, for example, the highly prized white coat with neat, round, black spots occurs only in heterozygotes. When two dogs with this ideal coat pattern are interbred, one-quarter of the offspring have coats that are almost entirely white, and one-quarter have very large irregular black patches.

BLENDING INHERITANCE I.

SNAPDRAGONS★
RED FLOWER_A
WHITE FLOWER_B
CROSS-POLLINATION_C

PINK FLOWER_D
SELF-POLLINATION_E

P₁ PHENOTYPE★

F₁ PHENOTYPE★

F₂ PHENOTYPE★   1 :2 :1
★ ★ ★

GENES★
RED GENE_{A¹}
WHITE GENE_{B¹}

SEGREGATION_F

P₁ GENOTYPES★ $C_R C_R$ $C_W C_W$

P₁ GAMETES★   ALL C_R C_W ALL C_R C_W

F₁ GENOTYPES★ $C_R C_W$

F₁ GAMETES★ $\frac{1}{2}$ C_R $\frac{1}{2}$ C_W
★ ★

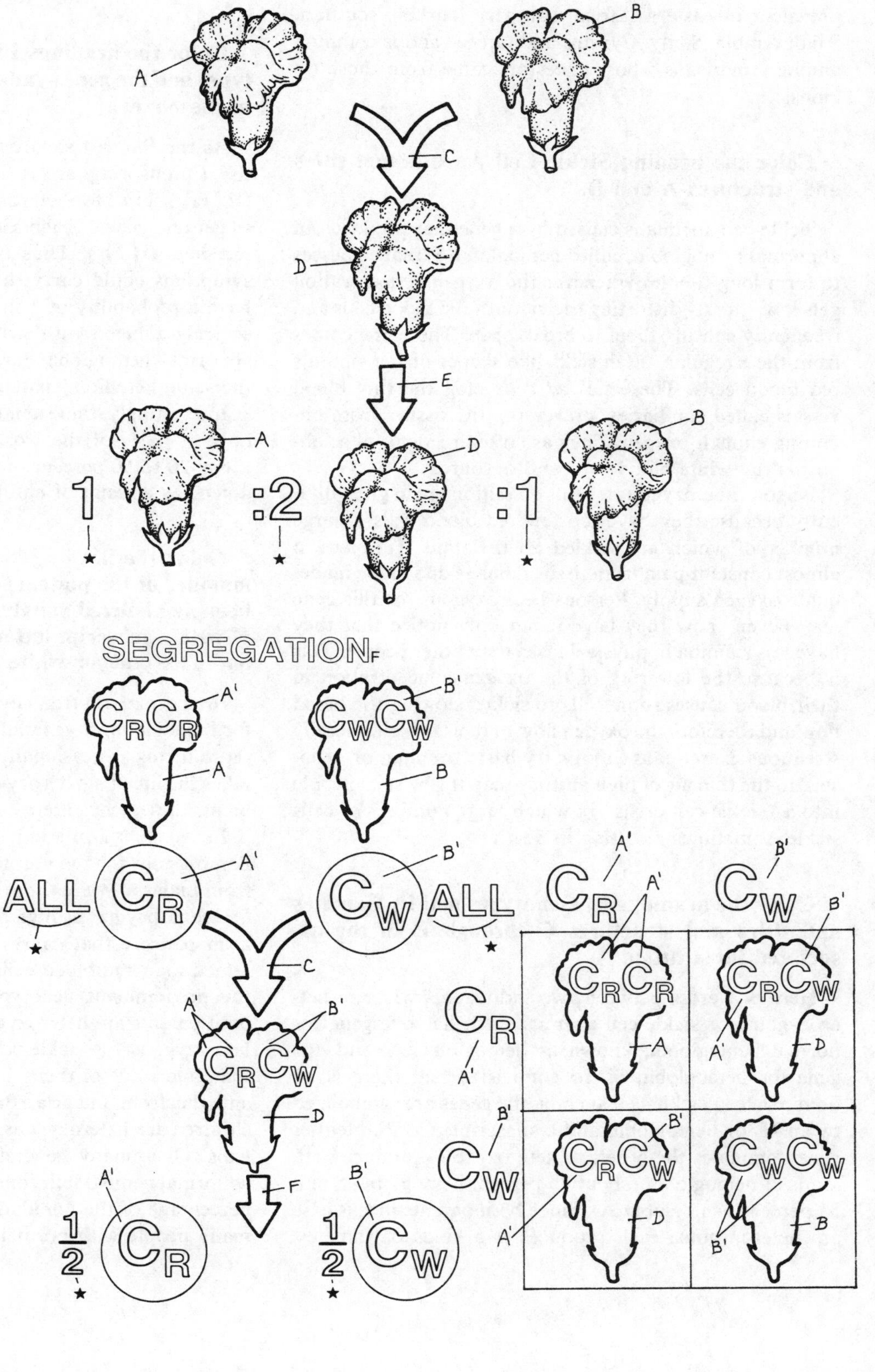

BLENDING INHERITANCE II

Blending inheritance in humans is seen in the genetic defect known as sickle-cell anemia, which is particularly prevalent in eastern Africa, southern Turkey, southern Saudi Arabia, Sicily, Cyprus, and Greece and is common among Americans whose ancestors came from those regions.

Color the heading Sickle-Cell Anemia and titles and structures A and B.

Sickle-cell anemia is caused by a gene that produces an abnormal hemoglobin, called hemoglobin S, that coalesces to form long tubules whenever the oxygen concentration gets low, grossly distorting the victim's red blood cells and frequently causing them to break open. The name comes from the irregular, often sicklelike shapes of the victim's red blood cells. These *sickled cells* clog the tiny blood vessels called capillaries, preventing the tissues from obtaining enough oxygen. They are often recognized as abnormal by white blood cells and devoured.

Persons homozygous for this condition are in great difficulty because they have too few red blood cells, a large number of which are sickled all the time. They are in almost constant pain from tissue damage due to the inadequate oxygen supply. Persons heterozygous for this gene may never know they have it but only notice that they have less stamina in physical tasks than other people. This is because the lowering of the oxygen concentration in their blood causes some cells to sickle, slowing the blood flow and therefore the oxygen flow to their muscles. Really strenuous exercise, as in military basic training, or exposure to the thin air of high altitude may throw such people into a "sickle-cell crisis" in which large numbers of cells sickle, sometimes resulting in death.

Color the headings P_1 Genotypes and P_1 Gametes and titles and structures C through E in the associated illustration.

Here is a cross between two individuals who are heterozygous for sickle-cell anemia. Each has one gene for normal hemoglobin, known as hemoglobin A, and one gene for hemoglobin S. To emphasize that there is no dominance in sickle-cell anemia, the genes are symbolized by an H for hemoglobin and a subscript A or S. Neither gene dominates the other, so heterozygotes produce both kinds of hemoglobin (about 45 percent hemoglobin S and 55 percent hemoglobin A). Since both parents in this case are heterozygous, each produces two kinds of gametes,

one carrying the gene for hemoglobin A, the other the gene for hemoglobin S.

Color the headings Punnett Square and F_1 Genotypes and the gene symbols comprising the genotypes in the boxes.

As the Punnett square shows, the probabilities among the F_1 offspring are 1 in 4 for homozygous normal ($H_A H_A$), 1 in 2 for heterozygous, since there are two ways to get one gene of each kind, and 1 in 4 for homozygous recessive ($H_s H_s$). Thus two parents who may have no symptoms could carry the defect in recessive form and have a probability of 1 in 4 that any given child will be severely afflicted with sickle-cell anemia. Other kinds of abnormal hemoglobins are also known, and they follow the same hereditary pattern.

Interestingly, these abnormal hemoglobins are common only in parts of the world where malaria is prevalent; there, 20 to 50 percent of the population may carry such a gene as a result of natural selection.

Color the heading Natural Selection and the remainder of the plate. (To save space, the genes have been symbolized simply by A and S, rather than by H with a subscript letter.) Use red for F, pink for G, and pale pink or white for H.

In the malaria-free environment, people homozygous for hemoglobin S generally die early in life, often before reproducing. Occasionally, a sickling crisis prematurely takes the life of a heterozygote as well. In such an environment, these genes are a significant disadvantage.

In a malaria-infested environment, however, people homozygous for the normal hemoglobin A frequently die from malaria, while those heterozygous for hemoglobin S find that they have a high resistance to malaria. The sporozoan parasite that causes malaria is simply not able to attack their red blood cells as effectively, so they become the predominant genotype in their population. Where both parents are heterozygous, some of their children are homozygous for sickle-cell anemia and die young from that, and some of their children are homozygous normal and die from malaria. But approximately half of their children are heterozygous and usually survive both problems. Over many generations, people heterozygous for abnormal hemoglobin come to make up a larger and larger percentage of the population because malaria kills off so many people with normal hemoglobin.

BLENDING INHERITANCE II.

SICKLE-CELL ANEMIA.★
NORMAL RED BLOOD CELL_A
SICKLED RED BLOOD CELL_B
HEMOGLOBIN A GENE_C

HEMOGLOBIN S GENE_D
SEGREGATION_E

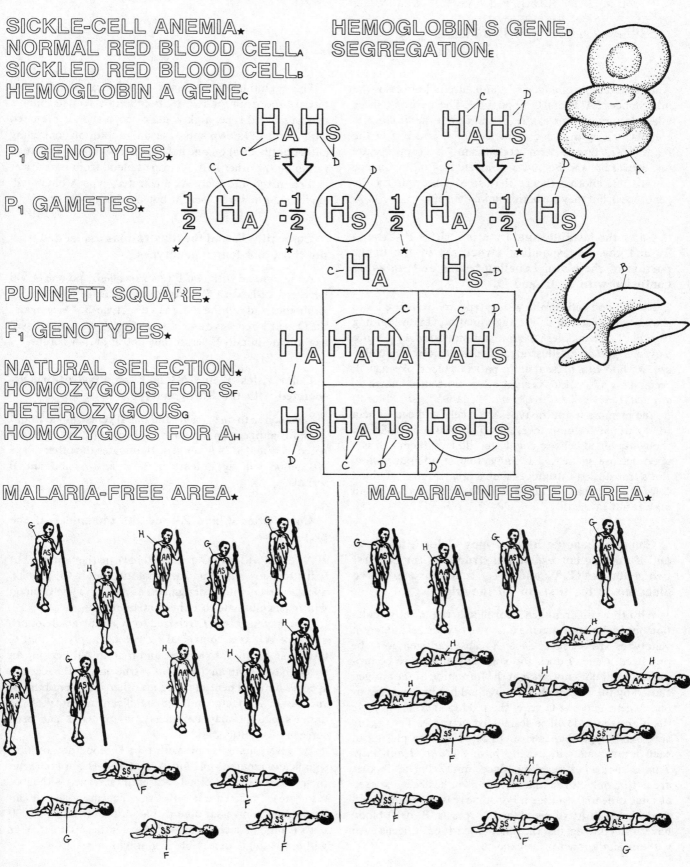

P_1 GENOTYPES.★

P_1 GAMETES.★

PUNNETT SQUARE.★

F_1 GENOTYPES.★

NATURAL SELECTION.★
HOMOZYGOUS FOR S_F
HETEROZYGOUS_G
HOMOZYGOUS FOR A_H

MALARIA-FREE AREA.★

MALARIA-INFESTED AREA.★

Another example of a lack of dominance in humans is the inheritance of A, B, and O blood types. There are only three alleles (alternative genes), A, B, and O, but the situation is complicated by the "codominance" of A and B over the recessive O. Here the term "codominant" is an appropriate one because A and B show a complete lack of dominance toward one another (hence the use of subscripts for the genotypes), but they are both dominant over O.

Color the heading Agglutination, titles P through T, and the corresponding structures in the upper portion of the plate. Labeling begins at P to avoid confusion with A, B, and O.

If red blood cells of the wrong type are transfused into a patient, a reaction called agglutination (Latin: "gluing together") occurs. Here the transfusion of *type A cells* into a *type B patient* is illustrated. Defense molecules called *anti-A antibodies* present in the patient's blood plasma will recognize and stick to certain other molecules known as *antigen A molecules* (Greek: *anti,* "against"; *gen,* "born") on the membranes of the type A red cells and cause them to agglutinate (clump together). This makes it easier for the white blood cells to engulf and destroy them and is a good defense against most things that invade our bodies. But a transfusion introduces such a huge number of blood cells at once that agglutination clogs blood vessels and may result in death.

Color the heading Blood Types. Color titles U, V, and X and the corresponding structures for the first two genotypes ($I_A I_A$ and $I_A I_O$) and their associated diagrams in the first row of the table.

With three different genes for blood type in the population but each person having only two of them, six different genotypes are possible. *Gene A* causes antigen A to be produced. *Gene O* does not cause any antigen to be produced, but it does not prevent the formation of the antigen caused by the A gene. If an individual has two A genes or one A gene and one O gene, the red blood cells will have the A antigen and will be designated as type A. The O gene acts like a typical recessive. In addition, the plasma of such individuals will usually have the *anti-B antibody*. This is somewhat puzzling, since most other antibodies are not manufactured unless the person has been exposed at some time to the antigen. For some reason yet undiscovered, the majority of people with types A, B, or O blood have antibodies against the other blood-type antigens even without any known prior exposure.

The agglutination reactions in a typical blood-typing test are shown in the two right columns. To determine a person's blood type, a glass microscope slide is prepared with a drop of known anti-A serum (a solution containing anti-A antibodies) on one half and a drop of known anti-B serum on the other half. A drop of blood from the patient is then mixed into each. As illustrated, type A blood will agglutinate in anti-A serum but not in anti-B serum.

Color title W and the illustrations associated with the third and fourth genotypes.

An individual with two *B genes* or one B and one O will have red cells with B antigen and plasma with anti-A antibodies and will be designated as type B. Once again, the O gene is recessive and has no effect. Type B blood will agglutinate in anti-B serum but not in anti-A serum.

Color titles Y and Y^1 and the illustrations associated with the fifth genotype.

A person with one A gene and one B gene will have both A and B antigens and neither antibody and will be designated as *type AB.* Neither gene dominates the other. Type AB blood will agglutinate in both anti-A and anti-B serum.

Color titles Z and Z^1 and the remainder of the plate.

A person with two O genes will have neither antigen but will often have anti-A and anti-B antibodies in the plasma. Such a person will be designated as *type O.* Type O blood will not agglutinate in either antiserum.

You may hear type O referred to as a "universal donor" and type AB as a "universal recipient," implying that O blood can be given to anyone and that an AB person can accept blood from anyone. This is true only within certain specific limits. There must be no other incompatibilities, such as in Rh factor, and no significant quantity of the donor's plasma may be transfused unless it has been screened for antibodies.

However, many people with type O blood do not have significant amounts of anti-A and anti-B antibodies in their plasma. If their blood is carefully screened with laboratory tests to be sure it is free of other antibodies, it can be used as "universal donor" blood. More than 100,000 units (pints) of such blood were used in the Vietnam War without a single detectable agglutination reaction.

CODOMINANCE AND BLOOD TYPES.

AGGLUTINATION.★
TYPE A RED BLOOD CELLS$_P$
  TYPE$_{P^1}$
TYPE B PATIENT$_Q$/CELL$_{Q^1}$
  TYPE$_{Q^2}$
ANTIGEN A$_R$
ANTI-A ANTIBODIES$_S$
ANTIGEN B$_T$

BLOOD TYPES.★
TYPE A GENE$_U$
TYPE O GENE$_V$
TYPE B GENE$_W$

ANTI-B ANTIBODIES$_X$
TYPE AB RED BLOOD CELL$_Y$/ TYPE$_{Y^1}$
TYPE O RED BLOOD CELL$_Z$/ TYPE$_{Z^1}$

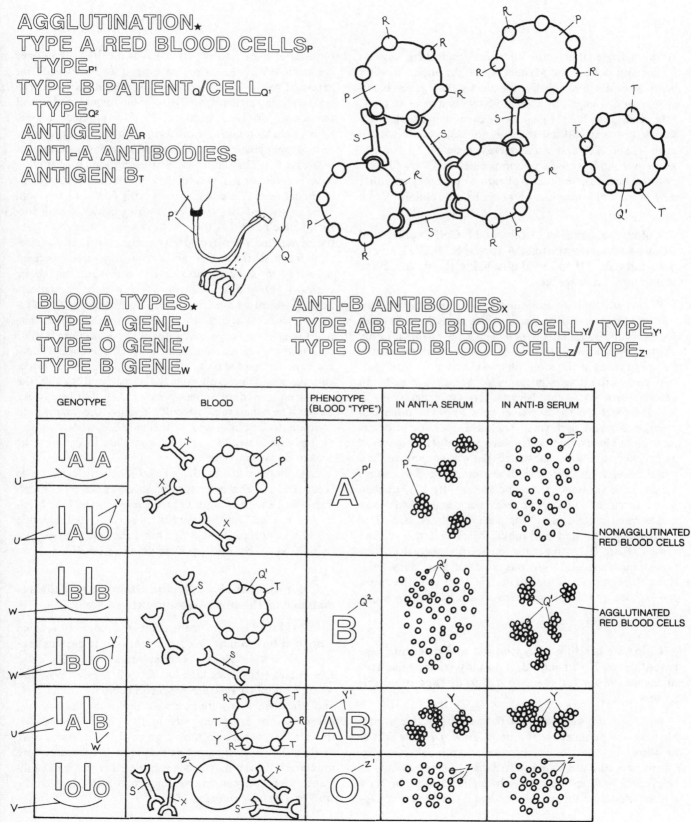

GENOTYPE	BLOOD	PHENOTYPE (BLOOD "TYPE")	IN ANTI-A SERUM	IN ANTI-B SERUM
I_AI_A / I_AI_O		A$^{P^1}$		NONAGGLUTINATED RED BLOOD CELLS
I_BI_B / I_BI_O		B$^{Q^2}$		AGGLUTINATED RED BLOOD CELLS
I_AI_B		AB$^{Y^1}$		
I_OI_O		O$^{Z^1}$		

DNA DISCOVERY

As knowledge of genetics increased and there was no longer any doubt that Mendel was right about heredity being determined by the discrete units we call genes, biologists naturally began to wonder just what a gene was like. It became clear that the genes were carried on the chromosomes, but that still left the questions of exactly what a gene looked like and what it was made of. Ultimately, genes were discovered to be molecules of DNA (deoxyribonucleic acid), and that realization opened up an entire new branch of study known as molecular biology.

Color the heading Analysis of Cell Nuclei and titles and representations A through D[1]. Use a very pale color for D[1] to avoid obscuring the words. Save bright red for later use.

We usually think of molecular biology as a very modern field of study, and it is true that the analytic techniques necessary for major advances have been available for only the last few decades. But the attempt to understand the nature of genes at the molecular level actually started just two years after the publication of Mendel's results. In 1868 Friedrich Miescher began a series of experiments in which he isolated the *nuclei* of cells from the other cell components and performed chemical analyses to determine what kinds of molecules were present. He discovered a previously unknown substance that was *not protein,* had a *high phosphorus content,* and was *mildly acid* in nature. He gave this substance the name *nuclein.* By 1889 nuclein had come to be known as nucleic acid, and R. Altmann had worked out a method for purifying it and had identified the purine and pyrimidine bases and the pentose sugars (Plate 22) comprising it. Shortly thereafter, one type of nucleic acid, DNA, was established as characteristic of all cell nuclei, although no one knew how the bases, the pentose, and the phosphate were arranged in the molecule.

Color the heading Alkaptonuria and titles and representations F[1] through J. You may wish to use two different colors for the two different functional enzymes (F[1], F[2]).

In 1908 a different piece of the puzzle became available with the publication of a book, *Inborn Errors of Metabolism,* by the English physician Archibald Garrod. Garrod became interested in metabolic errors, defects in body chemistry that produce dramatic changes in body tissues, blood, and urine. He studied patients with alkaptonuria, a disease in which molecules called alkaptones are excreted in the urine, causing it to turn black on exposure to air. Black pigments also accumulate in cartilage, causing arthritis, and can even turn the ears and the whites of the eyes black. He also studied patients with phenylketonuria, a disease in which the accumulation of phenylketones in the blood causes severe mental retardation. (The name refers to excretion of phenylketones in the urine, where they are easily detected by chemical tests.) In collaboration with Reginald Bateson, Garrod studied the families of these patients and correctly concluded that these defects were inherited as simple Mendelian recessives. He hypothesized that each defect resulted from the absence of a single *enzyme* necessary to carry out the next step in some metabolic pathway (Plate 52). The recessive genes for the defects he studied must have been unable to produce a *functioning enzyme,* so the pathway would operate up to the point where the enzyme was missing, but no farther. The *substrate* for the missing enzyme would then *accumulate* continuously and start to overflow into the blood, urine, and tissues. The result was the same as if one of the robots on a modern robot-operated assembly line broke down. The other robots wouldn't know this and would continue to function, and a great pile of partially assembled products would start to accumulate in front of the broken-down machine.

Garrod pointed out that it was likely that all genes controlled the characteristics of an organism by controlling which enzymes or other proteins are made. Like Mendel before him, Garrod found that the rest of the scientific world was not quite ready for his ideas, and it was more than 30 years before they were taken seriously.

Color the heading Specific Staining and the remainder of the plate. Use bright red for K.

In 1914 the German chemist Robert Feulgen developed a method for staining *DNA* a bright red. It did not produce this color with any other substances found in cells. Eventually (ten years later, in fact) he announced that when he applied this DNA-specific *stain* to intact *cells* and when the cells were examined under the microscope, the red color appeared only in the chromosomes. Although this was by no means proof that the genes were made of DNA, it did support that idea. If the genes were found only in the chromosomes, then any chemical substance found only in the chromosomes just might be the stuff that genes are made of.

DNA DISCOVERY.

ANALYSIS OF CELL NUCLEI.★
NUCLEI_A
CHEMICAL SEPARATION_B
PROTEIN_C
MYSTERY SUBSTANCE (NUCLEIN)_D
  CHEMICAL TESTS_E
  PROPERTIES_D'

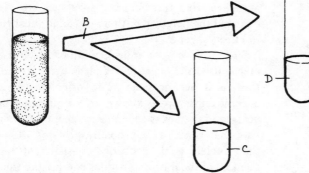
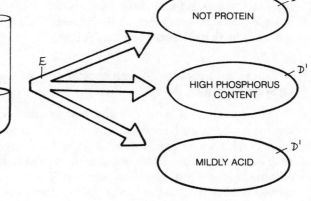

NOT PROTEIN _D'

HIGH PHOSPHORUS CONTENT _D'

MILDLY ACID _D'

ALKAPTONURIA.★
FUNCTIONAL ENZYME_F1, F2
REACTING COMPOUND_G1,G2
REACTION_H
ACCUMULATING COMPOUND_I
NONFUNCTIONAL ENZYME_J

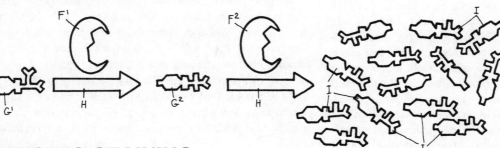

SPECIFIC STAINING.★
FEULGEN STAIN_K
CELLS_L
STAINED DNA_K1

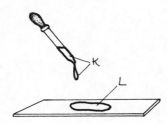

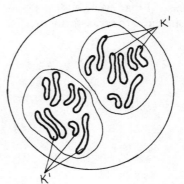

MAGNIFIED CELLS

DNA AND TRANSFORMATION

In 1928 another thread of evidence for the chemical nature of the gene came to light when the English bacteriologist Frederick Griffith reported his discovery of something peculiar in pneumococci, the bacteria that cause the most common form of pneumonia, as you might have guessed from the first half of their name. The second half of their name is the bacteriologist's term for bacterial cells that are spherical in shape: cocci (singular, coccus).

Color the heading Pneumococcus and titles and structures A and B.

The normal, disease-causing strain of pneumococci is known as the S strain because it forms very smooth-looking colonies when grown in a culture dish. The smooth appearance is due to a thick, gelatinous *capsule* surrounding the bacteria, which occur mostly in pairs. Griffith also had a mutant strain that could not form a capsule and was relatively harmless. It was called the R strain because the colonies it formed in a culture dish had a rough appearance.

Color the heading Injection and titles and structures C through F.

Griffith inoculated mice with various combinations of the two strains of bacteria, including cells of the S strain that had previously been killed with heat. The *live S strain bacteria* invariably killed the *mice* in a short time, while those of the *R strain* generally did not. When he inoculated some mice with *S strain bacteria* that had been *killed with heat,* the mice suffered no effect at all. But when he inoculated mice with a *mixture* of cells of the *R strain* and *heat-killed cells of the S strain,* the mice invariably died, and live S strain bacteria could be found in large numbers in the tissues of the dead mice. Over the next three years, other scientists found that this "transformation," as it came to be called, could be brought about even in a culture dish, without requiring a host animal to infect. In 1933 the American James Alloway reported that even cell-free extracts of S strain bacteria could cause the transformation.

In all of these cases, the S strain bacteria that resulted continued to reproduce thereafter as S strain.

These results made it clear that some substance from the dead S strain bacteria not only gave the R strain bacteria the ability to form a capsule and to cause pneumonia but also changed them genetically so they actually became S strain bacteria and remained so from then on. Whatever that transforming substance was, it obviously changed the heredity of those cells and must therefore be, in effect, a gene.

The discovery of transformation, of course, immediately raised the question, What is the chemical nature of this substance or gene? The principal components of chromosomes were known to be proteins and nucleic acids, and there was general agreement among most scientists that genes were probably proteins. It was thought that nucleic acids probably consisted of repeating sequences of the four nucleotides comprising them and thus were too simple to occur in enough different forms to account for the immense variety of known genes. However, 16 years after Griffith's work, a new discovery led to a reconsideration of nucleic acids.

Color the heading The Transforming Substance and the remainder of the plate as you read.

In 1944, after nearly ten years of painstaking work, three American biologists, Owen Avery, Colin McCarty, and Maclyn McLeod, announced that they had *isolated and purified* the various types of molecules found in S strain pneumococci and tested each one for its ability to accomplish the transformation discovered by Griffith. The only substance that would *transform* R strain cells into S strain cells was a particular kind of nucleic acid, deoxyribonucleic acid, better known today by its abbreviation, *DNA.*

This was not enough to convince everyone that genes were DNA, but it convinced some. DNA began to be taken more seriously, and investigations into its nature increased greatly.

DNA AND TRANSFORMATION.

PNEUMOCOCCUS.★
  CELL$_A$
  CAPSULE$_B$
MOUSE$_C$/DEAD MOUSE$_{C÷}$

INJECTION.★
LIVE S STRAIN$_D$
LIVE R STRAIN$_E$
HEAT-KILLED S STRAIN$_F$

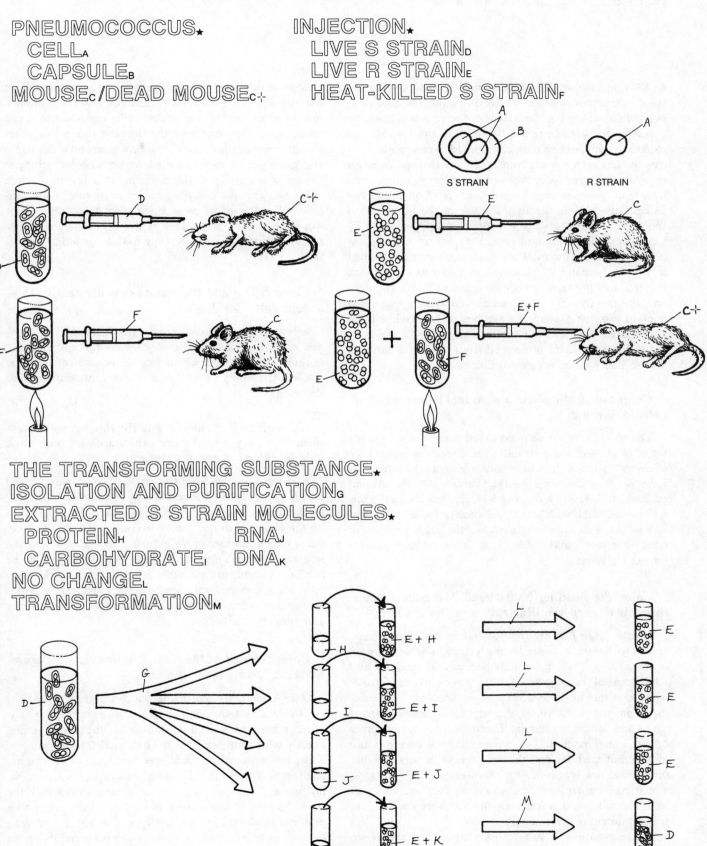

THE TRANSFORMING SUBSTANCE.★
ISOLATION AND PURIFICATION.$_G$
EXTRACTED S STRAIN MOLECULES.★
  PROTEIN$_H$ RNA$_J$
  CARBOHYDRATE$_I$ DNA$_K$
NO CHANGE$_L$
TRANSFORMATION$_M$

DNA IN *NEUROSPORA*

In 1941, just before the announcement that DNA was the substance responsible for transforming pneumococci, another discovery about the nature of the gene was made by two Americans, George Beadle and Edward Tatum, who experimented with the pink bread mold, *Neurospora*. *Neurospora* has even more advantages than *Drosophila* as an experimental organism. Not only does it reproduce even faster and produce even larger numbers of offspring, but its cells are haploid throughout most of the life cycle, so the effect of a recessive gene cannot be easily masked by a corresponding dominant gene. It grows readily in a test tube on a simple fluid culture medium, commonly called a "minimal medium," consisting of a solution of glucose, mineral salts (nitrates, phosphates, and sulfates), and only one vitamin (biotin). This means that *Neurospora* must contain enzyme systems to synthesize for itself all the amino acids and all the vitamins except biotin and does not need to get them preformed from its diet as humans and so many other organisms do.

Color the heading *Neurospora* and titles and structures A through E.

The mold first grows a mass called a *mycelium,* consisting of branching white strands that spread out in the slice of *bread* or other growth medium to draw in nutrients. Later it grows *stalks* up from the surface into the air, and on each stalk grows a *fruiting body* containing numerous *spore cases,* each with eight pink-orange spores in it. The spores are asexual reproductive cells, each capable of starting a new mold colony if it lands on any suitable nutrient material.

Color the heading Nutritional Mutants, titles F through F², and the illustration in the top row.

Beadle and Tatum exposed numerous *Neurospora* spores to X-rays in order to produce mutations (changes in the genes). These mutations occurred at random, and while some of them rendered the spores incapable of growing at all, other mutations left them incapable of synthesizing a particular enzyme necessary for manufacturing a single amino acid or vitamin. Those spores could not grow on a minimal medium, but they could grow on a mixture of minimal medium and the amino acid or vitamin that they could not synthesize for themselves. Such mutants, which have nutritional requirements that are different from the wild strains from which they were derived, are called "nutritional mutants."

The steps in Beadle and Tatum's experiment are shown in simplified form here. *Spores* of *Neurospora* were irradiated with *X-rays,* and then each spore was placed in

a separate tube of *"enriched medium,"* a growth medium consisting of water, several sugars, mineral salts, numerous vitamins, and all the amino acids necessary to make proteins. Any *mutant spore* that lost the ability to synthesize one or more necessary vitamins or amino acids could still grow on the enriched medium because everything it needed was already in the medium. In a few days, each spore had grown a complete *colony* in its tube, with hundreds of spores of its own. Since all of these new spores were the direct descendants of the single haploid spore that had started the colony, they were all genetically identical.

Color title J and the second-row illustration. Use a light color for J.

A few spores from each tube were then tested on *minimal medium.* If a spore didn't grow on minimal medium, the entire colony from which it came consisted of cells that had lost the ability to synthesize one or more nutrients for themselves.

Color titles K, L, and M and the third-row illustration. You may wish to use crosshatching with each pair of colors in the test tubes.

Additional spores from the mutant colony were placed in three other tubes, one containing *minimal medium plus all the amino acids,* one containing *minimal medium plus all the known vitamins,* and one containing *minimal medium plus a number of different sugars.* In this way the researchers could determine the general category of the needed nutrient; for example, a mutant that could grow only when supplied with amino acids must have a mutation that prevents it from synthesizing one or more of the amino acids for itself.

Color the rest of the plate, continuing with crosshatching if you used it earlier.

Once it was determined what kind of nutrient a particular strain could not synthesize, more spores from that same strain were placed in additional tubes to determine exactly which nutrient was the one needed. For example, if it grew only on minimal medium plus all the amino acids, spores were placed in numerous tubes, each containing minimal medium plus one particular amino acid. If the spores grew into mold colonies only in a tube containing *minimal medium plus,* for example, the amino acid *arginine,* it was clear that a mutation had occurred in a gene controlling the synthesis of that compound, arginine. The next plate shows some of the results.

DNA IN *NEUROSPORA*.

BREAD_A÷
NEUROSPORA★
MYCELIUM_B
STALK_C
FRUITING BODY_D
SPORE CASE_E
NUTRITIONAL MUTANTS★
SPORE_F
X-RAYS_G
X-RAYED SPORE_{F¹}
ENRICHED MEDIUM_H
NEUROSPORA COLONY_I
MUTANT SPORE_{F²}

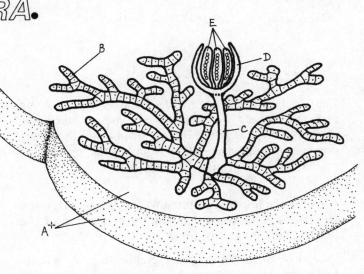

MINIMAL MEDIUM_J
WITH: ALL AMINO ACIDS_K
ALL VITAMINS_L
VARIOUS SUGARS_M
GLYCINE_N
ORNITHINE_O
CITRULLINE_P
ARGININE_Q

NO GROWTH ON MINIMAL MEDIUM MEANS MUTANT SPORES. THEY CANNOT SYNTHESIZE ALL NUTRIENTS.

GROWTH ON MINIMAL MEDIUM PLUS AMINO ACIDS MEANS MUTANT CANNOT SYNTHESIZE ONE OR MORE AMINO ACIDS.

GROWTH ON MINIMAL MEDIUM PLUS ARGININE MEANS MUTANT CANNOT SYNTHESIZE ARGININE.

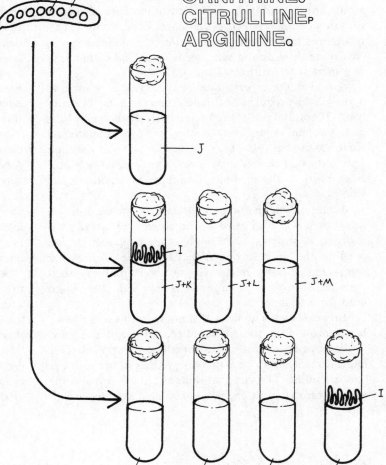

GENE FUNCTION

Continuing the experiments described in the preceding plate, Beadle and Tatum discovered many "nutritional mutants" of *Neurospora*. This plate illustrates three such mutants with different abilities to utilize the three amino acids ornithine, citrulline, and arginine.

Color the heading Related Mutants and titles and structures A through F in the upper portion of the plate. Use pale colors for A, B, C, and D.

None of the mutants in this group could grow on *minimal medium* alone. Mutant I could grow only on minimal medium plus *arginine.* Mutant II could grow on minimal medium plus *citrulline* or *arginine* but not ornithine. Mutant III could grow on minimal medium plus *ornithine* or *citrulline* or *arginine.* Although many other kinds of mutants were discovered, none were found with any other combination of requirements for these substances. For example, no mutant was found that required ornithine but would not grow if citrulline or arginine was supplied instead. Any mutant that required citrulline could also grow if arginine was supplied in its place. Beadle and Tatum recognized that the only explanation that made sense was that these three amino acids were sequential compounds in a metabolic pathway (Plate 52).

Moreover, the experimental results made it possible to figure out the sequence of these compounds in the pathway. (If you haven't looked too closely at the rest of the plate yet, you might want to stop here for a minute and see if you can figure out the sequence for yourself.) Mutant I gives the first clue. Mutant I won't grow unless arginine is supplied in the medium. Ornithine or citrulline just won't do.

Mutant II, on the other hand, will grow with citrulline, but it will also grow in the absence of citrulline if arginine is supplied. Although other explanations are possible, the simplest one is that citrulline is an earlier compound in some particular metabolic pathway than arginine, so as long as arginine is supplied, there is no need for citrulline.

Mutant III adds additional support to this hypothesis. It will grow if ornithine is supplied, but it will also grow in the absence of ornithine if citrulline is supplied. This indicates that ornithine is probably earlier in the pathway than citrulline. The fact that mutant III will grow in the absence of both ornithine and citrulline if arginine is supplied gives further support to the idea that arginine is later in the pathway than the other two.

Color the heading One Gene, One Polypeptide, titles G¹ through E³, and structures B, C, and D in the lower part of the plate. You may wish to use a different color for each gene and each enzyme to emphasize that three different genes and three different enzymes are involved.

Beadle and Tatum recognized that these results supported the hypothesis put forth by Archibald Garrod 33 years earlier (Plate 76), that genes were responsible for the functioning or nonfunctioning of enzymes catalyzing individual steps in metabolic pathways. Each of the mutant strains of *Neurospora* had to have a defective *enzyme* for a different step in the pathway. Mutant I must have a defect in some enzyme involved in converting citrulline into arginine. From the evidence of this experiment alone, we can't tell whether that occurs in a single step or in several, but biochemists today are convinced that each of these conversions is a single reaction, so we can confidently label this enzyme as enzyme 3. Mutant II must have a defect in a different enzyme, which we label as enzyme 2, responsible for converting ornithine into citrulline. Mutant III must have a defect in enzyme 1, which converts some unidentified precursor molecule into ornithine.

Since these enzyme deficiencies were demonstrated to be hereditary, there had to be something wrong with the *genes,* and Beadle and Tatum proposed that the principal way in which genes exert their influence on living organisms must be that each gene controls the production of a specific enzyme. Thus a defect in gene 1 results in a defect in enzyme 1. Similarly, defects in gene 2 and gene 3 result in a defects in enzyme 2 and enzyme 3. This idea became known as the "one gene, one enzyme" hypothesis.

After years of additional tests by numerous other scientists with all kinds of living organisms, that hypothesis is widely accepted— with one small modification. Today we know that nearly all large proteins are made up of more than one kind of polypeptide chain and that each polypeptide is controlled by a different gene. We also know that genes control the production of all proteins, not only enzymes. Beadle and Tatum's hypothesis has therefore been modified to "one gene, one polypeptide."

GENE FUNCTION.

RELATED MUTANTS.*   ARGININE_D

MINIMAL MEDIUM_A   *NEUROSPORA* COLONY_F

ORNITHINE_B

CITRULLINE_C

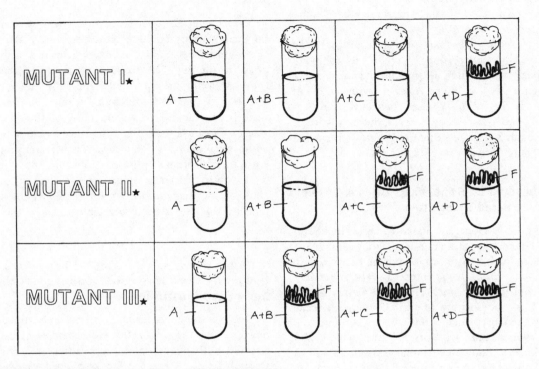

ONE GENE, ONE POLYPEPTIDE.*

GENE_{G¹, G², G³}

ENZYME_{E¹, E², E³}

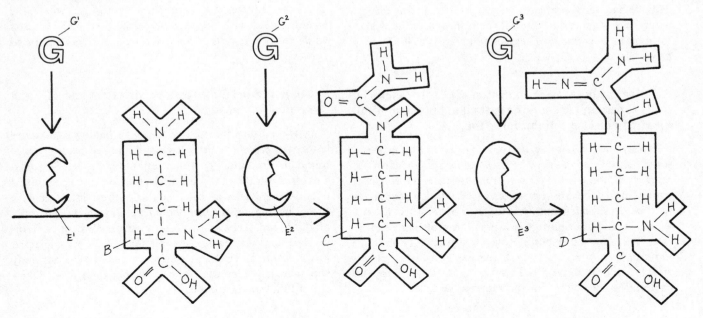

DNA IN BACTERIOPHAGE

The next major milestone in the search for the molecular nature of the gene came in 1952, when Alfred Hershey and Margaret Chase published the results of their experiments with bacteriophage, a type of virus that parasitizes bacteria (Greek: *phagein,* "to eat").

Viruses are incredibly tiny particles, so small that they can't be seen in a light microscope. They live as "obligate intracellular parasites," which means that they are unable to live independently but must enter a living cell and parasitize it from the inside. Viruses themselves have no cells, are unable to reproduce or carry out any sort of metabolism outside a host cell, and are generally regarded as not really being "alive."

Color the heading Bacteriophage, titles A through D, and the associated illustrations.

The virus known as bacteriophage (phage, for short) has a peculiar shape, consisting of a head, stalk, and arms. The arms appear to be used in attaching the phage to the bacterial cell *(bacterium)* to be infected. When Hershey and Chase began their experiment, little was known about the structure and composition of bacteriophages except that they consisted of protein and DNA. Nothing was known about which part was protein and which was DNA. It was known that a bacteriophage would attach to a bacterial cell, inject its *core* into the cell, leaving its outer *coat* as a "ghost" attached to the outside of the cell, and in less than an hour, 100 or more *new bacteriophage* particles would be formed inside the cell and the bacterium would disintegrate, releasing the phages to infect new cells. Whatever the phages were injecting, it was clearly acting like a gene, carrying all the information necessary for making new phage. The question was, Was it DNA or protein?

Color the heading Preparation of "Hot" Bacteria, titles E through H, and the associated portion of the illustration. Use a light color for F.

Hershey and Chase began their experiment by growing some *bacteria* of the species *Escherichia coli* (a harmless resident of almost everyone's large intestine) on *nutrients* containing *radioactive phosphorus (^{32}P),* which the bacteria took up and incorporated into their own molecules just as they would nonradioactive atoms. They also grew some other bacteria of the same species on nutrients containing *radioactive sulfur (^{35}S),* which the bacteria also took up and incorporated into their own molecules. Thus the bacteria became "hot" in the radioactive sense.

Color the heading Preparation of "Hot" Phage, titles I through H^2, and the associated portion of the illustration.

Next bacteriophages *(phage culture)* were inoculated into each of the bacterial cultures, where they invaded the bacterial cells and consumed them in the production of more phages. The *radioactive atoms of the bacteria* thus became incorporated into the newly formed phages, so that two different strains of *"hot" phages* were prepared: one in which many of the *sulfur atoms* were *radioactive* and the other in which many of the *phosphorus atoms* were *radioactive.* Hershey and Chase knew that protein contained sulfur but no phosphorus, while DNA contained phosphorus but no sulfur, so they could follow the radioactivity and tell whether the protein, the DNA, or both were injected into the bacterial cell.

Color the heading "Hot" Phage Plus "Cold" Bacteria, titles J through M, and the associated portion of the illustration.

Finally, some bacterial cultures were infected with phage containing radioactive sulfur and some others were infected with phage containing radioactive phosphorus. Enough time was allowed for the bacteriophages to inject their cores into the cells, but not enough for more phages to be produced. Next the cultures were run in a *blender* just long enough to knock the empty phage coats off the bacterial cells but not long enough to disrupt the cells. Then the cultures were spun in a *centrifuge,* where the bacterial cells with the *phage cores* settled to the bottom while the outer *coats of the bacteriophages* remained suspended.

Color the heading Results, titles G^3 and H^3, and the magnifications.

In the culture infected with phage having *radioactive sulfur (in protein),* virtually all of the radioactivity was found in the liquid portion containing the *phage coats.* That meant that nearly all of the phage protein remained outside the bacterial cell. In the culture infected with phage having *radioactive phosphorus (in DNA),* virtually all of the radioactivity was found in the sediment containing the bacterial cells, showing that DNA was injected *(phage core)* and protein was not injected. This strongly supported the idea that, in bacteriophage at least, DNA could function as a gene.

DNA IN BACTERIOPHAGE.

BACTERIOPHAGE★
 COAT_A
 CORE_B
NEW PHAGE_C
BACTERIUM_D

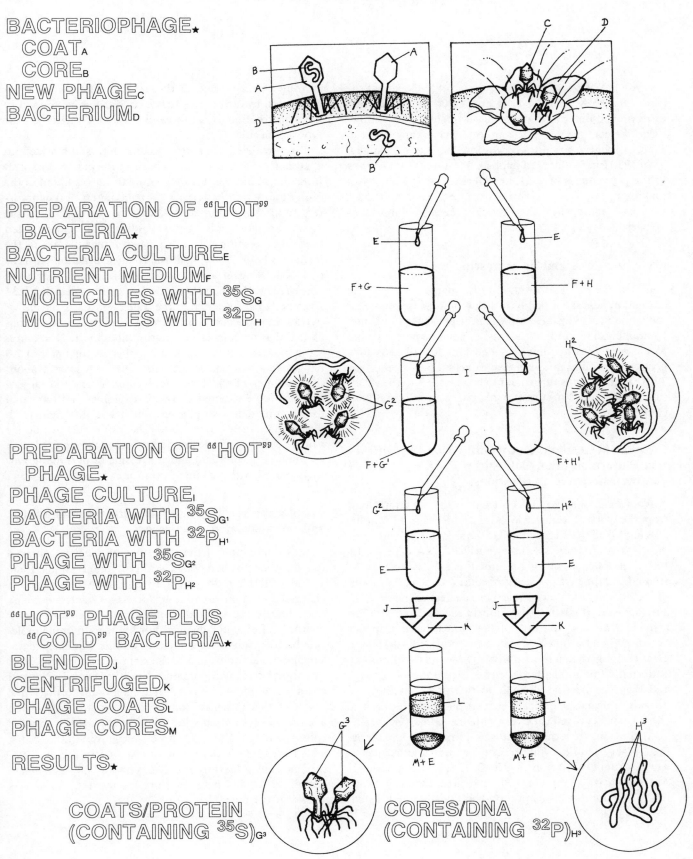

PREPARATION OF "HOT"
 BACTERIA★
BACTERIA CULTURE_E
NUTRIENT MEDIUM_F
 MOLECULES WITH ^{35}S_G
 MOLECULES WITH ^{32}P_H

PREPARATION OF "HOT"
 PHAGE★
PHAGE CULTURE_I
BACTERIA WITH ^{35}S_{G1}
BACTERIA WITH ^{32}P_{H1}
PHAGE WITH ^{35}S_{G2}
PHAGE WITH ^{32}P_{H2}

"HOT" PHAGE PLUS
 "COLD" BACTERIA★
BLENDED_J
CENTRIFUGED_K
PHAGE COATS_L
PHAGE CORES_M

RESULTS★

COATS/PROTEIN
(CONTAINING ^{35}S)_{G3}

CORES/DNA
(CONTAINING ^{32}P)_{H3}

DNA BASE PAIRING

As evidence began to mount in support of the idea that the gene was DNA, more investigators began to concentrate on trying to determine the structure of the DNA molecule. The components of the molecule were known (recall Plate 22), but it was not known how they were arranged.

In 1950 Erwin Chargaff published results of his analyses of the percentages of the four different nitrogenous bases making up the DNA of various species and announced a peculiar relationship that quickly became known as "Chargaff's rule."

Color title A and its representation.

Chargaff's rule states that in any sample of DNA, the amount of adenine is equal to the amount of thymine and the amount of cytosine is equal to the amount of guanine. Chargaff also found that the ratio of adenine plus thymine to cytosine plus guanine was constant for all cells in a given species but can vary from one species to another. These relationships turned out to be key clues to the structure of DNA and to how it could carry an informational code.

Color the headings X-ray Diffraction and Diffraction Pattern on Film. Color titles B, D, E, F, and G and the associated illustrations.

Additional clues to DNA structure and function came from the work of Maurice Wilkins and Rosalind Franklin at King's College, Cambridge University, in England, who did X-ray diffraction studies of purified DNA crystals. In this technique, *X-rays* are beamed at the *crystals* to be studied and the rays that are *diffracted* (bent) by the atoms of the crystal are allowed to expose a sheet of *X-ray film.* The angles at which the X-rays are diffracted depend on the arrangements and spacings of the atoms making up the crystal. When the film is developed, it shows light and dark areas that together comprise what is known as a diffraction pattern. Diffraction patterns are not the same as an image, and they can be very difficult to interpret, but they do provide information on spacing of repeating structures in a crystal, which proved to be very valuable in this case.

At the same time that Wilkins and Franklin were doing their X-ray studies, King's College had another team working on DNA: Francis H. C. Crick and a visiting American with a brand new Ph.D. degree, James D. Watson. Their approach was purely theoretical and consisted entirely of examining all the experimental evidence accumulated by other investigators and trying to apply it to the construction of a three-dimensional model of the DNA molecule.

They realized that the relationship summarized in Chargaff's rule had to have a basis in the molecular structure, and so did the three repeating distances found in the X-ray diffraction patterns: 0.34 nanometers, 2.0 nanometers, and 3.4 nanometers. Just two years earlier, the American biochemist Linus Pauling had demonstrated that many portions of most protein molecules are in the form of a helix. Looking at the X-ray diffraction patterns of DNA, Watson and Crick recognized that they were consistent with a helix shape. They also noticed that 0.34 nanometers is just about the thickness of a purine or pyrimidine molecule (both of which are flat), so they concluded that the purines and pyrimidines in the DNA molecule must be closely stacked together. A diameter of 2.0 nanometers was consistent with what was known about the diameter of the DNA molecule. It was too large a distance to be spanned by two pyrimidines and too small to accommodate two purines, but it was just right for a purine to join a pyrimidine. Chargaff's rule would be neatly explained if adenine (a purine) always paired with thymine (a pyrimidine) and if cytosine (a pyrimidine) always paired with guanine (a purine).

Color the heading Base Pairs, the remaining titles, and the associated illustration.

An examination of the structures of the bases showed that only adenine had polar (weakly charged) atoms (Plate 11) in just the right places to line up with oppositely charged atoms on thymine to form two *hydrogen bonds* (customarily represented by three dots as shown here). Neither cytosine nor guanine nor another thymine molecule has the right arrangement to do this. Similarly, cytosine had polar atoms in just the right places to form three hydrogen bonds with guanine. No other combinations of bases would allow for good hydrogen bond formation.

All of this evidence pointed to a structure in which adenine always paired with thymine and cytosine always paired with guanine and those purine-pyrimidine pairs, which are quite flat in shape, stacked up in the center of the molecule, leaving the deoxyribose and phosphate groups to form a helix twisting around the outside, as shown in the next plate.

DNA BASE PAIRING.

CHARGAFF'S RULE$_A$

X-RAY DIFFRACTION★
X-RAY SOURCE$_B$
CRYSTAL OF DNA$_D$
DIFFRACTED X-RAYS$_E$
X-RAY FILM$_F$

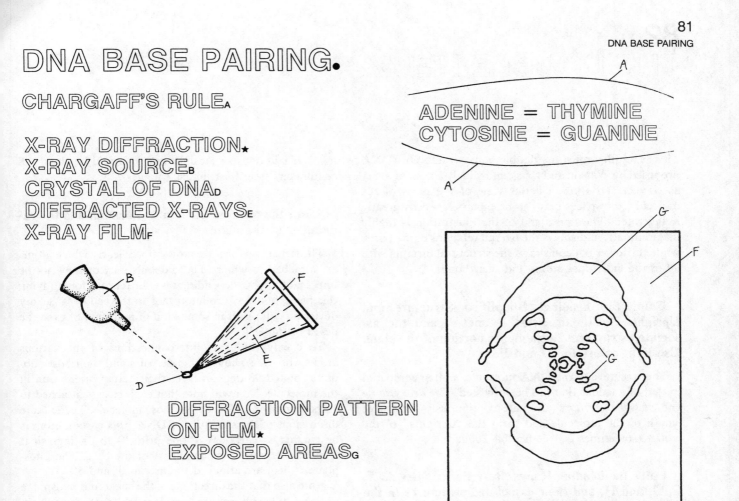

ADENINE = THYMINE
CYTOSINE = GUANINE

DIFFRACTION PATTERN
ON FILM★
EXPOSED AREAS$_G$

BASE PAIRS★
CARBON$_C$
NITROGEN$_N$
HYDROGEN$_H$
OXYGEN$_O$
HYDROGEN BONDS$_I$

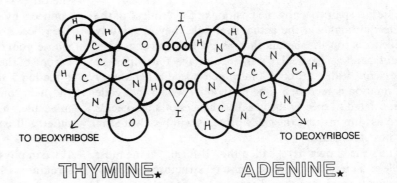

TO DEOXYRIBOSE

TO DEOXYRIBOSE

THYMINE★   ADENINE★

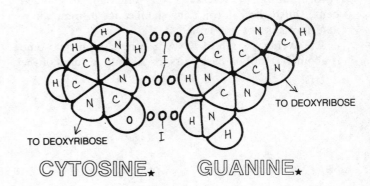

TO DEOXYRIBOSE

TO DEOXYRIBOSE

CYTOSINE★   GUANINE★

This plate illustrates the double helix structure of DNA proposed by Watson and Crick and widely accepted today as correct. To allow a better view of the parts of the molecule, the spaces between base pairs has been greatly exaggerated. The upper end of the illustration is highly diagrammatic and shows the overall relations of the parts, while the lower portion shows the structural formula with all of the individual atoms and their bonds.

Color the headings Simplified Structure and Uprights/Backbone, titles D and P, and the associated structures in the upper portion of the plate. Use light colors for D and P.

The structure of the DNA molecule is often compared to that of a ladder that has been twisted. The *deoxyribose* and *phosphate groups* alternate continuously the whole length of the molecule and form the "uprights" of the ladder (sometimes called the "backbone").

Color the heading Rungs/Base Pairs, titles A, T, C, G, and H, and their associated structures in the upper portion of the plate. Use light colors for A, T, C, and G.

The base pairs occupy the position of the "rungs" of the ladder, although in the actual molecule they are tightly packed on top of one another as no ladder rungs ever would be. The particular sequence of the four different bases constitutes a "code" in which specific hereditary information is recorded. The method by which that code is translated to specify the exact sequences of amino acids to be used in making the cell's proteins will be covered in the next few plates.

The helix shown here, called the "B-form," is the most stable and therefore the most common structure for DNA. In recent years it has been discovered that local regions of DNA may form a slightly more open helix, called the A-form, and in some cases a very different, left-handed helix, called the Z-form. (The significance of these alternate forms is not known.)

The average length of DNA in a human chromosome is about 140 million base pairs, or 14 million turns of the helix. If laid out in a straight line, it would be about 4.8 centimeters long (just under 2 inches).

Color the heading Structural Formula and the remainder of the plate.

The structural formula shows more clearly which atoms are attached to which. Those details may or may not be important to you, depending on your reasons for studying this plate, but they are important to the cell because any deviation will result in some kind of mutation or even the death of the cell.

To clarify the exact interconnections of the various atoms, this view shows the base pairs and the ribose subunits rotated 90 degrees from their actual orientation in the molecule. You will note that each base is attached to carbon number 1 of its deoxyribose molecule. To facilitate discussions of the structure of DNA, this carbon atom is designated as carbon 1' ("one prime") to distinguish it from the carbon atom number 1 of the base. The phosphates, then, are attached to carbons 3' and 5'.

Note also that the directions of the sugar and phosphate uprights or backbones are "antiparallel"; that is, the chain on one side runs in the opposite direction to the chain on the other side. On one side, the 5' carbon of each ribose connects by way of a phosphate group to the 3' carbon of the ribose above. On the other side, the 5' carbon of each ribose connects by way of a phosphate group to the 3' carbon of the ribose below. Thus the chains progress in the direction 5' to 3' up the helix on one side and 5' to 3' down the helix on the other side. At each end of the DNA molecule, then, one strand will end with a 3'-OH and the other will end with a 5'-phosphate.

In 1958 Watson, Crick, and Wilkins received the Nobel Prize in physiology and medicine for this discovery of the structure of DNA. It was an extremely important achievement because, as Watson and Crick pointed out in their paper announcing the discovery, not only could such a structure carry genetic information coded in the varying sequence of the bases, but there was also an obvious mechanism by which the molecule could be self-replicating, so that exact copies could be supplied for each daughter cell in cell division.

THE DOUBLE HELIX.

SIMPLIFIED STRUCTURE★
UPRIGHTS/BACKBONE★
DEOXYRIBOSE_D
PHOSPHATE_P
RUNGS/BASE PAIRS★
ADENINE_A
THYMINE_T
CYTOSINE_C
GUANINE_G
HYDROGEN BOND_H

STRUCTURAL FORMULA★

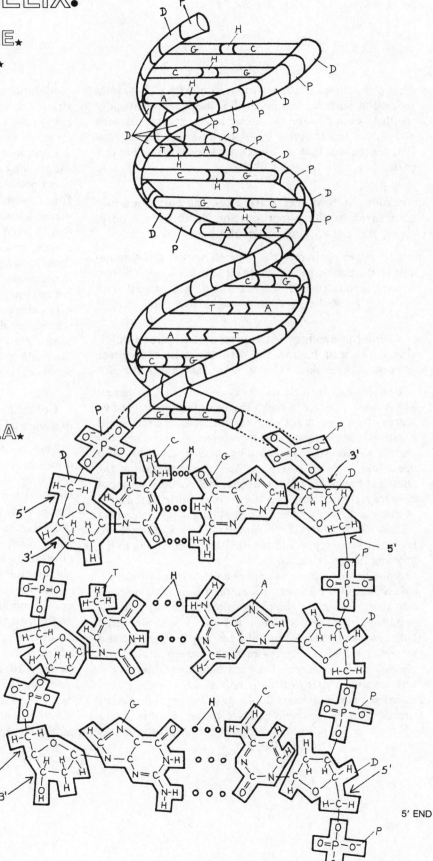

3' END

5' END

DNA REPLICATION

This plate shows how the structure of DNA provides a method of exact replication. Although we cannot actually see these events at the molecular level, experiments with radioactive tracer atoms have provided very strong support for the idea that the events occur as illustrated in this plate.

Color the first seven titles and the corresponding structures in the upper portion of the plate only, using the same colors as in the previous plate.

The upper portion of the plate shows the DNA molecule in the double helix form prior to replication. The two strands of this "parent" DNA are held tightly together by *hydrogen bonds* between the bases in each pair.

Color the heading Unzipping and Replication, titles B, E, and F, and the structures in the center portion of the plate. Use a very light color for B.

For replication to occur, the two strands of the parent DNA must separate in much the same fashion as the two halves of a zipper unzip, exposing the bases in an unpaired condition. An enzyme, *DNA polymerase,* adds new nucleotides from the surrounding nucleoplasm to form "daughter" strands of DNA. Since the formation of the chemical bonds joining the nucleotides requires an input of energy, the cell must provide the nucleotides in the form of nucleoside triphosphates (base + sugar + three phosphates). Two of the three *phosphates* are *hydrolyzed* to provide the energy to add the nucleotide remaining to the growing daughter strand.

Since only *adenine* fits *thymine* and has opposite electrical charges in the correct places, only adenine is inserted into the growing daughter strand wherever a thymine is present in the parent strand, and vice versa. Similarly, only *guanine* is inserted into the growing daughter strand wherever a *cytosine* is present in the parent strand, and vice versa. This assures that each daughter strand is exactly complementary to the parent strand on which it is assembled and is an exact duplicate of the opposite parent strand. The two complete DNA molecules that result when the entire process is completed are exact duplicates of the original DNA molecule.

Molecules of DNA polymerase move along the parent strands, assembling new daughter strands as they go. They always assemble the new strand in the 5' to 3' direction, forming a covalent bond between the phosphate of the new nucleotide and the 3' oxygen atom of the deoxyribose on the growing end of the strand. This means that replication is in opposite *directions* on the two sides of the molecule. One strand, called the leading strand (left side of the plate), is assembled in a direction toward the "replication fork" (point of unzipping). The other strand, called the lagging strand (right side in the plate), is assembled in a direction away from the replication fork and must be done in segments. The segments are then connected by a different enzyme, DNA ligase, which also repairs DNA when it is damaged. Actually, numerous replication forks are started simultaneously at various points along a chromosome, usually in pairs running in opposite directions, so the leading strand also ends up being made in segments and needing DNA ligase to join the segments together.

Color the heading Duplicate Strands and the remainder of the plate.

The consequence of all these steps is that two exact duplicate strands of DNA are produced, each of them consisting of one of the original parent strands and a new daughter strand. The cell can now undergo cell division and provide each daughter cell with a complete copy of this DNA molecule.

In recent years one of the intriguing discoveries (not illustrated) has been that while one part of the DNA polymerase molecule functions as a polymerase, attaching nucleotides, another portion of the same molecule acts as an "exonuclease" (nucleotide-cutting-out enzyme) and performs a "proofreading" function. It is estimated that about one time in 10,000 to 100,000 the wrong base is added to a growing DNA strand. Somehow the exonuclease portion of the DNA polymerase molecule recognizes nearly all such mistakes and removes each erroneous base as fast as it is added so that another attempt to add the correct one can be made. The result is that there is an estimated error rate of only one in one billion base pairs copied.

DNA REPLICATION.

DEOXYRIBOSE D
PHOSPHATE P
ADENINE A
THYMINE T
CYTOSINE C
GUANINE G
HYDROGEN BOND H

UNZIPPING AND
   REPLICATION ★
DNA POLYMERASE B
DIRECTION OF
   SYNTHESIS E
HYDROLYSIS F

NUCLEOSIDE
TRIPHOSPHATES

DUPLICATE
STRANDS ★

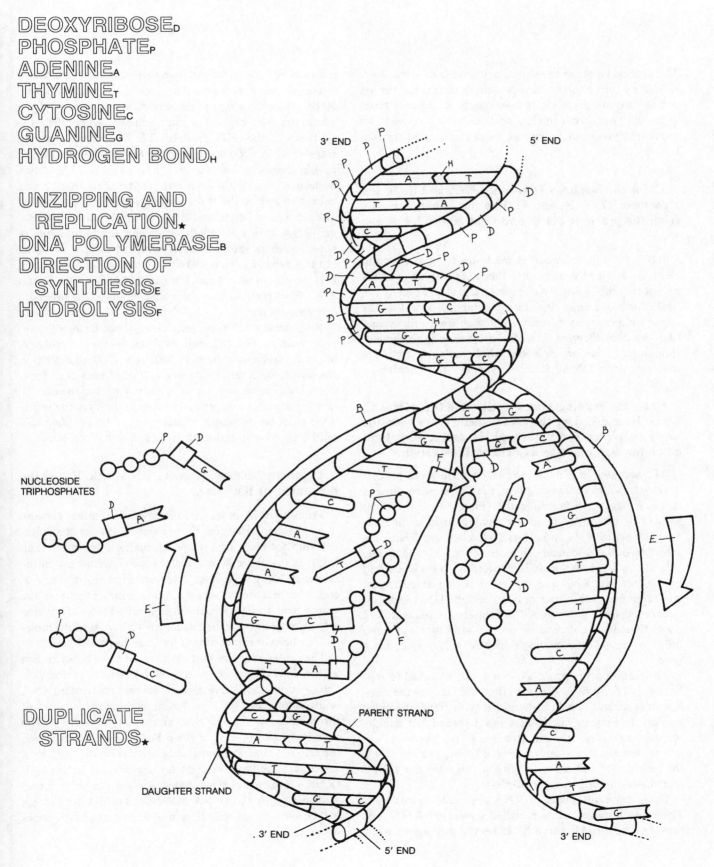

3' END   5' END

B

PARENT STRAND

DAUGHTER STRAND

. 3' END

5' END   3' END

DNA TRANSCRIPTION

The process in which the hereditary code carried by DNA is used by the cell to control protein synthesis has turned out to be quite complex. It involves three different types of a slightly different nucleic acid, called RNA, and two sequential processes known as transcription and translation.

Color the headings DNA and RNA and titles and structures D, T, R, and U. Use the D and T colors from the previous plates and light colors for R and U.

RNA (ribonucleic acid) is made up of numerous nucleotides assembled in exactly the same way as in DNA except that RNA is mostly single-stranded and mostly not in the form of a helix. It differs in composition in that the sugar component is *ribose,* rather than *deoxyribose,* and that the base *thymine* is replaced by *uracil,* a different, though quite similar, molecule. The other three bases are the same as in DNA: adenine, cytosine, and guanine.

Color the heading Transcription and titles P, A, C, G, H, B, and E. Use the established colors from the previous plates for P through H and a very light color for B. Color the associated illustration.

The first step in utilizing the DNA code is the process of transcription. In transcription, the DNA unzips, just as if it were going to be replicated (Plate 83), except that instead of DNA polymerase attaching to it, a different enzyme, called *RNA polymerase,* attaches, synthesizing a molecule of RNA instead of a molecule of DNA. Only one side of the DNA molecule is transcribed. This is assured by the fact that RNA polymerase is not attracted to just any stretch of DNA but only to certain DNA base sequences, called "promoters." Promoters are sequences of bases that do not determine protein structure but serve only to convey the message "RNA polymerase, start here."

The transcription process is essentially identical to replication. The differences are that the complementary daughter strand is being assembled with ribonucleotides instead of deoxyribonucleotides and that the RNA daughter strand does not remain attached to the parent DNA strand. Instead it separates from the DNA, and the DNA then zips back together. The RNA migrates out of the nucleus of the cell to the cytoplasm.

Three different classes of RNA are made in this way. The most abundant class is called messenger RNA (abbreviated mRNA) because it carries the message of what amino acids are to be put together in what sequence to make the cell's proteins. The second class is ribosomal RNA (rRNA), which is an important component of the ribosomes, the organelles that actually accomplish the synthesis of the cell's proteins. The third class is called transfer RNA (tRNA) because it transfers the amino acids to the ribosome for assembly into proteins. The RNA molecule being synthesized in the center of this plate could belong to any of the three classes of RNA. They are all alike except in length and the sequence of bases. Messenger RNA varies in length according to the number of amino acids in the protein for which it carries the code, but it is typically from 900 to 1500 nucleotides in length. It is mostly linear, although it can fold back on itself, and a few short sections may form a helix where the bases are complementary.

Ribosomal RNA takes only certain specific lengths, approximately 120, 1500, and 3000 nucleotides in prokaryotic cells and approximately 120, 160, 2000, and 5000 in eukaryotic cells. It is extensively folded back and forth upon itself, because it forms a framework for the attachment of a number of protein molecules to form the somewhat globular ribosome, which is slightly more than half rRNA by weight and slightly less than half protein.

Color the heading Transfer RNA, title F, and the remainder of the plate.

Transfer RNA deserves some special attention because of its peculiar structure. There must be at least one different kind for each amino acid (actually a few more than that), all are about 80 nucleotides in length, all end in the sequence CCA (cytosine, cytosine, adenine) on their 3' ends, that end always serves as the attachment point of the amino acid to be transferred, and all tRNAs are folded into a complex "hairpin" structure with most of the molecule in helix form but three loops of unpaired bases.

The center loop has a set of three unpaired bases known as the *"anticodon"* (see next plate), which serves as a "recognition code" and assures that that particular tRNA is attracted only to a particular complementary set of three bases on the mRNA, known as a "codon." The unpaired bases on one of the other loops serve to attach the tRNA to the ribosome, and the bases on the third loop serve as a recognition code for the specific aminoacyl-tRNA synthetase enzyme that attaches a particular amino acid to a particular tRNA molecule. The correct protein will be synthesized only if each of these recognition codes is correct.

DNA TRANSCRIPTION.●

DNA.★
DEOXYRIBOSE_D
THYMINE_T

RNA.★
RIBOSE_R
URACIL_U

TRANSCRIPTION.★
PHOSPHATE_P
ADENINE_A
CYTOSINE_C
GUANINE_G
HYDROGEN BOND_H
RNA POLYMERASE_B
DIRECTION OF
    TRANSCRIPTION_E

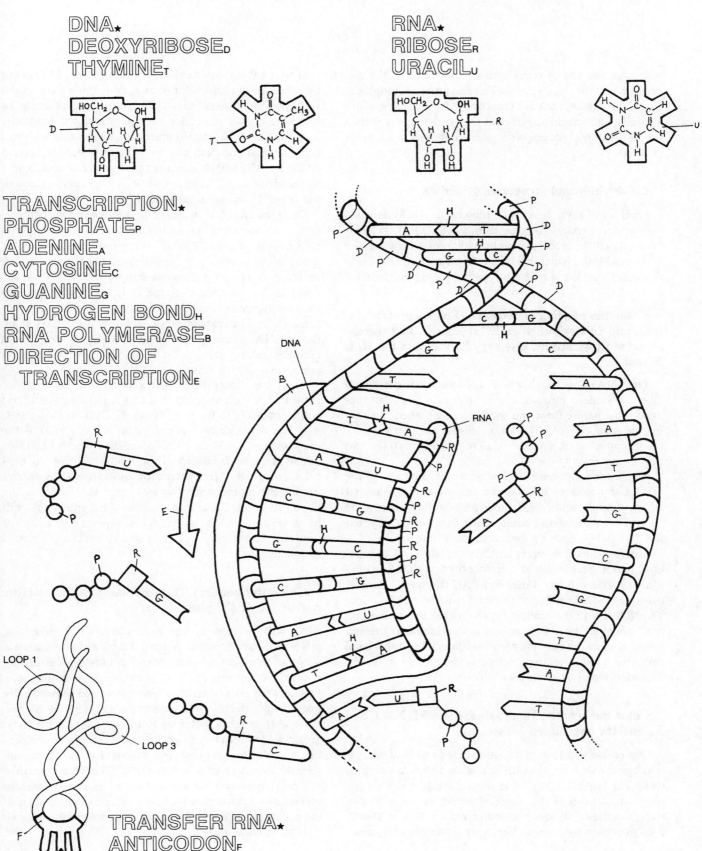

TRANSFER RNA.★
ANTICODON_F

PROTEIN SYNTHESIS: TRANSLATION

After the process of transcription, messenger RNA migrates out of the nucleus of the cell and into the cytoplasm. There the genetic code is "translated" by a ribosome into a specific amino acid sequence in the synthesis of a protein (or at least the polypeptide portion if it is a complex protein).

Color titles and structures B and D.

Each set of three bases on a messenger RNA molecule constitutes a *codon* for one amino acid. Each codon is "recognized" by a complementary *anticodon* on a transfer RNA molecule, which brings the correct amino acid into position for addition to the polypeptide being synthesized.

Color the headings Ribosome, Phases of Translation, and Initiation, titles E through L, and the associated illustrations. Use very light colors for H, I, J, and K.

The ribosome is roughly half protein and half ribosomal RNA, organized into *small* and *large subunits.* The subunits are separate from one another except when translating messenger RNA. The large subunit has two separate binding sites for tRNA, known as the *P* (peptidyl) *site* and the *A* (aminoacyl) *site.*

The initiation phase begins with the binding of the *mRNA* to the small subunit of a ribosome. Next the first *tRNA,* with its amino acid, binds to the mRNA. Then the large ribosomal subunit binds, doing so in such a way that the first tRNA ends up bound to the P site.

The first codon of every mRNA is always AUG (adenine, uracil, guanine), and therefore the first tRNA to bind is always one with the anticodon TAC (thymine, adenine, cytosine), which is complementary to the AUG codon. The tRNA with that anticodon always has the amino acid *methionine* attached to it, so methionine is always the first amino acid in the new polypeptide chain. (In prokaryotic cells, the closely related N-formyl methionine is used.) The methionine is often removed later.

Color the heading Elongation, titles M, N, O, and J', and the associated illustration.

Immediately following initiation, the process of elongation begins, with the binding of a second tRNA, carrying its specific amino acid, to the second binding site on the ribosome, known as the A site. (Remember, each kind of tRNA carries only one particular amino acid of the 20 used in protein synthesis.) The A site is immediately adjacent to the P site, so the tRNA binding to it is always the one that has an anticodon that is complementary to the very next three bases (codon) on the mRNA molecule. In this plate, the second codon is GCU, which will bind only with the tRNA having the anticodon CGA, which always carries the amino acid *alanine.* The fact that the second codon is GCU will thus assure that the second amino acid in the chain being formed will be alanine and not one of the other 19 amino acids used to make proteins.

Once the tRNA is in place, an enzyme (peptidyl transferase) detaches the first amino acid, methionine, from its tRNA on the P site and joins its carboxyl end to the amino group of the amino acid (alanine) attached to the tRNA on the A site to form a *peptide bond* (Plate 18). Then the first tRNA is released from the P site to go back out into the cytoplasm for another amino acid molecule to be attached to it. The ribosome moves exactly three bases along the mRNA, moving the second tRNA to the P site (shown), taking along with it what is now a dipeptide (composed of two amino acids).

The elongation process then repeats over and over. A third tRNA with an anticodon (AAA in this illustration) complementary to the third codon (UUU) binds to the A site. The dipeptide constructed so far is transferred by the enzyme from the second tRNA on the P site to the third amino acid (on the tRNA at the A site) to form a tripeptide. Because the tRNA with the anticodon AAA always carries the amino acid *phenylalanine,* the fact that the third codon is UUU assures that the third amino acid will be phenylalanine. The same process repeats over and over until the entire polypeptide specified by the mRNA is complete.

Color the heading Termination, title P, and the remainder of the plate.

As will be seen in the next plate, three particular mRNA codons do not code for particular amino acids but serve as termination codons (also commonly called *stop codons*). When one of them, such as UGA shown here, reaches the A site, it attracts a protein called *release factor* instead of a tRNA. This causes the peptidyl transferase enzyme to break a water molecule to obtain a hydroxyl group, add that hydroxyl group to the end of the polypeptide chain to make a complete carboxyl group there, and release the completed polypeptide. The ribosome then separates into two subunits. The mRNA may bind to several more ribosomes and make several more copies of the polypeptide, but before long it is broken down by an enzyme, ribonuclease.

PROTEIN SYNTHESIS: TRANSLATION.

CODON_B
ANTICODON_D
RIBOSOME★
SMALL SUBUNIT_E
LARGE SUBUNIT_F
P SITE_H
A SITE_I

PHASES OF TRANSLATION★
INITIATION★
mRNA_J
tRNA_K
METHIONINE_L

ELONGATION★
ALANINE_M
PEPTIDE BOND_N
PHENYLALANINE_O
STOP CODON_{J1}

TERMINATION★
RELEASE
   FACTOR_P

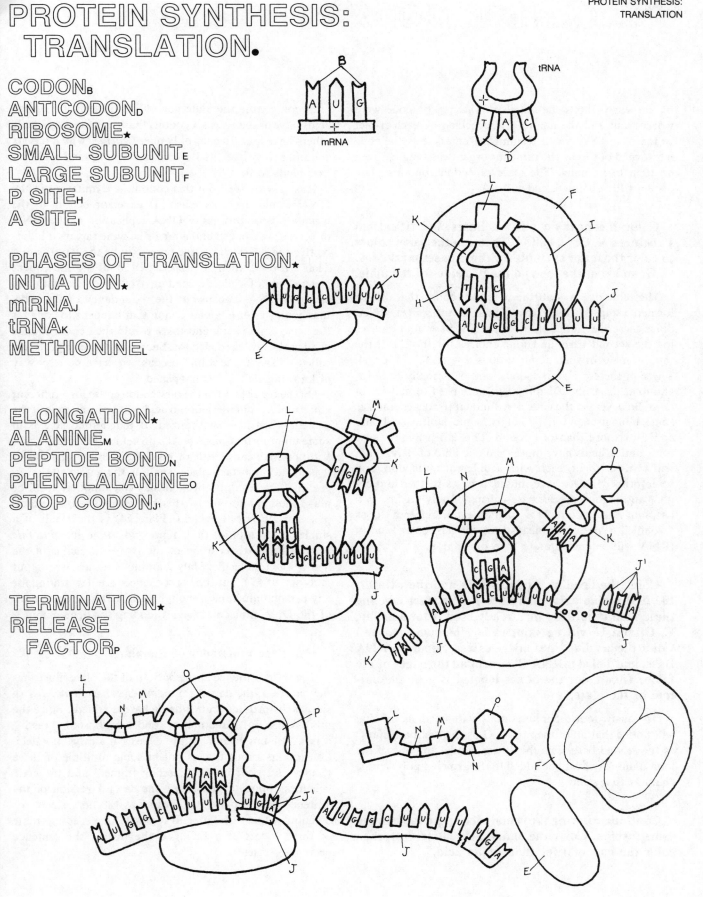

THE GENETIC CODE

By the early 1960s the details of the genetic code were worked out and the amino acid specified by each triplet codon of mRNA was determined. Remarkably, the code has turned out to be the same for organisms ranging from bacteria to humans. The table of codons on this plate displays the complete code.

Color the heading mRNA Bases and titles and structures A, C, G, and U. Then use the same colors to color the letter symbols for those bases marked A, C, G, and U at the top and the two sides of the table.

The table displays all the possible codons that can be formed using the four bases found in messenger RNA. Since there are four choices for the first base, four choices for the second base, and four choices for the third, the total number of such combinations is 4 × 4 × 4, or 64. Three of these are stop codons, and 61 actually code for amino acids. Only 20 amino acids are used to make proteins, however, so the code is redundant (or degenerate, as some biologists put it). That is, some amino acids are coded by more than one codon. This is feasible because some amino acids have more than one kind of tRNA that will transfer them, each with a different anticodon, and some other tRNAs have modified bases in their anticodons and require precise base-pairing of only the first two bases of the codon. This phenomenon, referred to as "wobble" in the code, means that one particular kind of tRNA will recognize more than one codon.

Color the heading Amino Acids and the title for the first amino acid, alanine (B). (We start labeling these with B because we already used A for adenine. C, G, and U will be skipped for the same reason.) Color the heading First mRNA Base, Second mRNA Base, and Third mRNA Base around the edges of the table. Then color the boxes labeled B near the bottom of the plate.

Obviously, four different codons code for *alanine.* You will notice that all of them have G as the first base and C as the second base. The third base can be any one of the four; alanine will still be added to the growing polypeptide chain at that point.

Continue coloring the remainder of the table in the same fashion. Color the amino acid name first, then color the codon(s) for that amino acid.

In completing the plate you will find that some amino acids have as many as six codons that code for them while others have four, two, or only one. *Leucine,* which has six codons, is very likely to have more than one kind of tRNA that binds to it.

It is easy to see from this code how a mutation in the DNA would exert its effect. If an error occurs in the usually precise process of DNA replication or if a DNA molecule is damaged and a repair enzyme repairs it incorrectly, one or several of the bases will be the wrong ones. That error will then be perpetuated in all the subsequent generations of cells descending from the one with the changed DNA. Because of the redundancy of the code, that could result in a new codon that happens to code for the same amino acid, and there would then be no change at all in the protein coded for. Such a mutation is generally called a "silent mutation" because we have no easy way of knowing that it has happened.

On the other hand, if the new codon codes for a different amino acid, a different amino acid will be inserted into the protein when it is synthesized. If that amino acid is in some unimportant place on the protein molecule, the substitution may cause little or no change in function. But if it is at an important place, such as the active site of an enzyme, the resulting protein may not function at all or may function improperly. It is well established, for example, that sickle-cell anemia (Plate 74) is the result of a single substitution of the amino acid *valine* for *glutamic acid* in the sixth position of the two beta chains of the hemoglobin molecule. Only 2 amino acids are wrong out of a total of 574, but the consequences are devastating for any person unfortunate enough to have this error on both of the DNA molecules (genes) coding for hemoglobin.

Color the remainder of the plate.

The short sentence at the bottom of the plate illustrates the nature of the damage done when a base is *deleted* (a similar distortion occurs when a base is added). Since the mRNA has no "punctuation" and the ribosome "reads" it in a "reading frame" of three bases at a time, any deletion (or insertion) that is not in some multiple of three results in a shift in the "reading frame," and the code becomes complete nonsense. The second version of the sentence on the plate is the same as the first except for deletion of the sixth letter. If it is read in a reading frame of three letters at a time, all the rest of the sentence becomes nonsense.

THE GENETIC CODE.

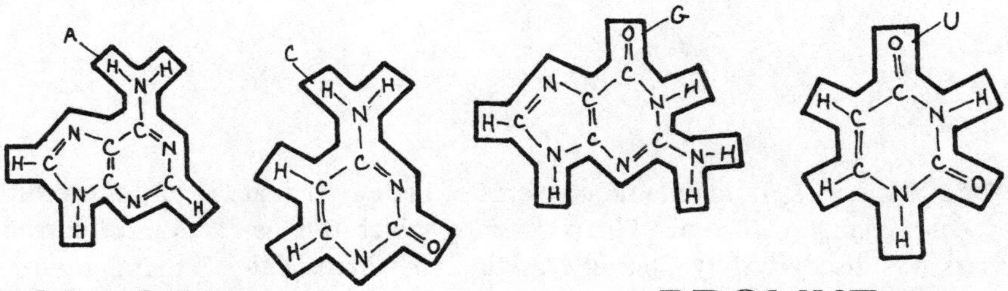

mRNA BASES★
ADENINE$_A$
CYTOSINE$_C$
GUANINE$_G$
URACIL$_U$
AMINO ACIDS★
ALANINE$_B$
ARGININE$_D$
ASPARAGINE$_E$
ASPARTIC ACID$_F$
CYSTEINE$_H$
GLUTAMINE$_I$
GLUTAMIC ACID$_J$

GLYCINE$_K$
HISTIDINE$_L$
ISOLEUCINE$_M$
LEUCINE$_N$
LYSINE$_O$
METHIONINE$_P$
PHENYLALANINE$_Q$

PROLINE$_R$
SERINE$_S$
THREONINE$_V$
TRYPTOPHAN$_W$
TYROSINE$_X$
VALINE$_Y$

SECOND mRNA BASE★   **STOP CODON$_Z$**

FIRST mRNA BASE★ / **THIRD mRNA BASE★**

	U $_{U'}$		C $_{C'}$		A $_{A'}$		G $_{G'}$		
U $_{U'}$	UUU	Q	UCU	S	UAU	X	UGU	H	U $_{U'}$
	UUC	Q	UCC	S	UAC	X	UGC	H	C $_{C'}$
	UUA	N	UCA	S	UAA	Z	UGA	Z	A $_{A'}$
	UUG	N	UCG	S	UAG	Z	UGG	W	G $_{G'}$
C $_{C'}$	CUU	N	CCU	R	CAU	L	CGU	D	U $_{U'}$
	CUC	N	CCC	R	CAC	L	CGC	D	C $_{C'}$
	CUA	N	CCA	R	CAA	I	CGA	D	A $_{A'}$
	CUG	N	CCG	R	CAG	I	CGG	D	G $_{G'}$
A $_{A'}$	AUU	M	ACU	V	AAU	E	AGU	S	U $_{U'}$
	AUC	M	ACC	V	AAC	E	AGC	S	C $_{C'}$
	AUA	M	ACA	V	AAA	O	AGA	D	A $_{A'}$
	AUG	P	ACG	V	AAG	O	AGG	D	G $_{G'}$
G $_{G'}$	GUU	Y	GCU	B	GAU	F	GGU	K	U $_{U'}$
	GUC	Y	GCC	B	GAC	F	GGC	K	C $_{C'}$
	GUA	Y	GCA	B	GAA	J	GGA	K	A $_{A'}$
	GUG	Y	GCG	B	GAG	J	GGG	K	G $_{G'}$

ORIGINAL MESSAGE$_{AA}$

THE BIG DOG BIT TED AND RAN OFF

DELETION AND FRAME SHIFT$_{BB}$

THE BID OGB ITT EDA NDR ANO FF

↑
DELETION

One of the most fundamental concepts of biology is that species change with time. The process by which this occurs was discovered by Charles Darwin and Alfred Wallace. They announced their results together (in 1858), but each made the discovery independently, and Darwin is given the greater credit, not only because he made the actual discovery first but also because he spent 28 years meticulously collecting, reviewing, and organizing a vast array of data and wrote a lengthy book explaining it all (*The Origin of Species,* published in 1859).

Darwin was born in England in 1809. His father and grandfather were wealthy physicians, and Charles himself studied medicine for several years but quit because he hated the sight of blood. He then studied for the clergy, but after completing those studies he decided that a career as a clergyman wasn't for him either. By a remarkable stroke of good luck there was a position open for a ship's naturalist on a voyage of exploration, and through the influence of one of his professors, who had noticed that he was a keen and ardent observer of nature, Darwin obtained that position. He sailed in 1831 on a ship called the *Beagle* and spent five years observing animals, plants, geologic formations, and fossils, primarily on the two coasts of South America.

Color title A and color over the the associated line tracing Darwin's voyage on the *Beagle*.

When Darwin departed on that voyage, he was convinced, like most people of his time, that species were fixed and unchanging. But the science of geology had been advancing rapidly. Many fossils had been found, and the science of paleontology had gotten its start. Although the fossil record was obviously incomplete, some groups of fossils clearly displayed a gradual change of some species over time. On the voyage, Darwin read a new book by his friend Charles Lyell, *Principles of Geology,* which presented convincing evidence that the earth was much older than people had thought, that it had been slowly changing over immense periods of time, and that it was still changing. At the same time, Darwin was discovering many fossils himself and seeing in the geology of South America more evidence to support Lyell's views. He was struck by the fact that many fossils resembled living forms, yet were different enough that they had to be regarded as different species. Of all the areas visited by Darwin, it was the Galápagos Islands, 600 miles off the west coast of Ecuador, which

had the most profound influence on his thinking. These almost barren islands had so much exposed volcanic rock that Darwin reasoned they must have been thrust up from the ocean floor quite recently, as geologic changes go. Darwin was struck by the fact that only a few of the species he had seen on the South American mainland were represented here. In particular, all the small birds present were finches, but they had differentiated into 14 species with remarkably different beaks, each especially well adapted to a particular food source.

Color the heading "Galápagos Finches" and the heading "Food Source."

Darwin knew that small land birds are often blown far to sea by storms and that plants are sometimes carried great distances by ocean currents, frequently with small animals as passengers. He theorized that the finches arrived soon after the islands were formed and, over a period that could have been several thousand years, gradually differentiated into 14 separate species, each adapted to a different "niche" in the environment. Biologists today call this process "adaptive radiation."

Color titles B through G and the remainder of the plate as each finch is discussed.

Geospiza fortis is a ground finch and feeds on *seeds*. Its beak is somewhat short and stoutly built to enable it to break those seeds open. *Geospiza scandens* is sometimes called the cactus finch because its principal food is the prickly pear cactus. Its beak is long and nearly straight for reaching into the *flowers of the cactus* to feed on nectar and the soft tissues. *Camarhynchus crassirostris* lives high in the trees in dense forests. It feeds primarily on buds, flowers, and *fruit* but does eat some seeds as well, so it has a strong beak for crushing. *Certhidea olivacea* is often called the warbler finch because it so greatly resembles a warbler that even Darwin was fooled at first. It eats *insects* and has a thin, sharp beak for capturing them. *Camarhynchus pallidus* is the most unusual of all. It has a spadelike beak, much like a woodpecker's, and feeds as a woodpecker does on insects and grubs it digs out from under bark. But it lacks the woodpecker's long tongue, so it uses a cactus spine or a twig as a *tool* to dig out its prey—the only known case of a tool-using bird.

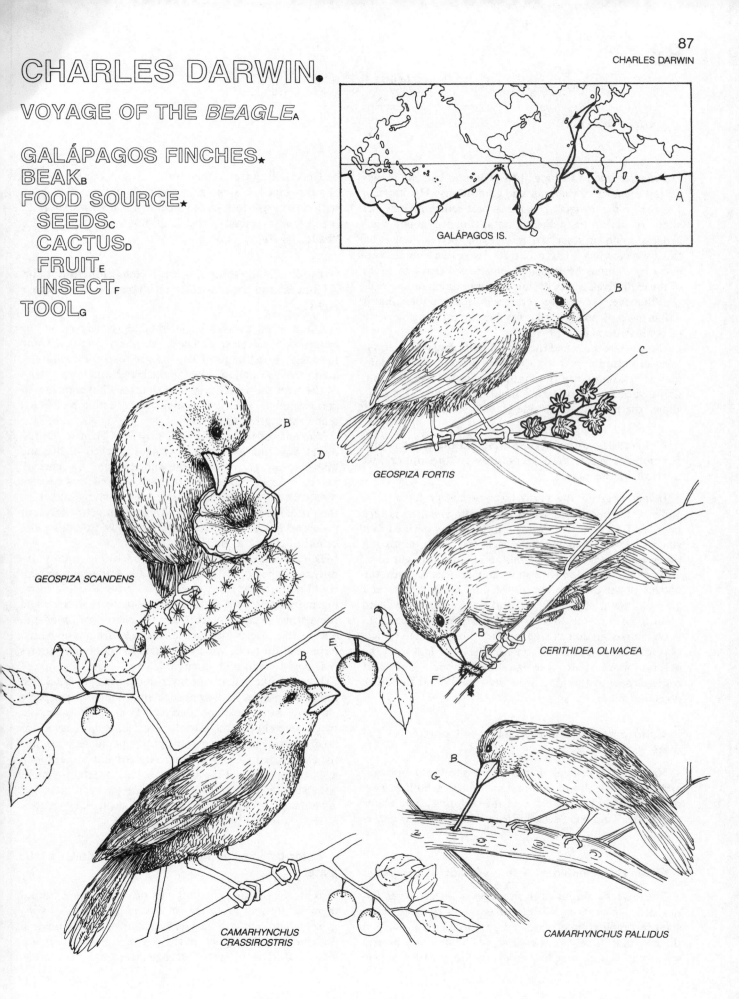

CHARLES DARWIN.

VOYAGE OF THE *BEAGLE*A

GALÁPAGOS FINCHES.★
BEAKB
FOOD SOURCE.★
  SEEDSc
  CACTUSD
  FRUITE
  INSECTF
TOOLG

GALÁPAGOS IS.

A

GEOSPIZA FORTIS

GEOSPIZA SCANDENS

CERITHIDEA OLIVACEA

CAMARHYNCHUS CRASSIROSTRIS

CAMARHYNCHUS PALLIDUS

ARTIFICIAL SELECTION AND NATURAL SELECTION

Darwin returned to England in 1836 and became an accepted member of the scientific community. His journal, *Voyage of the Beagle,* became a best-seller, and he set about reviewing his collected data, thinking about what process could produce the changes in species he was by this time convinced had occurred. He read a now famous essay by Thomas Malthus, which warned (back in 1798) of the explosive growth of human population, and he realized that every species reproduces in such numbers that it will grow explosively until it reaches the carrying capacity of its environment.

Darwin suddenly had the insight that new species developed in nature by exactly the same process that humans had been using since time immemorial to develop more useful varieties of domestic animals and plants. This plate shows the two processes side by side for comparison.

Color the heading Artificial Selection, titles A and B, and the associated structures in the upper left portion of the plate.

In many parts of the world, farmers have problems with badgers stealing their chickens, so hunting badgers is very popular. Centuries ago, farmers in Germany decided that since selective breeding had been useful in improving other kinds of domestic animals and plants, it ought to be able to develop a dog with short enough legs to chase the badger down into its burrow, strong enough paws and claws to dig after the badger, and large enough jaws and teeth to be a match for the badger when it was caught.

Of course, no dog existed with all these characteristics, but of the dogs available, some had shorter legs than the others, some had stronger paws, and so on, so dogs that came closest to the ideal were separated from the other dogs and mated.

Color arrow C and the center left portion of the plate.

Naturally, the ideal dog was not produced immediately, but with the recombination that is a built-in part of the genetic process, a few of the offspring came closer to the ideal and were selected to produce the next generations.

Color the remainder of the left side of the plate.

The farmers continued this process of selection generation after generation. With the passage of *time,* the constant recombination of genes, an occasional mutation, and the constant "selective pressure" of the *breeders* always separating out as breeding stock the dogs with the best

badger-hunting traits, the ideal breed for the job eventually developed. It was half a dog high and two dogs long, with short legs, stout paws and claws, and strong jaws: the breed known today as the dachshund (German: *Dachs,* "badger"; *Hund,* "dog").

Color the heading Natural Selection and titles A^1 and B^1 and the associated illustration at the upper right.

Giraffes had always fascinated biologists as outstanding examples of adaptation. Their extremely long necks and long legs, with the front legs longer than the hind ones, adapt them so well to reaching the leaves and tender twigs of the trees on which they feed that they have no serious competitors for that food source. The question was, What made them grow that way?

The answer proposed in 1809 by Jean Baptiste de Lamarck was that as ancestral giraffes reached higher and higher to eat, their bodies responded to an "inner need" and their necks grew longer. These longer necks were then passed on to their offspring, who kept reaching higher, so their necks grew still longer, and so on. Unfortunately, our present-day knowledge of heredity shows this otherwise beautiful theory to be false.

Darwin correctly recognized that variation occurs naturally in all species and that some factor in the environment could perform the same role as a selective breeder, selecting individuals with certain variations to reproduce and preventing, or at least reducing, the reproduction of others. In the case of the giraffe, the *selective factor* would have been the trees, which provided food only for giraffes that could reach high enough.

It seems likely that whenever ancestors of giraffes first found themselves in their present environment, they probably had short necks too, and as long as some leaves and twigs were close enough to the ground, a short neck was no disadvantage. But when the giraffe population reached the carrying capacity of its environment and a food shortage developed—as it always will, sooner or later—giraffes with slightly longer than average necks had a *survival* advantage. If they weren't the only ones to survive, they at least survived in larger numbers.

Color title and arrow C^1 and the remainder of the plate.

With the passage of thousands of years in an environment where any mutation or recombination of existing genes would confer a survival advantage if they made it possible to reach higher on the tree, the giraffes that survived had to be the sort of strange creatures we see today.

ARTIFICIAL SELECTION AND NATURAL SELECTION.

ARTIFICIAL SELECTION★
BREEDER A
DESIRED PARENTS B
TIME C
RESULT D

NATURAL SELECTION★
SELECTIVE FACTOR A¹
SURVIVOR B¹
TIME C¹
RESULT D¹

NATURAL SELECTION
WE CAN SEE

Since Darwin's exposition of the principles of natural selection, biologists have found numerous examples of natural selection occurring in periods of time much shorter than the many thousands of years usually required. One outstanding example of this is what is known as industrial melanism (Greek: *melas,* "black"), the turning black of certain species in areas that were blackened with the soot of the coal-burning factories that sprang up in great numbers during the Industrial Revolution.

Color titles A, A¹, and B and the associated structures in the upper illustration. Use light gray or gray-green for A and A¹. Use black for B.

One of the best-studied cases of industrial melanism is the change in color of the peppered moths in the vicinity of Manchester, England. Nature study has been popular for centuries in England, so there are records of observations and insect collections going back several hundred years. In the vicinity of Manchester, a certain rather large moth *(Biston betularia)* was well known. It was called the *peppered moth* because it resembled a white moth on which pepper had been sprinkled. It was nocturnal in its habits and spent all the daylight hours resting on the trunks of trees, where it blended in almost perfectly with the *lichens* covering the tree trunks, since the lichens had the same "peppered" coloration. Only occasionally was a rare *black* member of this species spotted.

Color title B¹ and structures A, B, and B¹ in the lower illustration. Use black for B¹.

In the second half of the nineteenth century, however, more and more black moths began to show up. That change in the moths corresponded exactly with the progress of the Industrial Revolution. In that coal-burning part of England, the amount of *soot* put into the air by factories was so great that it *covered the tree trunks* in industrial areas, killing the lichens and turning the tree trunks black. Under those conditions, of course, the black moths were as well camouflaged as the peppered ones had been on the lichen-covered trunks. Eventually 98 percent of the moths of this species in industrial areas were black. This same change was observed in many dozens of other species of moths in similar industrial areas in England and the United States, wherever forests became polluted with soot. In unpolluted forests, the moths retained their light coloration.

The question then arose of just how this change was occurring. One hypothesis was that it must just be the results of natural selection, due to birds or other predators eating moths that didn't blend in with the background. But that left unanswered the question of how the first black moths appeared originally. Another hypothesis was that the moths ate or absorbed something in the soot that turned them black. Yet another hypothesis was that the moths had some sort of built-in reflex that caused them to turn themselves black whenever they sensed that their principal background had turned black. The English biologist H.B.D. Kettlewell decided to go out into the forests to investigate this problem thoroughly.

Color the remainder of the plate.

Kettlewell captured equal numbers of black and peppered moths, put identifying paint marks on their undersides where the marks wouldn't show when the moths were resting on a trunk, and released one set in an area with blackened tree trunks and another in an unpolluted area with trunks still covered with lichens. When he came back to recapture the moths, he recovered only half as many of the peppered moths as he did black moths in the soot-blackened forest, and in the light-colored forest he recovered only half as many black moths as he did peppered moths. He also examined the stomach contents of birds known to feed on the moths and found that in blackened forests they ate a disproportionately large number of light-colored moths and in light-colored forests the reverse was true. He also set up movie cameras and captured on film what is summarized in the illustrations of this plate. When a bird is zooming in toward a tree trunk looking for lunch, it is much more likely to see and therefore capture a black moth on a lichen-covered tree trunk or a peppered moth on a blackened tree trunk.

Kettlewell also found that geneticists had already established that the coloration of these moths was determined by a single pair of genes, with the peppered coloration recessive to black. Clearly, then, this was a case of natural selection in action. There was no "battle for survival" according to the "law of the jungle," as is sometimes mistakenly assumed to be a requirement for evolution. Survival or nonsurvival may depend on something as simple as the color of the background. Evolution, then, is simply the process of heredity, with all its lotterylike characteristics, extended over a long period of time, with the environment selecting which survive and which do not.

NATURAL SELECTION
WE CAN SEE.

PEPPERED MOTH_A
BLACK MOTH_B•
LICHEN_{A¹}
SOOT-COVERED TRUNK_{B¹}•

PREDATOR_C
CAPTURE PATH_{C¹}

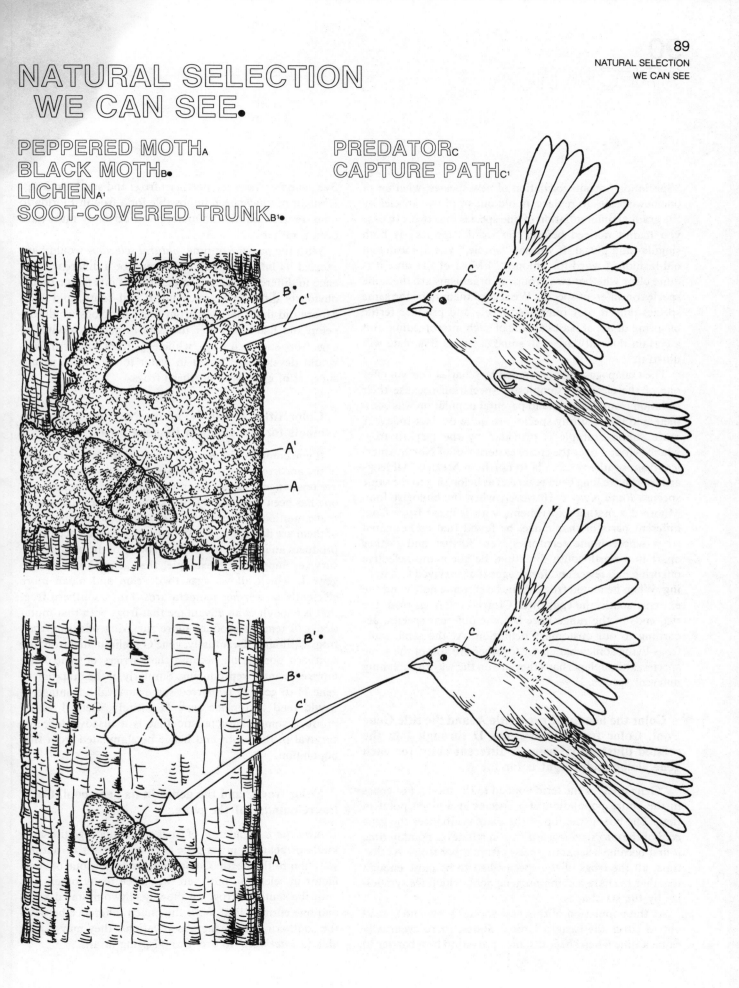

"Speciation" means formation of new species, whether of one new species to replace an old one or of two species by the gradual differentiation of one species into two. (In case you hadn't noticed before, the word "species" is both singular and plural. If you say "specie," you are using an old-fashioned word for "money.") Most of the time it is quite clear whether two groups of organisms are the same species or different species. We regard them to be the same species only if they interbreed freely and produce fertile offspring or are at least capable of such interbreeding. But a certain difficulty arises in some cases, as this plate will illustrate.

The Galápagos finches studied by Charles Darwin offer one of the clearest examples of speciation because their isolation from the mainland parental population was complete and the resulting species are quite distinct today. A more subtle example is provided by the leopard frog, which is found over the entire eastern half of North America, from southern Canada to northern Mexico. All leopard frogs have long been regarded as belonging to the same species, *Rana pipiens.* However, when the biologist John Moore did mating experiments with leopard frogs from different parts of that range, he found that as he mated frogs whose home territories were farther and farther apart in a north-south direction, he got many defective offspring and fewer and fewer eggs that survived to hatching. When he mated frogs from the extreme north and the extreme south, the eggs died. Clearly, the frogs from the two ends of the range have become different species, according to our customary definition. At the same time, frogs living somewhat closer together are still of the same species, although perhaps they are on the way to becoming different species.

Color the heading Frogs, title A, and the title Gene Pool. Color frog A and genes D through I in the related illustration, using a different color for each gene shape and a light color for A.

"Gene pool" is the term applied to the total set of genes in the entire population of a species at a given point in time. The box at the top of the plate symbolizes the *gene pool* of the *ancestral leopard frog* at whatever point in time it first became a separate species from other frogs. At that time, all the frogs of the species had to be close enough together to share a common gene pool, which we symbolize by the six shapes.

As the population of this new species grew and spread out to cover the eastern United States, there eventually came a time when sheer distance proved to be a barrier to free gene flow between northern frogs and southern frogs. It might not have been impossible for a frog in Florida to migrate to Minnesota to mate, but it certainly must have been a rather rare event.

With the passage of time, *natural selection* would have exerted its influence, and that would be a different influence in different places. Temperature is one factor that is obviously different between north and south, and John Moore did find that eggs from northern frogs would develop and hatch in water too cold to permit survival of eggs from southern frogs, while eggs from southern frogs would develop and hatch in water too warm to permit survival of eggs from northern frogs.

Color titles B and N, arrow N, frog B, and the symbols for the genes in its gene pool.

If you compare the genes of the *southern frog* with those of the ancestral frog, you will see that three kinds of genes are unchanged, two of them are somewhat changed, and one has been lost entirely. We know from genetic studies that mutations do occur in genes from time to time. Most of them are damaging, but occasionally one can occur that produces an improvement in the ability of the organism to survive. Suppose, for example, that gene G mutated to gene J, which allows eggs to develop and hatch more efficiently at warmer temperatures. For a southern frog, that is enough of an advantage that frogs with that mutation will tend to produce more surviving offspring than frogs without it. Eventually, gene G is eliminated from the southern population because the carriers of gene J are more efficient at reproducing. Similarly, if the mutation of gene H to gene K conferred some survival advantage, it would tend to replace gene H entirely. Gene I would produce some characteristic that is a disadvantage for survival in the south and hence be eliminated from that population.

Color frog C and the genes in its gene pool. Color the remainder of the plate.

A *northern* population of frogs would undergo the same kinds of changes in its gene pool as a result of natural selection in the north. Cooler temperature is one obvious factor in selection, but there would be many others. As with the southern frog, we show here two mutated genes and one entirely lost. With this much difference between the southern and northern populations, they may not be able to *interbreed* to produce offspring at all.

SPECIATION.

FROGS★
  ANCESTRAL_A
  SOUTHERN_B
  NORTHERN_C
GENE POOL_D–M
NATURAL SELECTION_N
INTERBREEDING_O
DEAD EGGS_P

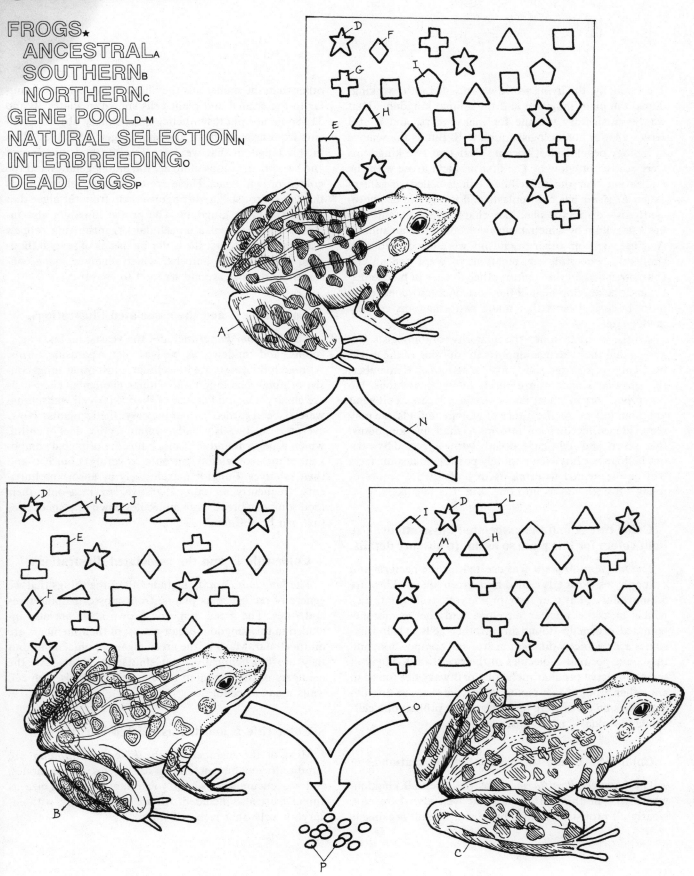

KINGDOMS OF THE LIVING WORLD

Traditionally, the living world was divided into two kingdoms, the plant kingdom and the animal kingdom. That was a satisfactory scheme for many years, and it still serves a lot of nonbiologists quite well. But as the science of biology developed, it became clear that two kingdoms were really not enough. The first problem arose with the realization that the unicellular (single-celled) organism called *Euglena* has chloroplasts and is capable of photosynthesis—distinctly plantlike characteristics—but also has a flagellum by which it can swim about like an animal. As time went on, other organisms were discovered with similar disconcerting combinations of plant and animal characteristics. In this century, the advance of biochemistry and the development of the electron microscope led to more biological facts that made two kingdoms entirely inadequate.

At the present time, the majority of biologists are agreed that the evidence supports the division of the world of living organisms into five kingdoms. Remember, though, that kingdoms are purely human inventions. No other living organism, as far as we can tell, cares a bit what kingdom it is in. As the future of biology unfolds, we may see evidence for division into more than five kingdoms. (Six, seven, and eight have already been suggested by various biologists.) It is even remotely possible that some facts will be discovered to bring us to unify two or more of today's five kingdoms into one. But it is not likely.

Color title A and the associated illustrations. Use light colors for this plate so as not to obscure details.

The *kingdom Monera* was created for the bacteria and cyanophytes (formerly called blue-green algae). They are all unicellular, but they have prokaryotic cells (Plate 32), which lack the nuclear envelope and the membrane-bounded organelles found in all other cells. Their ribosomes are distinctly different in size and composition, and they have peculiar molecules in their cell walls. Many of them also have peculiar metabolic pathways not found in any other group. Most biologists believe that the first living things on earth were probably single prokaryotic cells similar to today's monerans.

Color title B and the associated illustrations.

The *kingdom Protista* serves as the catch-all kingdom where we put things that don't fit well anywhere else. Nearly all protists are unicellular and, like all organisms other than the monerans, they have eukaryotic cells similar to the animal and plant cells shown in Plates 30 and 31. Some are photosynthetic, some simply absorb nutrients from around them, and some ingest solid food. It is in this kingdom that we place *Euglena* and its relatives and two other groups of algae, the dinoflagellates and the golden brown algae. These are commonly referred to as the "algal protists" to distinguish them from the algae that are in the plant kingdom. The protist kingdom also includes *Paramecium,* a unicellular organism that propels itself rapidly through the water by means of several thousand cilia, and the amoeba, which changes shape constantly and flows around its food to engulf it.

Color title C and the associated illustrations.

The *kingdom Fungi* includes the yeasts, molds, mushrooms, and mildews, as well as many parasitic forms. Although the yeasts are unicellular, most other fungi consist of numerous long multicellular filaments. Their cells are eukaryotic, and because of their thick cell walls, fungi were once regarded as nonphotosynthetic plants. However, their cell walls usually contain a great deal of chitin, which is never found in plants but is the principal component of the exoskeleton (the outer covering) of insects and their relatives. Fungi live exclusively by absorbing nutrients, frequently secreting digestive enzymes onto their food source to break large molecules down into small ones that can be absorbed.

Color title D and the associated illustrations.

The *kingdom Plantae* includes all of the things that we generally recognize as plants, from mosses to pine trees and roses. The green, red, and brown algae are also included in this kingdom because most of their members are multicellular. Many of the green algae are unicellular, but they are biochemically most similar to plants. All the members of this kingdom have eukaryotic cells with cell walls made of cellulose.

Color title E and the associated illustrations.

Most of the members of the *kingdom Animalia* are familiar to you, but such things as sponges, sea anemones, and sea cucumbers, which you might not recognize as animals, are also included. All are multicellular, with eukaryotic cells that lack cell walls.

KINGDOMS OF THE LIVING WORLD.

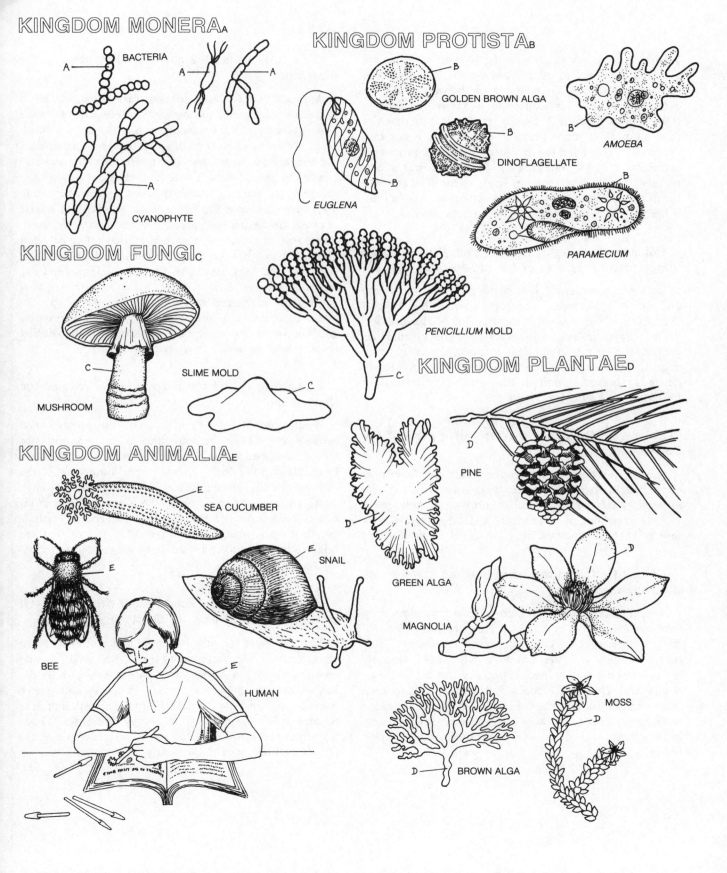

KINGDOM MONERA_A

A — BACTERIA

A ⎯ A

CYANOPHYTE

A

KINGDOM PROTISTA_B

B — GOLDEN BROWN ALGA

B — DINOFLAGELLATE

AMOEBA

B

B

EUGLENA

PARAMECIUM

B

KINGDOM FUNGI_C

PENICILLIUM MOLD

C

MUSHROOM

C

SLIME MOLD

C

KINGDOM PLANTAE_D

D — PINE

D

GREEN ALGA

D

MAGNOLIA

D

KINGDOM ANIMALIA_E

E — SEA CUCUMBER

E

BEE

E — SNAIL

E — HUMAN

D — BROWN ALGA

MOSS

D

No living organism exists entirely by itself. It is always profoundly influenced by its environment. The branch of biology that studies the relationships between living organisms and their environments is known as ecology (Greek: *oikos,* "house"). Ecologists concentrate much of their study on communities and ecosystems. A community is defined as all of the organisms living in a given area and interacting with one another. An ecosystem is a community plus all of the nonliving components of its environment. This plate shows some of the components of a typical biological community and their relationships.

Color the heading "Trophic Levels," title A, and the corresponding part of the illustration.

Within a community, organisms are categorized into different "trophic levels" according to how they nourish themselves (Greek: *trophe,* "nourishment"). The most important organisms are the *producers,* the green plants that capture the energy of sunlight to make the energy-rich organic molecules on which all the rest of the community depends. (In some communities, algae or even certain bacteria may be the important producers.)

Color title B and the corresponding parts of the illustration.

Feeding directly on the producers are the *herbivores* (Latin: *herba,* "grass"; *vorare,* "to devour"), also known as *primary consumers.* Familiar members of this group include grasshoppers, butterflies, and other herbivorous insects, rabbits, squirrels, mice, and seed-eating birds.

Color title C and the corresponding parts of the illustration.

Animals that feed on the herbivores are called *primary carnivores* (Latin: *caro,* "flesh"). They are also called *secondary consumers.* It's unfortunate that they are "primary" one time and "secondary" another, but both naming systems are widely used. If you think about what the words actually mean, it really isn't too difficult to keep them straight. Included among the primary carnivores are such animals as foxes, owls, frogs, insectivorous (insect-eating) birds, and predatory insects such as the praying mantis.

Color title D and the corresponding parts of the illustration.

Animals that feed on primary carnivores are called *secondary carnivores* (or *tertiary consumers*). A snake that eats a frog is a secondary carnivore. So is a hawk that eats an insectivorous bird. Nature, of course, does not entirely cooperate with our desire for nice, neat categories. A fox may eat a frog, becoming a primary carnivore in the process; it may then eat a snake, becoming a tertiary carnivore in that process. Similarly, a mouse may eat an occasional insect, becoming thereby a primary or even a secondary carnivore, depending on what kind of insect it eats. Some animals, such as humans, baboons, and rats, routinely feed at all levels and are called omnivores (Latin: *omni,* "all"). Recognizing that the various categories of carnivores are oversimplifications, ecologists still find them useful, and carnivores and omnivores are traditionally assigned to the highest trophic level at which they feed.

Color title E and the corresponding part of the illustration.

Feeding on all the other levels is the group called *decomposers.* (They are sometimes called reducers, but they do not reduce things in the chemical sense; they live by oxidation.) We don't apply the term "omnivore" to the members of this group, bacteria and fungi, because they do nearly all their feeding on dead organisms. The decomposers break down the dead remains of all species (including their own) into small, inorganic molecules that are released into the soil and water to be recycled as nutrients for the producers.

Color the heading "Food Web," titles F and G, and the associated parts of the illustration.

The pattern of the flow of energy and matter within a community is often referred to as a "food web." In the community illustrated here, that pattern is shown by the arrows, which indicate the transfer of energy and matter from one organism to the next. Only the direction of flow is shown, not the quantity of energy or matter. Those quantities are customarily shown by means of ecological pyramids, illustrated in the next plate.

COMMUNITIES.

TROPHIC LEVELS.★
PRODUCER.A
HERBIVORE (PRIMARY CONSUMER).B
PRIMARY CARNIVORE (SECONDARY CONSUMER).C
SECONDARY CARNIVORE (TERTIARY CONSUMER).D
DECOMPOSER.E
FOOD WEB.★
CONSUMPTION.F   DECOMPOSITION.G

In trying to understand communities, ecologists find it useful to determine certain numeric values and convert them into graphs that give a pictorial representation of the relationships. Some of the most valuable of these are ecological pyramids. This plate shows the three kinds of pyramids in common use.

Color titles A through D, the heading Pyramid of Numbers, and the structures in the two pyramids in the first section.

One kind of ecological pyramid is the pyramid of numbers. The organisms in each trophic level are actually counted, where possible, or estimated from representative samples. In a very small forest, for example, it is entirely possible to count all the trees. Counting all the individual plants in even a tiny meadow would be a different matter.

A sort of pyramid is then constructed, making the area of each box proportional to the number of individuals in that community. The decomposers are usually not shown separately in ecological pyramids but are included as part of each level of consumer. If they were shown separately in a pyramid of numbers, they would overwhelm the other trophic levels. One cubic centimeter of soil often contains more than a million bacteria, for instance. (How many cubic centimeters of soil are there in a small forest?)

A pyramid of numbers will take different shapes according to the sizes of the producers. In a grassland, each *producer* is very small, so their numbers are very considerable. An equal area in a forest will contain only a few large trees, so a pyramid of numbers for a forest will show a very small area for producers, although the trees might support just as many consumers as the grass does in the grassland.

Color the heading Pyramid of Biomass and the trophic levels in the two pyramids in that section.

"Biomass" means the actual mass (weight) of living matter in the organisms in each trophic level. Collecting the data from which to build this pyramid is even more tedious than for a pyramid of numbers, but it has been done for many communities. In a typical terrestrial community, a pyramid of biomass has the conventional pyra-

midlike shape, with a large base to represent the mass of plants necessary to support a smaller mass of *herbivores,* which in turn support a smaller mass of *primary carnivores,* and so on. However, since a pyramid of biomass shows the biomass at one particular point in time, the proportions can be distorted if one trophic level has a peculiar reproductive rate. This often happens in aquatic communities, where the producer level is dominated by algae that reproduce so rapidly that they replace the ones that are eaten as fast as they are consumed. At any given time, there is a smaller biomass of algae than of organisms feeding on them, but if we were to make a pyramid of the biomass produced over an extended period of time, that pyramid would closely resemble the pyramid of energy described below.

Color the heading Pyramid of Energy and structures A through D in the remaining pyramid.

A pyramid of energy displays the total amount of energy captured and stored in the biomass of each trophic level over one year. (The energy is measured in kilocalories—what nutritionists call Calories, with a capital "c"—or in joules, a unit of energy from physics.) A pyramid of energy takes very nearly the same shape for every community. Each trophic level captures only about 10 percent of the energy contained in the biomass of the level below it. The remaining 90 percent is unassimilated (since even the most efficient digestive system cannot digest and absorb everything) or is used and dissipated as heat in the activities of life. Thus a *secondary carnivore* eating a primary carnivore takes in only about 1 percent (10 percent of 10 percent) of the energy present in the original producers and converts only about 0.1 percent of that energy into its own body mass.

The pyramid of energy shows very clearly that if food for feeding people is scarce, we can feed far more people on plant foods than we can on meat from plant-eating animals. It also shows why in nature the largest number of trophic levels normally found is five, and then usually only in aquatic communities where the big fish eat the little fish who eat the littler fish who eat the almost microscopic organisms who eat the algae.

ECOLOGICAL PYRAMIDS.

PRODUCERS_A → PRODUCERS_A

PRODUCERS_A
HERBIVORES_B

PRIMARY CARNIVORES_C
SECONDARY CARNIVORES_D

PYRAMID OF NUMBERS.⋆

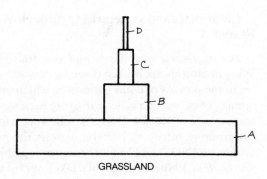

GRASSLAND

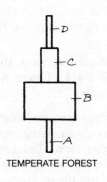

TEMPERATE FOREST

PYRAMID OF BIOMASS.⋆

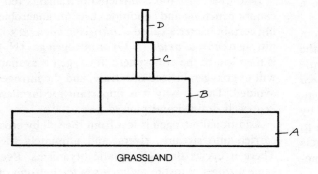

GRASSLAND

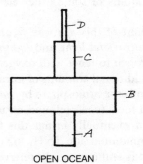

OPEN OCEAN

PYRAMID OF ENERGY.⋆

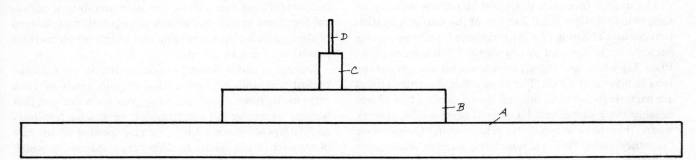

BIOGEOCHEMICAL CYCLES

As discussed in Plate 43, the flow of energy in the biosphere is in one direction only: from the sun, through living organisms, into the environment, and out into space. Matter, on the other hand, cycles constantly from organism to organism as well as to and from the environment, which acts as a reservoir. The cyclic pathways taken by various elements in passing through living organisms and the earth, its atmosphere, and its bodies of water are generally called biogeochemical cycles. This plate illustrates one of the most important of those cycles, the nitrogen cycle.

Color titles and structures A through J. Reserve a blue-green color for L. Leave A[1], E[1], F[1], J[1], and J[2] uncolored for now.

Proteins and nucleic acids are essential to life, and in order to make them, every living organism must obtain nitrogen in a suitable form. The earth's atmosphere is approximately 79 percent *nitrogen* and serves as a nitrogen reservoir. However, the nitrogen atoms in the air are tightly joined in diatomic ("two-atom") molecules (N_2), and no animal or plant known can separate them to use them. To be useful, the nitrogen must first be "fixed," that is, attached to atoms of some other elements to form a compound.

A small amount of this *nitrogen fixation* results from the passage of *ultraviolet light* and *lightning* through the air, causing nitrogen to react with oxygen to form *nitrate ions* (NO_3^-). Additional amounts of nitrate and *ammonia* (NH_3) are put into the atmosphere by *volcanoes,* by combustion of fossil fuels (coal, oil, and natural gas), and by forest fires. *Rain* eventually brings this fixed nitrogen to the soil—as ammonium ion (NH_4^-), once it contacts water, though it is still commonly referred to as "ammonia" to simplify discussion.

Color arrow E[1], titles K through P, and their corresponding parts of the illustration. Use blue-green for L. Color J[1] as well.

The abiotic processes described above are actually responsible for only a small fraction of the nitrogen fixation that occurs. The bulk of it is carried out by *nitrogen-fixing bacteria* in the soil and by *cyanophytes* (blue-green algae; Plate 32) which live mostly as a scum on submerged objects in lakes and ponds. The most efficient nitrogen fixers are bacteria found in nodules on the roots of certain *plants,* notably the legumes (alfalfa, beans, peas, lentils, clover), where they have a symbiotic relationship (Greek: *sym,* "together"; *bios,* "life") in which the bacteria obtain some nutrients from the plant but provide ammonia in return,

which the plant can use to make amino acids. Commercial processes for fixing atmospheric nitrogen to produce chemical *fertilizers* also add nitrogen to the soil (about 10 percent of what biological fixations add). *Animals* are totally unable to fix nitrogen or even to utilize inorganic nitrogen compounds. They must *consume* already formed amino acids in their food.

Color titles and structures Q through W, including F[1] and J[2].

The *excretions* of animals and the dead bodies of all living organisms are broken down in the soil by *decomposers* in the process of *ammonification,* which produces ammonia. Then bacteria called *nitrifying bacteria* convert the ammonia to *nitrite ion* (NO_2^-), following which a different group of nitrifying bacteria convert the nitrite ion to nitrate ion (NO_3^-). (The two processes together are called *nitrification.*) Nitrate is readily taken up by the roots of plants and utilized.

Color the remainder of the plate, including A[1].

If soil becomes too compacted or remains too wet, air cannot penetrate and conditions become anaerobic, allowing certain bacteria called *denitrifying bacteria* to convert nitrate to *nitrous oxide* (N_2O) or nitrogen gas (N_2), which is then lost to the atmosphere. If oxygen is available, they will use oxygen instead of nitrate, and the nitrogen loss is avoided. That is why it is important for farmland to be kept well drained and plowed.

Additional nitrogen is lost from the soil by erosion and carried into streams, rivers, and ultimately the ocean. There it cycles through aquatic organisms. Eventually, some nitrogen is lost to *sediments* at the bottoms of oceans or lakes too deep for the nitrogen to be recycled by currents. It will remain there until major upheavals of the earth's crust bring those sediments to the surface again. This does not mean that we are running out of nitrogen. While supplies of fossil fuels last, the opposite is true, and in some farming areas, the use of large amounts of chemical fertilizers sometimes results in population explosions of algae, which clog waterways, and health problems from pollution of drinking water.

Dozens of other elements cycle in similar ways. In the phosphorus and calcium cycles, certain kinds of rock serve as the reservoirs. Some ecologists are concerned that we may be mining too much phosphate rock for fertilizers and detergents, since we have greatly speeded up the rate of transfer of phosphate to deep ocean sediments, which will not return to the surface for millions of years.

BIOGEOCHEMICAL CYCLES.

NITROGEN CYCLE.★
ATMOSPHERIC NITROGEN$_{A, A^1}$
SUN$_B$
ULTRAVIOLET LIGHT$_C$
LIGHTNING$_D$
NITROGEN FIXATION$_{E, E^1}$
NITRATE ION$_{F, F^1}$
VOLCANO$_G$
FACTORY$_H$
RAIN$_I$
AMMONIA$_{J, J^1, J^2}$
NITROGEN-FIXING
  BACTERIA$_K$
NITROGEN-FIXING
  CYANOPHYTES$_L$

FERTILIZER$_M$
PLANT$_N$
ANIMAL$_O$
CONSUMPTION$_P$
EXCRETION$_Q$
DEATH$_R$
AMMONIFICATION$_S$
DECOMPOSERS$_T$
NITRIFICATION$_U$
NITRIFYING BACTERIA$_V$
NITRITE ION$_W$
DENITRIFICATION$_X$
DENITRIFYING BACTERIA$_Y$
NITROUS OXIDE$_Z$
LOSS TO SEDIMENT$_{ZZ}$

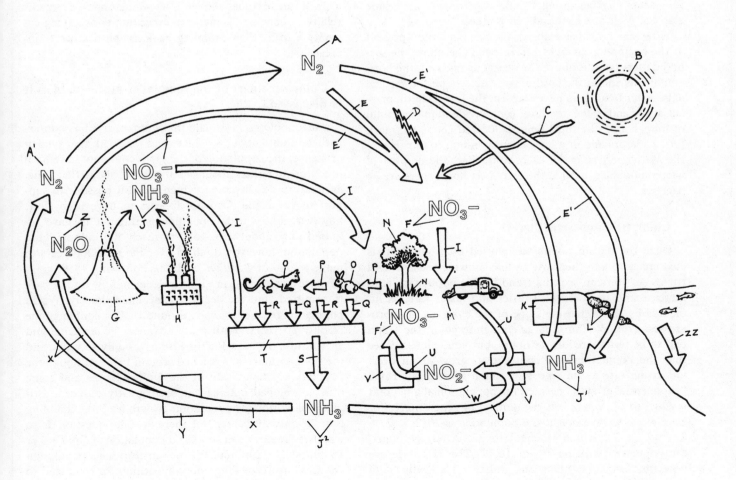

Another thing studied by ecologists is population growth, and certain common patterns are found regardless of the species studied. The upper part of this plate illustrates the population growth when a single bacterium is placed in a tube of nutrient broth at a suitable temperature. Bacteria reproduce rapidly, and it is not uncommon for some of them to produce a new generation every hour, as shown here.

Color the heading Exponential Growth, titles A and B, and their associated structures in the first graph. Then color title C and each datum point in the first graph as it is discussed in the text.

On the graph, each *datum point* marks the population size for that particular time. Population size is indicated by the distance of the point measured along the *vertical axis,* while elapsed time is indicated by the distance along the *horizontal axis.* The edges of the graph are labeled to indicate what each axis represents. The first datum point represents the beginning of the experiment, zero time elapsed, with one bacterial cell present.

After one hour has elapsed, the one bacterium present at the beginning divides to form two. Therefore, the second datum point is above the one-hour mark on the horizontal axis and at the level of two bacteria on the vertical axis. After two hours have elapsed, the two bacteria present at the end of the first hour each divide to make a total of four; hence, the third point is located above the mark for two hours and at a level corresponding to four bacteria. With the passage of each hour, then, the number of bacteria doubles, until at the end of six hours there are 64 bacteria.

Color title and structure D.

Once the datum points are plotted on a graph, it is customary to join them with a line showing their general trend. As you can see, the trend in this population is to increase and to do so at a faster and faster rate, as indicated by the increasing slope of the *curve* as time progresses. To make things large enough to color, the graph had to be terminated at six hours, but you can easily see that the population will continue to grow at an ever-increasing rate as time goes on. Such growth is termed "exponential growth" because the population at any point is equal to 2^n, where *n* is the number of generations. For example, after five hours, the population is 2^5 ($2 \times 2 \times 2 \times 2 \times 2$), which is equal to 32. After 24 hours, the population will be 2^{24}, or 16,777,216. (Bacteria are very tiny, and that many and more will actually fit in a test tube, but by 24 hours they will begin to run out

of nutrients and start to die from accumulated wastes.)

All populations, even those in which two parents produce only one offspring every few years, tend to grow exponentially. If conditions allow the population to grow at all, then with each generation there are more individuals to contribute to the next growth phase.

Color the heading Logistic Growth Curve, title E, and the associated illustration.

Exponential growth cannot continue indefinitely. Food, light, water, and space are never infinite; they combine with disease and the effect of predators to establish a relatively steady *carrying capacity* for any population in a given environment: a maximum population size that the environment can support indefinitely. Hence in nature we see a pattern known as logistic growth. This pattern will usually be followed by any population that moves into a suitable new environment. Growth is exponential at first, then it decreases as it approaches the carrying capacity, where the growth rate becomes zero and the population size remains relatively constant, sometimes exceeding the carrying capacity a little, then dropping back as the death rate increases.

Color each part of the remainder of the plate as it is discussed.

Anthropologists estimate that humans lived as hunter-gatherers until about 8000 B.C. and that with that type of existence, the population of the earth could not have been much greater than about 5 million. At about that time, agriculture was invented, which made it possible to support more people. Growth appears to have been slow, however, and world population is thought to have required until about A.D. 1650 to reach 500 million. That population, however, doubled over the next 200 years to make a total of *1 billion* by 1850. Following 1850, great advances were made in the knowledge of how diseases are transmitted, and the death rate declined sharply. World population reached *2 billion* around 1930, a doubling time of only 80 years. With more advances in medicine and agriculture, the next doubling time was only 45 years, and the *4 billion* mark was reached around 1975. Although the population growth rate in the United States and some other developed nations has been greatly reduced, estimates of worldwide population growth lead to a doubling time of only 39 years. But there is some question as to whether the earth can support a population of *8 billion,* or 16 billion, or 32 billion. We may instead see a population *crash,* as overcrowding and competition for food lead to epidemics and wars.

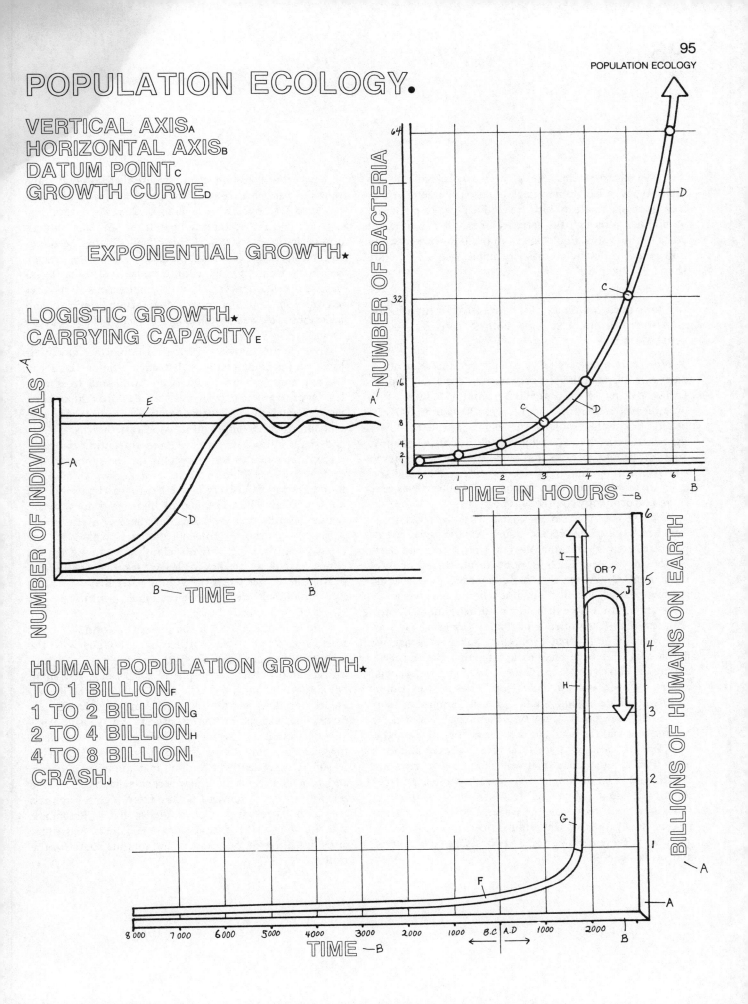

POPULATION ECOLOGY.

VERTICAL AXIS $_A$
HORIZONTAL AXIS $_B$
DATUM POINT $_C$
GROWTH CURVE $_D$

EXPONENTIAL GROWTH. ★

LOGISTIC GROWTH. ★
CARRYING CAPACITY $_E$

HUMAN POPULATION GROWTH. ★
TO 1 BILLION $_F$
1 TO 2 BILLION $_G$
2 TO 4 BILLION $_H$
4 TO 8 BILLION $_I$
CRASH $_J$

Terrestrial ecosystems are customarily classified into major types called biomes. Each biome is characterized by a distinctive climate and by distinctive types of plant and animal life, although the actual species differ in different parts of the world. Ecologists are not all in agreement on how many different biomes are appropriate. Eight are shown in this plate.

Color each biome as you come to it in the discussion that follows. Use light colors so as not to obscure details.

Tropical rain forests have warm temperatures and abundant rainfall, resulting in very dense vegetation and an immense variety of species of both plants and animals. Tall, broad-leaved trees form a high canopy 30 meters (100 feet) or more above the ground. Other plants, adapted to lower light intensities, live at various levels below the canopy. Some of them have leaves that hold small pools of water in which entire communities live out their lives. Typical animals are monkeys, reptiles, and insects of all varieties, tree frogs, and brilliantly colored birds.

Savanna is a tropical grassland with scattered trees, found in parts of Africa and South America where temperatures are high but rainfall is only moderate and there is a long dry season. Each supports a wide variety of grazing animals. In Africa they include giraffes, wildebeests, gazelles, zebras, impalas, buffaloes, lions, and hyenas.

Desert occurs where there is very little rainfall. Deserts with less than 2 centimeters of rainfall per year may support virtually no life, but with a little more than that, we find a wide variety of plants and animals that are highly adapted to conserve water. Plants have tough, waxy coats, and many, such as cacti, have thick leaves to store water. Animals include rodents, snakes, lizards, tarantulas, scorpions, and a few birds. Many of the animals burrow in the ground to avoid the heat and become active only at night.

Temperate deciduous forest is found where rainfall is abundant, summers are relatively long, and winters are rather cold. The dominant plants are broad-leaved trees that are deciduous (drop their leaves in the fall), although some evergreen trees grow there also. Typical animals are insects, small rodents, white-tailed deer, raccoons, foxes, and a very few remaining wolves, bobcats, mountain lions, and black bears.

Temperate grassland occurs where there is not enough rainfall to support trees and summers are warm and winters cold (as opposed to savannas, where the alternation is in rainfall rather than temperature). Grazing animals and periodic fires also make it difficult for tree seedlings to get a start. The dominant plants are, obviously, grasses, but there are numerous wildflowers as well. Like savannas, temperate grasslands support numerous herbivores such as bison, antelope, prairie dogs, ground squirrels, and jackrabbits, as well as foxes and coyotes, which prey on them.

Scrub forest, known as chaparral in North America, is found in maritime climates that have little or no rain in the summer. The dominant plants are small trees with leathery leaves and extreme resistance to drought and fire, which frequently sweeps through these areas. Typical animals are rabbits, chipmunks, wood rats, lizards, snakes, and small birds. Mule deer are present during the rainy season but move to cooler regions for the summer.

Temperate coniferous forest, also known as taiga, is found almost exclusively in the Northern Hemisphere in more northerly latitudes and at higher altitudes, where winter is quite cold and wet but there is a reasonably warm, though short, summer. Dominant plants are coniferous ("cone-bearing") trees such as pine, fir, and spruce, which have waxy needles for leaves to prevent water loss during the winter when the groundwater may be frozen. Typical animals are rabbits, porcupines, rodents, moose, elk, deer, and grizzly bears.

Tundra is the biome of the far north, forming a narrow band between the northernmost boundary of the taiga and the perpetual ice of the arctic. There is no tundra in the Southern Hemisphere except for a few small patches at high altitudes. The sun's rays are at so low an angle that they provide little heat and significantly reduced light energy for photosynthesis. The subsoil exists all year round as "permafrost"—frozen solid—and the topsoil thaws only during the six- to eight-week growing season. Consequently, the dominant plants are mosses and lichens along with a few perennial herbs and some greatly dwarfed birch and willow trees. Typical animals are caribou, reindeer, wolves, foxes, hares, lemmings, and the snowy owl. Geese, ducks, and numerous other birds nest there in the summer, but migrate south for the winter.

BIOMES.

TROPICAL RAIN FOREST_A

SAVANNA_B

DESERT_C

TEMPERATE
DECIDUOUS FOREST_D

TEMPERATE GRASSLAND_E

SCRUB FOREST_F

TEMPERATE
CONIFEROUS FOREST_G

TUNDRA_H

AQUATIC ECOSYSTEMS

Aquatic ecosystems are more complex than terrestrial ones and thus do not lend themselves to regular classification and are highly interdependent. Organisms in aquatic ecosystems are classified as plankton, nekton, or benthos. Plankton are the organisms that either have no locomotion or are too tiny or too feeble to swim against the current. The photosynthetic plankton are called phytoplankton (Greek: *phyton,* "plant") and include the algae and the photosynthetic protists. Phytoplankton are eaten by zooplankton (Greek: *zoon,* "animal"), such as protists and tiny crustaceans. Nekton are the organisms that swim well enough to move against the current, such as squid, fish, whales, and dolphins. Benthos are the organisms that live attached to, burrowed into, or crawling on the bottom.

Color the heading Marine Ecosystems, titles A through D, and the corresponding parts of the illustration. Reserve your light colors for later.

The entire ocean is essentially one huge ecosystem. Although we can divide it into a number of regions or zones based on water depth, they interact much more than two adjacent terrestrial biomes would, and a number of their organisms travel back and forth quite regularly. Some fish, of course, leave the marine environment entirely and enter fresh water to spawn.

In most parts of the world, there is a *continental shelf* where the depth increases very gradually. At the end of this shelf, the bottom drops away more steeply to form the *continental slope,* which levels off at a depth of about 4000 to 6000 meters. In some places there is no continental shelf and the bottom slopes steeply away from the shore. At the bottom of the slope is the broad, somewhat level *abyssal plain,* with submerged mountain ranges (not shown) and a few extremely deep *trenches,* also known as *deeps,* with depths up to more than 10,000 meters.

Color the headings Neritic Province and Pelagic Province, titles E through J, and the associated parts of the illustration.

In the neritic province of the ocean, light penetrates to the bottom, a depth of about 200 meters (650 feet) where the water is clear. Most of the continental shelf is included in this province. The *littoral region* is the region shallow enough to be regularly mixed by wave or tidal action, thus keeping the nutrients suspended and the water well oxygenated. The section between high and low tide is called

the intertidal zone. A large variety of littoral communities are found in different places, depending on whether the shore is sandy or muddy, requiring organisms to put down roots or to burrow, or rocky, requiring that they cling tightly to or crawl under the rocks. In the tropics, coral reefs have their own immense variety of living organisms. The *sublittoral region* is also rather rich in life, since there is enough light throughout for photosynthesis.

The pelagic province is the open ocean. The uppermost layer is the *epipelagic region,* where there is enough light for photosynthesis. Life is not nearly as abundant here because the remains of dead organisms sink below this region and their nutrients are not directly recycled back to producers, which here are exclusively phytoplankton.

Next below is the *mesopelagic region,* where photosynthesis is impossible and the dim light fades out with increasing depth until the total darkness of the *bathypelagic region* is reached at around 1000 meters. Most of the organisms in these regions, including the very deep *hadal region,* are scavengers or decomposers, although some of the strangest predators are found at great depths.

Color the remainder of the plate as you read.

Standing-water ecosystems vary widely, from temporary puddles to lakes of all sizes and depths. A lake or pond has three primary regions. The *littoral region,* where light reaches the bottom, is the richest in life, with numerous rooted plants, floating plants, algae, small insects and insect larvae, and small fish. The *limnetic region* consists of open water with sufficient light penetration for photosynthesis. The only photosynthesizers in this region are the phytoplankton.

Below the limnetic region is the *profundal region.* Here there is insufficient light for photosynthesis, and virtually all the organisms are dependent on debris sinking down from above. There are a few fishes that live here as scavengers and numerous fungi, bacteria, worms, and other decomposers.

Flowing-water ecosystems come in an overwhelming variety of sizes, bottom types, velocities, and depths. To survive in flowing water, organisms must be able either to swim against the current, to cling firmly to rocks or to the bottom, or to burrow under rocks or into the bottom. Nutrients are well suspended and the water well oxygenated. The principal producers are aquatic plants that are rooted in the bottom and algae that glue themselves to rocks and other objects.

AQUATIC ECOSYSTEMS.

MARINE ECOSYSTEM★
BOTTOM★
CONTINENTAL SHELF_A
CONTINENTAL SLOPE_B

ABYSSAL PLANE_C
TRENCH/DEEP_D

NERITIC PROVINCE★
LITTORAL REGION_E
SUBLITTORAL REGION_F

PELAGIC PROVINCE★
EPIPELAGIC REGION_G
MESOPELAGIC REGION_H
BATHYPELAGIC REGION_I
HADAL REGION_J

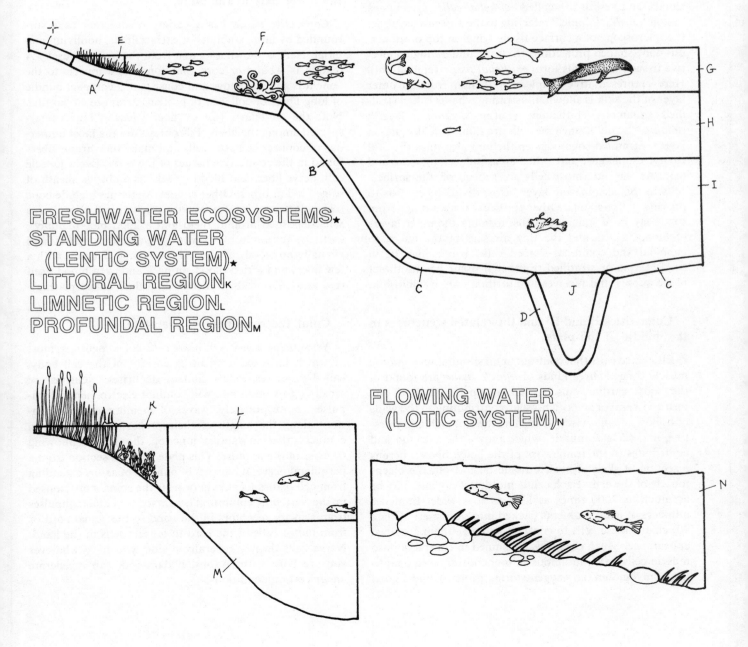

FRESHWATER ECOSYSTEMS★
STANDING WATER
(LENTIC SYSTEM)★
LITTORAL REGION_K
LIMNETIC REGION_L
PROFUNDAL REGION_M

FLOWING WATER
(LOTIC SYSTEM)_N

CELLS AND TISSUES

With this plate we begin an examination of how cells are put together to form tissues, organs, and organ systems. A tissue is defined as a collection of cells with a common function, usually similar in shape and attached together. Four major categories of tissues are distinguished.

Color titles A and A¹ and the related structures in the upper part of the plate.

Cells that line the surface of the body, cavities, ducts, canals, and vessels are called *epithelial cells* (Greek: *epi,* "upon"; *thele,* "nipple," referring to the appearance under the microscope of a surface tissue lying on top of numerous cone-shaped projections from the underlying connective tissue). Such cells form *epithelial tissue* (epithelium). Here a representative sample of epithelium from the outer layer of the skin is shown, illustrating what is called stratified squamous epithelium (Latin: *stratum,* "layer"; *squama,* "scale"), since the cells are thin, scalelike, and in layers. Stratified squamous epithelium also lines the oral cavity, esophagus, and other areas subject to wear and tear. As the outermost cells are rubbed off the surface, cells in the germinating layer below divide by mitosis to provide replacements. Other epithelial tissues range from single layers of squamous cells, forming the inner layers of blood vessels and the tiny air sacs of the lungs, to cuboidal and columnar-shaped cells, which may be in single layers or stratified. Epithelial tissue has no direct blood supply and receives its nutrition only by diffusion.

Color titles B and B¹ and the related structures in the middle of the plate.

Elongated *cells* that contract upon stimulation compose muscle tissue. Three kinds of *muscle tissue* are found in the body: cardiac muscle, making up the walls of the heart; visceral or smooth muscle, contributing to the walls of hollow organs (viscera) such as the stomach and intestine; and skeletal muscle, which moves the skeleton and contributes to the framework of the body. Shown here is a section of skeletal muscle tissue taken from the biceps muscle of the arm. Each cell is multinucleate and may be as much as 2000 times as long as it is wide. Immense numbers of microfilaments packed into sarcomeres (Plate 40) enable these cells to contract and give them a striped appearance. Much oxygen is consumed in contraction, so muscle cells have numerous mitochondria, large quantities of myoglobin (an oxygen-storing protein), and a good

blood supply. They also have extensive endoplasmic reticulum and numerous ribosomes for synthesizing the needed proteins. Cardiac muscle also has a striped appearance due to its sarcomeres, but the cells are extensively branched and interconnected so that if one contracts, they all contract. Smooth muscle has its sarcomeres more randomly organized and hence lacks the striped appearance.

Color titles C and C¹ and the related structures in the lower part of the plate.

Connective tissue has a variety of different cells surrounded by large amounts of extracellular, nonliving material. Here is shown a tendon from the biceps muscle. A tendon is a band or cord that connects a muscle to the bone it is going to move and consists of a compact bundle of long fibers of collagen (a protein) arranged in parallel. Note the *fibroblasts* (*fibro,* "fiber"; *blast,* "bud") interspersed among the fibers. Fibroblasts are the most numerous of connective tissue cells and make the various fibers found in the connective tissues of the body. Every muscle cell, nerve fiber, and blood vessel has a fibrous sheath of some kind or other. Other connective tissues include bone (osteoblasts and osteocytes with calcium and phosphorus salts deposited around them and a good blood supply), cartilage (chrondroblasts with a gelatinous matrix and virtually no blood supply), adipose tissue (fat cells with a few fibers and a rich blood supply), and even blood itself (red and white cells immersed in a fluid called plasma).

Color the remainder of the plate.

Nerve tissue consists of *nerve cells* called neurons, most of which have extravagant extensions of the cell body, called processes, axons, and/or dendrites. Neurons are sensitive to stimuli and will conduct electrochemical impulses (more precisely, waves of membrane depolarization) along the length of their processes. Each neuron contacts others in a special junction, the synapse, allowing transfer of the impulse. This plate shows a section from a peripheral nerve; it consists of bundles of axons extending from cell bodies located in or near the spinal cord housed in the vertebral column. These axons conduct impulses from sensory receptors in the hand to the spinal cord or from motor cells in the cord to muscle cells in the hand. Nerve cells do not generally divide; you have whatever you are born with. Axons, if damaged, can regenerate under certain conditions.

CELLS AND TISSUES.

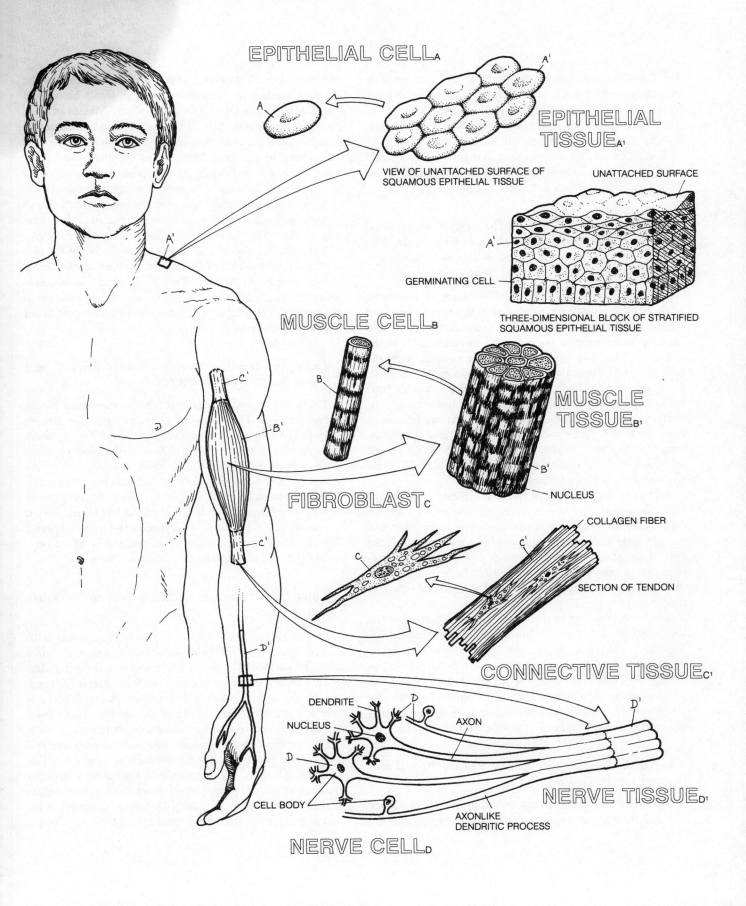

EPITHELIAL CELL A

EPITHELIAL TISSUE A'

VIEW OF UNATTACHED SURFACE OF
SQUAMOUS EPITHELIAL TISSUE

UNATTACHED SURFACE

A'

GERMINATING CELL

THREE-DIMENSIONAL BLOCK OF STRATIFIED
SQUAMOUS EPITHELIAL TISSUE

MUSCLE CELL B

MUSCLE TISSUE B'

NUCLEUS

FIBROBLAST C

COLLAGEN FIBER

SECTION OF TENDON

CONNECTIVE TISSUE C'

DENDRITE

NUCLEUS

AXON

CELL BODY

AXONLIKE
DENDRITIC PROCESS

NERVE TISSUE D'

NERVE CELL D

The four kinds of tissues of the body are combined in various ways to make functional structures called organs, which are always composed of at least two different tissues. Most organs have a working surface of epithelium, usually some amount of muscle tissue for movement or control of blood flow, nerve tissue to regulate activity, and connective tissue to hold it all together. This plate shows a section of the wall of a typical organ, the stomach.

Color titles A and A¹ and the related structures. Note the glands labeled A in the lamina propria. The vessels, composed of four tissues themselves, are not to be colored.

The stomach is lined with a single layer of columnar *epithelial tissue,* which produces a significant layer of mucus on the free surface of the tissue to protect it and ease the movement of the stomach contents. This is the working layer, the first to come into contact with the outside world (food, bacteria, water, chicken bones, glass, toys, and so on). The epithelial layer displays numerous invaginations (inward growths) into the underlying connective tissue layer, the lamina propria. At the bottom of each pit is the opening of one or several gastric glands, which extend even farther into the connective tissue and are themselves lined with columnar epithelia that secrete hydrochloric acid and digestive enzymes.

Although the epithelia of the small intestine absorb all kinds of things, the stomach epithelia appear to absorb only water, salts, alcohol, and a few other drugs. Alcohol and strongly acidic or alkaline mixtures can damage these epithelial cells; prolonged damage can produce serious structural changes (ulcers). Normally, the epithelia reproduce rapidly and regularly, creating a virtually new epithelial layer every 20 to 30 hours.

The outer covering of the stomach (at the bottom of the illustration) is a single layer of squamous or cuboidal cells. This layer, along with its supporting connective tissue fibers (not shown), is called the *serosa* (visceral peritoneum). It produces a watery secretion that lubricates the surface of the stomach so that it does not chafe on the other internal organs (viscera) as it goes through its churning contractions to mix food with digestive juices.

Color the heading Connective Tissue, titles B and B¹, and the related layers.

Just below the epithelial lining of the stomach is a layer of connective tissue known as the *lamina propria* (Latin:

lamina, "layer"; *propria,* "one's own"). It supports the gastric glands and contains numerous small blood and lymphatic vessels as well as nerve fibers. Masses of lymphocytes (nodules), part of the immune system, may be seen here as well, ready to move into an area that is injured or invaded by bacteria. Many loose fibroblasts and scavenger cells (not shown) migrate through this tissue, ready to repair any damage. The lamina propria, its overlying epithelium, and its underlying muscle layer (to be colored shortly) are collectively called the mucosa (a term generally employed for such layers in all viscera with cavities).

The deeper layer of connective tissue (between the two muscle layers) is more fibrous and supports rather large vessels and nerve trunks. This layer is called the *submucosa,* since it lies deep below the mucosa. Ulceration to the depths of this layer will generally bring on significant bleeding (hemorrhagic ulcer).

Color the heading Muscle Tissue, titles C and C¹, and the related structures.

There are two layers of smooth muscle tissue in the stomach wall. The layer in contact with the lamina propria is known as the *muscularis mucosa.* Its rhythmic contractions enhance mixing of food with digestive juices in the gastric pits. The *muscle* layer *(tunic)* next to the serosa has fibers running in three different directions: longitudinal, circular, and diagonal. These fibers provide mixing on a larger scale. When the food has been digested to a semiliquid state, they move it on to the small intestine. The rhythmic contraction of these muscles is called peristalsis (Greek: *peri,* "around"; *stellein,* "to place").

Color title D and the related nerves and pressure receptor.

The epithelial layer of the stomach is supplied with sensory fibers *(nerve tissue)* conducting nerve impulses from the surface cells to the brain and spinal cord. Additional sensory fibers carry impulses from pressure receptors in the lamina propria and submucosa, and perhaps even the muscle layers. Much of the sensory input from the stomach wall contributes to an overall sense of wellbeing. For example, if you have just had a good meal, you usually feel good; but if your stomach is upset, you feel nauseated, listless, and incapable of taking an interest in much of anything. Motor fibers from the brain stimulate muscle peristalsis and glandular secretion, although secretion is primarily under hormonal control.

TISSUES AND ORGANS.

EPITHELIAL TISSUE A
SEROSA A¹
CONNECTIVE TISSUE ★
LAMINA PROPRIA B
SUBMUCOSA B¹
MUSCLE TISSUE ★
MUSCULARIS MUCOSA C
MUSCULAR TUNIC C¹
NERVE TISSUE D

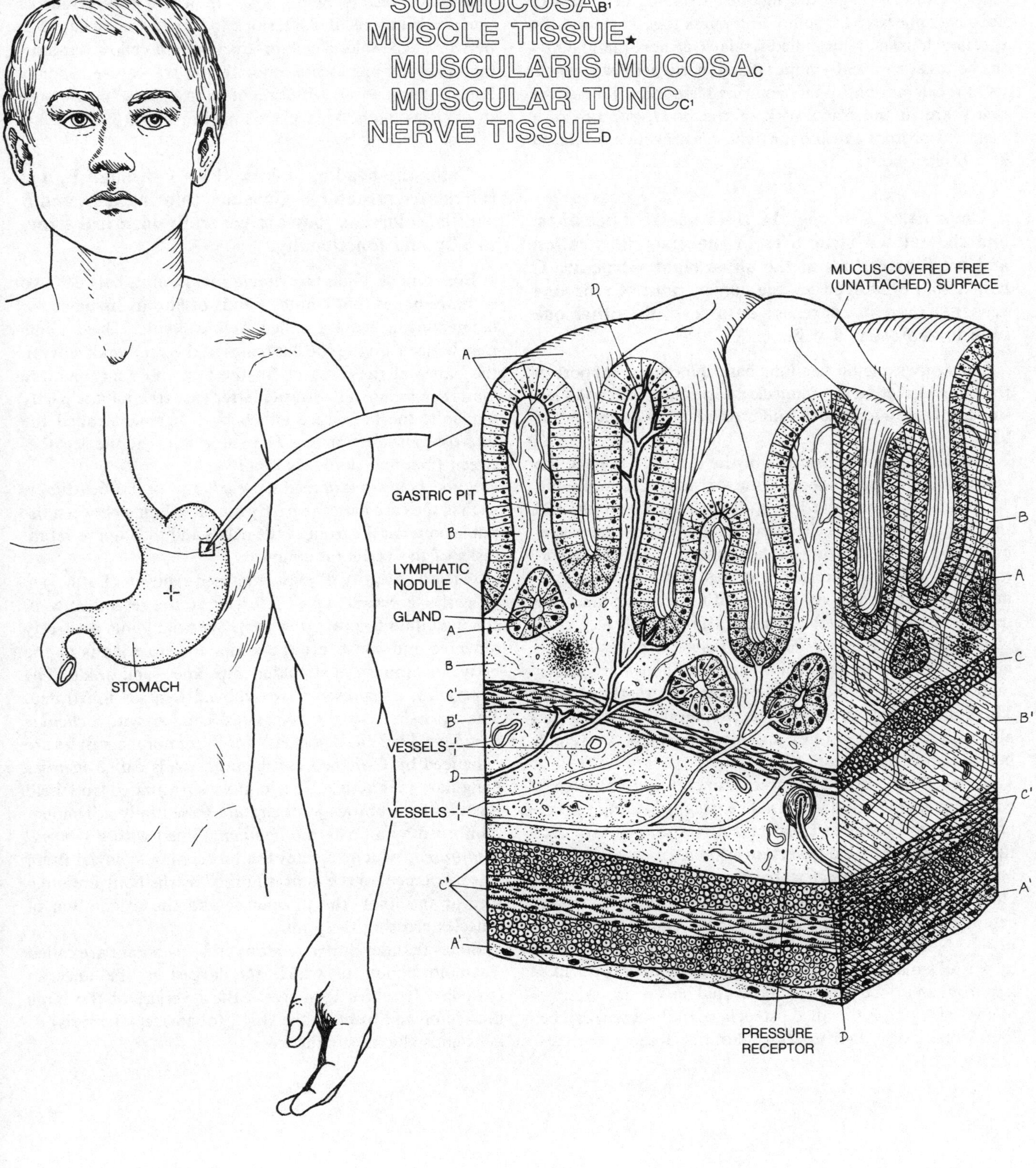
STOMACH
MUCUS-COVERED FREE (UNATTACHED) SURFACE
D
A
GASTRIC PIT
B
LYMPHATIC NODULE
GLAND
A
B
C'
B'
VESSELS
D
VESSELS
C'
A'
PRESSURE RECEPTOR
D
B
A
C
B'
C'
A'

The skeletal system consists of the bones, the articular (joint) cartilages at the ends of the bones, the costal cartilages of the rib cage, the ligaments (bands or sheets of fibrous connective tissue holding joints together), and the sheets of fibrous tissue called fasciae (see next plate). Only the bones are normally mineralized (hardened with deposits of a calcium phosphate complex). As important as the bones are in the framework of the body, rupture of a ligament or joint capsule may cause as severe a disability as a broken bone.

Color titles A through H, the associated headings, and the related structures in the main illustration and the illustrations at the upper right. Structure C can also be colored at the lower right. Color one vessel (F) red to represent an artery, the other one blue to represent a vein.

The bony skeleton has four basic functions: support of the body, attachment of muscles, formation of blood cells (in the marrow cavities), and storage of calcium for the blood.

The *axial skeleton* consists of the skull (22 bones), the vertebral column (26 bones), the sternum (2 bones), and 24 ribs. The cranium of the skull encloses the brain, and the facial part forms the framework of the face. The vertebrae have weight-bearing parts, arches to protect the spinal cord, and projections (processes) for muscle attachment. The upper 24 vertebrae (with one exception) are separated by intervertebral disks of fibrocartilage.

The *appendicular skeleton* consists of bones of the upper and lower limbs. The lower-limb bones are larger than those of the upper limbs to bear the greater weight. The bones of the hands and feet are similar in structure, although the ankle bones are more adapted to weight bearing and locomotion.

By weight, bone is 65 percent mineral and 35 percent cells, fibers, and blood vessels. The weight-bearing part of bone is called *compact bone,* and it is arranged into distinctive patterns called haversian systems or osteons. Here bone cells called *osteocytes* are arranged in concentric circles (lamellae) surrounded by a *mineralized matrix.* In the center of each set of lamellae is a *canal* that carries small *vessels.* Nutrients and oxygen reach the isolated osteocytes by way of small canals that radiate away from the cells like legs from an insect. Between individual haversian systems, bone-destroying cells called osteoclasts (not shown) can be found breaking down existing patterns, freeing the cal-

cium phosphate complexes to enter the blood as ions.

At the ends of bones and along the *marrow cavities* we find *cancellous* (spongy) *bone,* in which tiny beams of bone called trabeculae interlock to form a somewhat open network. Red (blood cell–producing) and yellow (fat-storing) marrow are located among the trabeculae. Spongy bone undergoes reorientation of its trabeculae in response to changes in the load placed on it.

Color the heading Joints, titles I through P, and the related structures. Consider coloring I, N, and P similar colors, as they are generally identical structurally and functionally.

Bones move at joints (articulations). Joints between two or more bones that employ bands of fibrous tissue as the connecting agent are called *fibrous* joints. These joints may be seen among the flat bones of the skull and between the bones of the forearm. In the first case (sutures), the joints are immovable. In the latter case, the joint is partly movable: the bones are joined by a ligament called the interosseous membrane. The same joint in the leg (between tibia and fibula) is immovable.

Some bones are joined by *cartilage* or fibrocartilage; such is the case with the partly movable joints between the pubic bones at the front of the pelvis and the intervertebral disks of the vertebral column.

A third category of joint is termed *synovial* (Latin: *syn,* "together"; *ovum,* "egg," referring to the resemblance of synovial fluid to raw egg white). Synovial joints are freely movable and constitute the major movable joints of the body, including the shoulder, hip, knee, and ankle. The knee joint, a representative synovial joint, is illustrated. The opposing bony surfaces are capped with *articular cartilage.* The tibial "sockets" for the femoral condyles are enhanced by C-shaped cartilaginous pads called *menisci* (singular, meniscus). The joint is surrounded and held together by a fibrous joint *capsule* (essentially a ligament reinforced by an external *ligament*) lined with a *synovial membrane,* which secretes the lubricating synovial fluid. The movement of the joint is limited by the bony architecture at the joint, the ligaments, and the orientation of muscles crossing the joint.

Bones that develop in tendons crossing a joint are called sesamoid bones, of which the largest is the kneecap (patella). It serves to increase the leverage of the large muscle on the front of the thigh (quadriceps femoris) as it extends the knee joint.

SKELETAL SYSTEM.

AXIAL SKELETON A

BONE ★
COMPACT C
CANCELLOUS C'
MARROW CAVITY D

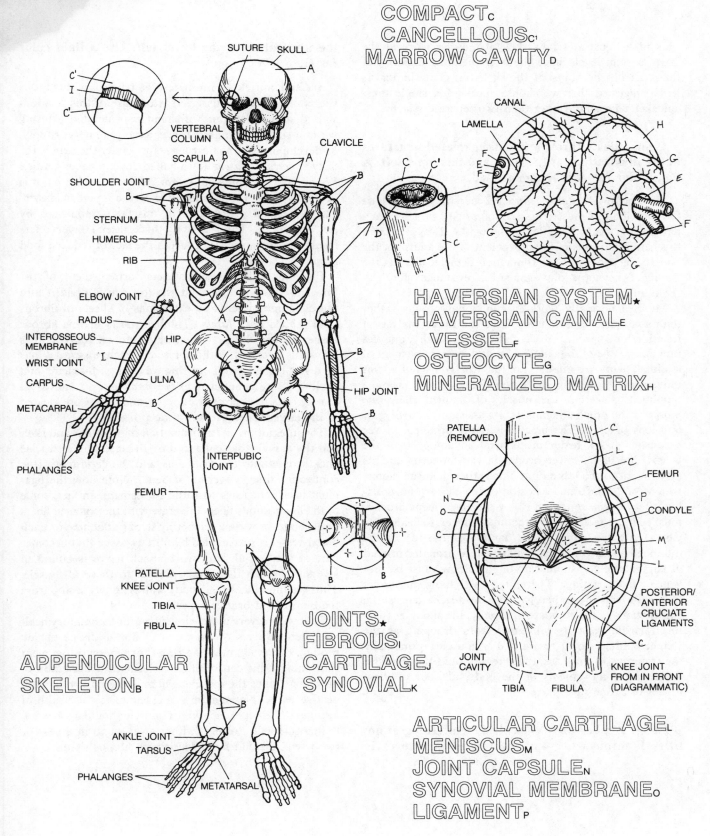

SUTURE   SKULL
A
C'
I
C'
VERTEBRAL
COLUMN
SCAPULA   B   A   CLAVICLE
SHOULDER JOINT   B
B
STERNUM
HUMERUS
RIB
ELBOW JOINT
RADIUS
INTEROSSEOUS
MEMBRANE   I
WRIST JOINT   HIP
CARPUS   ULNA
METACARPAL   B
HIP JOINT
PHALANGES   B
INTERPUBIC
JOINT
FEMUR
PATELLA
KNEE JOINT   K
TIBIA
FIBULA
B
A   B
B
I
J
B   B

CANAL
LAMELLA
C   C'
D   c
F
F
F
G
H
G
E
F
G

HAVERSIAN SYSTEM ★
HAVERSIAN CANAL E
VESSEL F
OSTEOCYTE G
MINERALIZED MATRIX H

PATELLA
(REMOVED)   C
P   L
N   C
O   FEMUR
C   P
CONDYLE
M
L
JOINT
CAVITY   POSTERIOR/
ANTERIOR
CRUCIATE
LIGAMENTS
TIBIA   FIBULA   C
KNEE JOINT
FROM IN FRONT
(DIAGRAMMATIC)

JOINTS ★
FIBROUS I
CARTILAGE J
SYNOVIAL K

APPENDICULAR SKELETON B

ANKLE JOINT
TARSUS
PHALANGES
METATARSAL
B

ARTICULAR CARTILAGE L
MENISCUS M
JOINT CAPSULE N
SYNOVIAL MEMBRANE O
LIGAMENT P

This plate illustrates how skeletal muscles work at the organ system level. Since the overwhelming majority of them act on the bones of the skeleton, many biologists have suggested that we should speak of a single musculoskeletal system rather than two separate systems.

Color titles A through E and the related structures in the illustrations at the right and the upper left. A light color for A is recommended.

Most action of *skeletal muscle* occurs at joints, where the muscle itself must cross the joint or be attached to a tendon that crosses the joint. Biologists always describe this action in terms of the movement of the joint, not the bone being moved. Thus the action depicted here is the flexing of the elbow joint against the resistance offered by the dumbbell.

Muscles normally work in functional groups rather than as isolated muscle units. This can make analysis of function for a specific muscle difficult. Generally, muscles function in one of four modes for any given movement of a joint. The mover muscles are the protagonists of a joint movement. In flexing the elbow joint, as in curling a dumbbell to exercise the muscles of the arm, the *prime mover* is the brachialis muscle. Its assistant, or *synergist,* is biceps brachii. (The brachialis is the principal mover because it has a better mechanical advantage than the biceps due to its lower origin on the humerus and its attachment to the more rigid ulna rather than the radius, which rotates.) To make flexion of the elbow joint possible, the *antagonist* to the elbow flexors, triceps brachii, must be inhibited from contracting. Further, in flexing the elbow with the palm of the hand over the top of the dumbbell instead of under it (elbow joint pronated instead of supinated), the supinators of the elbow must be prevented from acting. Therefore, the elbow pronators (pronator teres shown here) act as *neutralizers* to resist the contractions of the supinators. Finally, the shoulder (pectoral) girdle must be prevented from drooping as the weight is lifted, so there are *fixator* muscles, principally the trapezius in this case, that literally hold the shoulder up. In most movements of the body, all four of these modes of action take place.

Color the heading Musculoskeletal Integration, titles F through G³, and the related structures in

the illustration at the lower left. Use a light color for G.

Skeletal muscle is intimately associated with fibrous sheets of connective tissue. It is because of this relationship that muscles can withstand considerable bruising without rupture of muscle tissue, damage to nerve supply, or development of hemorrhage. Just under the skin is the fatty *superficial fascia* (also called subcutaneous tissue). When a person "puts on weight" above that desired, it is generally due to the addition of fat to the superficial fascia. The pattern of distribution of this fat is influenced by sex hormones. This variably thick fascial layer offers insulation and a source of fuel in the face of reduced food input.

Examining the musculoskeletal arrangement of the forearm, you will note that the forearm is divided into muscle compartments by extensions of sheets of fibrous tissue called *deep fascia*. Thinner sheets of this fibrous tissue further divide the muscle into smaller bundles known as fascicles. Within a fascicle, each *muscle cell* has its own fibrous *sheath.* In this way, nerve filaments and capillaries are conducted to the fascicles and individual muscle cells. Larger *arteries, veins,* and *nerves* are located in the larger sheets (septa) of deep fascia.

The integration and coordination of muscular and skeletal tissue is further enhanced by the insertion of fasciae and tendons into the *ligaments* and the *periosteum* (the connective tissue covering of bone). Note how the ligament itself integrates with the periosteum. In fact, some of the ligamentous fibers integrate with the collagen fibers in the bone, providing a superbly strong attachment. Each blood vessel is surrounded by fibrous tissue that becomes part of the fascial framework. Each nerve is bound in layers of fibrous tissue very similar to those of muscle. Thus the musculoskeletal body wall is largely secure from the bumps and bruises experienced in life.

Following severe muscle injury, postinflammatory healing often involves proliferation of fibrous tissue, which produces scarring, meaning fascial or tendon or ligament shortening. This can result in a permanent limitation of movement since the muscle is literally tied down by connective tissue. Nerve supply is essential to the health of muscle. If the nerve supplying a muscle should be severed, the muscle will atrophy (wither away), and in a year or two it will die and be replaced with fibrous tissue.

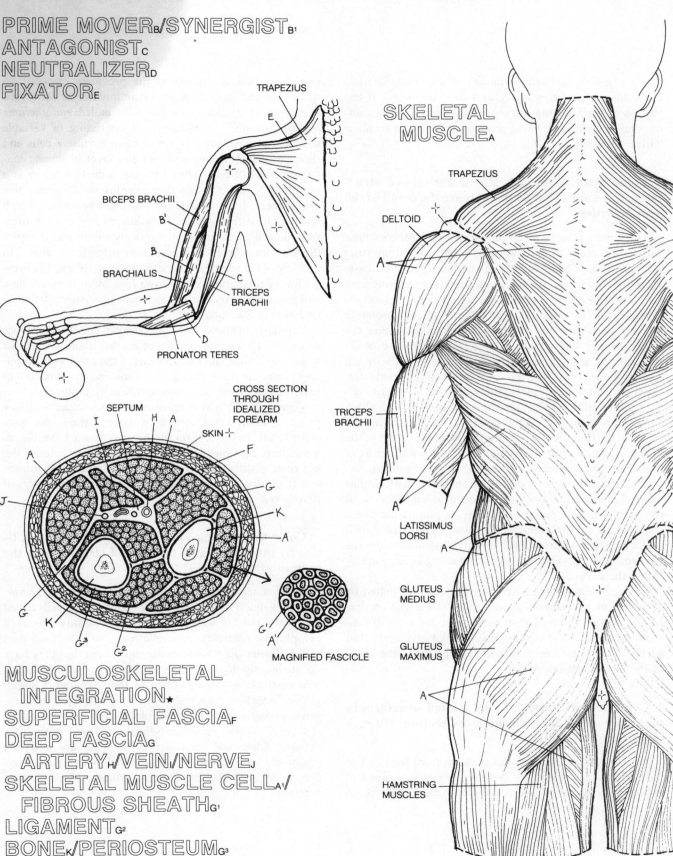

MUSCULOSKELETAL SYSTEM.

PRIME MOVER B/SYNERGIST B¹
ANTAGONIST C
NEUTRALIZER D
FIXATOR E

TRAPEZIUS

BICEPS BRACHII

BRACHIALIS

TRICEPS BRACHII

PRONATOR TERES

SKELETAL MUSCLE A

TRAPEZIUS

DELTOID

TRICEPS BRACHII

LATISSIMUS DORSI

GLUTEUS MEDIUS

GLUTEUS MAXIMUS

HAMSTRING MUSCLES

CROSS SECTION THROUGH IDEALIZED FOREARM

SKIN

SEPTUM

MAGNIFIED FASCICLE

MUSCULOSKELETAL INTEGRATION ★
SUPERFICIAL FASCIA F
DEEP FASCIA G
ARTERY H/VEIN/NERVE J
SKELETAL MUSCLE CELL A¹/
FIBROUS SHEATH G¹
LIGAMENT G²
BONE K/PERIOSTEUM G³

The cardiovascular system consists of the heart, arteries, veins, networks of capillaries, and the blood itself. It has the job of transporting oxygen, nutrients, hormones, antibodies, and soluble waste products from place to place within the body.

Color titles A through F and the related structures. Use colors that do not obscure the detail of the heart cavities.

The human *heart* is an organ with four chambers whose walls are formed of cardiac muscle cells and supporting fibrous tissue. The right side of the heart receives deoxygenated blood from the two great veins, the superior vena cava and the inferior vena cava, which collect blood returning from all over the body. These vessels discharge blood into a thin-walled muscular chamber called the *right atrium*. The atrium contracts to force the blood through the right atrioventricular valve into the much thicker-walled *right ventricle*. A fraction of a second later, the right ventricle contracts and pumps the blood through the right semilunar valve into the *pulmonary trunk,* which divides into right and left pulmonary arteries, carrying the blood to the lungs for oxygenation. Valves are a part of the fibrous skeleton of the heart and serve to prevent reflux of blood into the chamber from which it was pumped.

The left heart consists of similar chambers with similar names. The *left atrium* receives oxygenated blood from the lungs via four pulmonary veins. It pumps the blood through the left atrioventricular valve into the *left ventricle,* which then contracts to pump the blood through the left semilunar valve (not shown) into a single, large artery called the aorta.

The contractions of cardiac muscle are coordinated by a system of specialized muscle cells called the cardiac conduction system (not shown). Cardiac muscle will contract rhythmically in the absence of any nerve input, but nerves to the heart regulate the rate to adjust to the needs of the rest of the body.

Color titles G through L and related structures in the figure at the left and the cross sections through the artery and vein.

Arteries are vessels that convey blood from the heart to the body tissues. *Veins* are vessels that conduct blood to the heart from body tissues. The structure of these blood vessels is directly related to the pressures imposed on them. Arteries have three layers: an inner layer composed of simple squamous epithelial cells, called *endothelium* (*endo-,* "inside"), a middle layer consisting of variable numbers of concentric layers of *smooth muscle* cells and their fibrous envelopes, and an outer layer of *fibrous connective tissue*. Sandwiched between each two layers is a *lamina* of variable thickness containing *elastic* connective tissue fibers. In the aorta and its major branches, which receive blood from the heart in surges and at high pressure, muscle fibers are almost entirely absent and the middle and outer layers are predominantly elastic fibers. In medium-sized arteries, where the pressure is less, the muscle layers predominate, allowing regulation of blood flow to a given region by contracting (causing vasoconstriction) or relaxing (vasodilation).

Capillaries (not shown) are simple endothelial tubes supported by a small amount of fibrous tissue, connected to arteries at one end and to veins at the other end. Capillary walls are thin enough to allow ready diffusion of materials between the blood and nearby tissue cells.

Veins collect blood from the capillaries under very low pressure; many veins, especially those of the limbs, have endothelial valves to prevent reverse flow of the blood. Veins have an inner layer of *endothelium* as arteries do, but their middle layer is largely devoid of *smooth muscle* and the outer layer is thickest, characterized by layers of *fibrous connective tissue*.

Color the heading Formed Elements of Blood, titles M through P, and the related structures at the bottom of the plate.

Blood consists of red, oxygen-carrying cells called *erythrocytes,* a much smaller number of white blood cells called *leukocytes* (two of which are shown here), tiny masses of cytoplasmic particles called *platelets,* and the fluid part called *plasma*. Erythrocytes are unusual in that they have no nuclei, having lost them in the bone marrow during development. About one-third of their volume is hemoglobin, which is responsible for 85 percent of the oxygen-carrying capacity of the blood and gives blood its red color. White blood cells are associated with body defense. Some are scavenger cells, some are antibody producers, and others are associated with allergic hypersensitivity reactions. Platelets initiate blood-clotting reactions.

CARDIOVASCULAR SYSTEM.

HEARTₐ
RIGHT ATRIUM_B
RIGHT VENTRICLE_C
PULMONARY TRUNK_D
LEFT ATRIUM_E
LEFT VENTRICLE_F

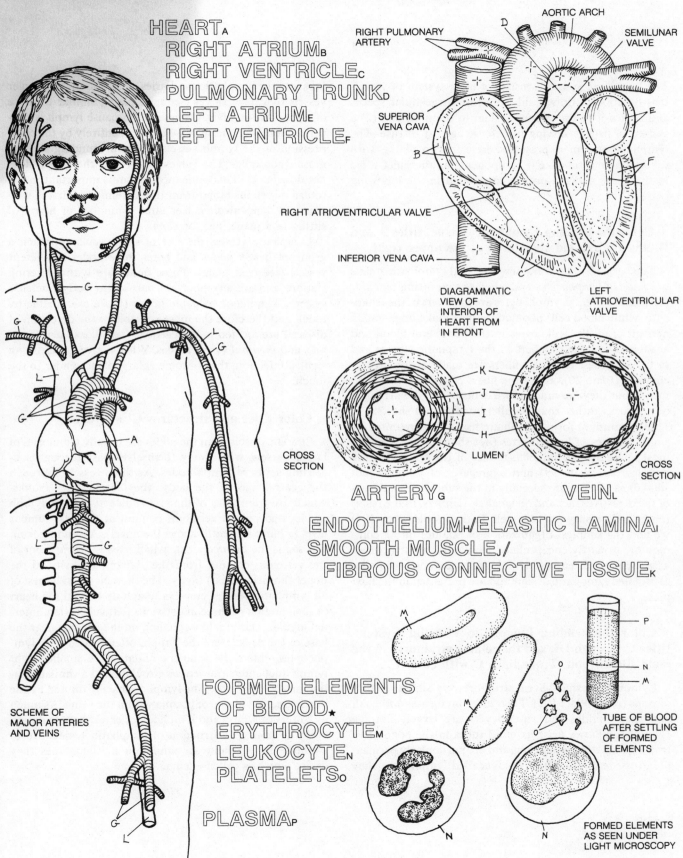

RIGHT PULMONARY ARTERY

AORTIC ARCH

SEMILUNAR VALVE

SUPERIOR VENA CAVA

RIGHT ATRIOVENTRICULAR VALVE

INFERIOR VENA CAVA

DIAGRAMMATIC VIEW OF INTERIOR OF HEART FROM IN FRONT

LEFT ATRIOVENTRICULAR VALVE

LUMEN

CROSS SECTION

CROSS SECTION

ARTERY_G

VEIN_L

ENDOTHELIUM_H/ELASTIC LAMINA_I
SMOOTH MUSCLE_J/
FIBROUS CONNECTIVE TISSUE_K

SCHEME OF MAJOR ARTERIES AND VEINS

FORMED ELEMENTS
OF BLOOD.★
ERYTHROCYTE_M
LEUKOCYTE_N
PLATELETS_O

PLASMA_P

TUBE OF BLOOD AFTER SETTLING OF FORMED ELEMENTS

FORMED ELEMENTS AS SEEN UNDER LIGHT MICROSCOPY

The lymphatic system consists of (1) a system of vessels that function to draw fluid from the extracellular spaces and return it to the venous system in the neck and (2) a system of organs serving in defense against disease. The lymphatic system is presented here in two plates, with the vascular ("tubelike") components as the chief topic in this plate and the organs considered in the following plate.

Color the heading Lymphatic Tissue, titles A and B, and the related structures at the upper right.

The basic structural framework of all lymphatic organs is a mass of *lymphocytes* and a bed of supporting *reticular* ("netlike") *fibers.* Lymphocytes are members of the leukocyte (white blood cell) population; they make up about 35 percent of all the leukocytes in the peripheral blood and are found in large numbers in the lymphatic system and in tissue spaces. Reticular fibers are made of the protein collagen (Plate 20), as are the fibers of tendons and ligaments, but they are much more delicate and are arranged in networks rather than parallel bundles and sheets.

In its simplest form, lymphatic tissue is represented by small colonies of lymphocytes (not shown) that roam almost all of the subepithelial tissues of the body. Solitary nodules (see Plate 99) and aggregates of nodules (not shown) can be found especially in the subepithelial tissues of the digestive tract and, in smaller numbers, in the respiratory and urinary tracts as well. In the tonsils (Plate 104) we find the simplest organization of nodules into a common organ partly encapsulated from adjacent structures. The thymus, spleen, and lymph nodes are completely encapsulated lymphatic organs and are presented in the next plate.

Color the heading Lymphatic Vascular System, titles C through D², and the related structures in the main illustration. Trace lines C with color C.

Lymphatic *vessels* have a structure very similar to that of veins (recall Plate 102). Their inner lining is endothelial, and their middle and outer layers are largely fibrous. Lymph capillaries arise as blind tubes in the connective tissue spaces near small veins and are particularly permeable to tissue fluids. These fluids, called lymph once they pass into the lymphatic capillaries, pass into slightly larger vessels that have valves (see next plate), similar to those of veins. These valves are essential because lymph is propelled along the lymph vessels almost entirely by the compression of the lymph vessels when surrounding skeletal muscle contracts. The valves assure that the lymph does not flow back. (This same mechanism is important in the return of venous blood from the extremities. That is why you may have swollen feet after a number of hours of sitting in a plane, bus, or car.)

Lymphatic vessels the size of small veins approach a group of *lymph nodes* and enter these nodes as afferent vessels (see next plate). These nodes are lymph-filtering stations and are an important part of the body's defense system. The major palpable nodes (those examinable by touch and therefore the most valuable in the detection of disease) are located in the *cervical* (neck), *axillary* (armpit), and *inguinal* (groin) areas. When active in defending against infection, they become enlarged and tender to the touch.

Color titles and structures C¹ and C².

Efferent vessels from the nodes are usually tributaries of larger vessels, which may themselves pass through succeeding series of lymph nodes. As these vessels approach the central axis of the body, they form larger trunks, which are tributaries of the *cysterna chyli,* a large lymph sac located on the vertebral column at the first lumbar level (a hand's breadth above the navel). Ascending from this sac is the *thoracic duct,* which passes just in front of the vertebral column from the abdominal cavity to the upper thoracic (chest) cavity. The thoracic duct, largest of all lymphatic vessels, conveys lymph up behind the heart to the junction of three major veins (left subclavian, internal jugular, and brachiocephalic), on the left side near the base of the neck. Here the lymph, often loaded with lymphocytes, enters the venous circulation. A smaller right lymph duct enters the venous circulation at a similar location on the right side. The lymph is sharply diluted by the much greater volume of plasma within the veins. As much as 2 liters of lymph and lymphocytes enter the circulation every 24 hours. Obstruction of lymphatic vessels causes swelling due to fluid accumulation in the tissues they drain, which is called edema.

LYMPHATIC SYSTEM I.

LYMPHATIC TISSUE★
LYMPHOCYTE A
RETICULAR FIBER B

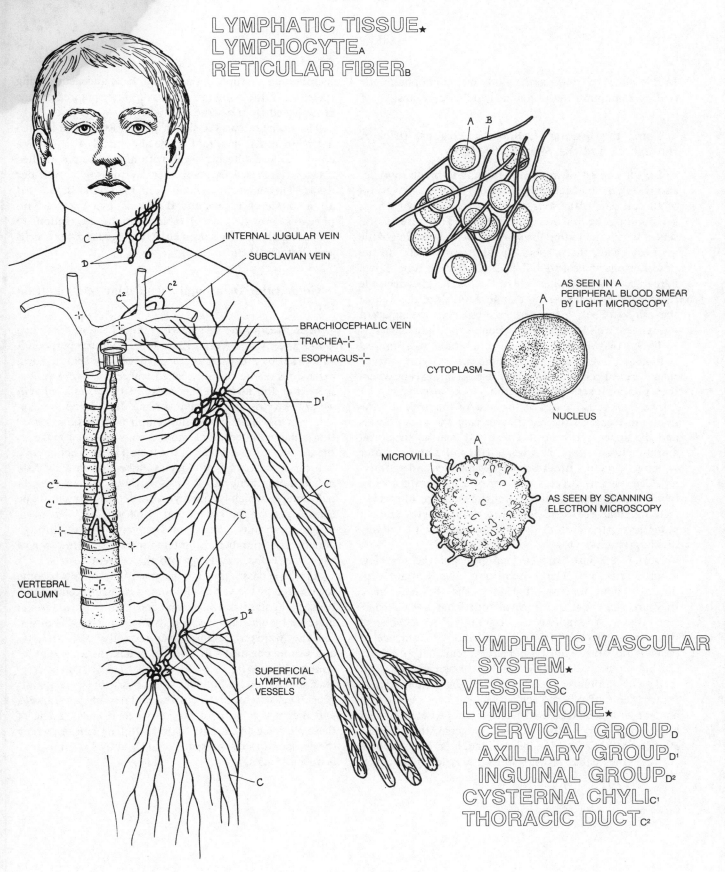

INTERNAL JUGULAR VEIN

SUBCLAVIAN VEIN

BRACHIOCEPHALIC VEIN

TRACHEA

ESOPHAGUS

VERTEBRAL COLUMN

SUPERFICIAL LYMPHATIC VESSELS

AS SEEN IN A PERIPHERAL BLOOD SMEAR BY LIGHT MICROSCOPY

CYTOPLASM

NUCLEUS

MICROVILLI

AS SEEN BY SCANNING ELECTRON MICROSCOPY

LYMPHATIC VASCULAR SYSTEM★
VESSELS C
LYMPH NODE★
CERVICAL GROUP D
AXILLARY GROUP D1
INGUINAL GROUP D2
CYSTERNA CHYLI C1
THORACIC DUCT C2

In this plate we continue the study of the lymphatic system by examining the principal lymphatic organs.

Color the heading Lymphatic Organs, titles A through C, and the related structures.

Tonsils consist of masses of lymphatic tissue (nodules and their germinal centers) under the epithelial coverings of the mucosa of various parts of the pharynx (throat) surrounding the oral and nasal cavities. There are several sets of tonsils, including the pair around the opening of the auditory tubes, the adenoids (pharyngeal tonsils) on the posterior pharyngeal wall, the palatine tonsils (one shown here in section) on either side of the junction between the mouth and pharynx, and the lingual tonsils at the base of the tongue. Each tonsil is separated from the adjacent pharyngeal wall by a capsule of connective tissue fibers. It is likely that tonsils play an important role in immunologic development during the preteen years, hence their surgical removal is no longer popular, except where they present a serious obstruction to the airway.

The *thymus* is located at the base of the neck and the uppermost part of the chest, overlying the great vessels and the upper portion of the heart. It is at its structural and functional peak between birth and puberty, after which its structure progressively degenerates and its activity declines. The thymus consists of a mass of grapelike lobules, each with a central core (the medulla) of specialized reticular cells and immature lymphocytes and an outer layer (the cortex) thickly populated with actively dividing lymphocytes.

During the last part of fetal life and the first few months after birth, large number of lymphocytes produced in bone marrow migrate to the thymus, where they are somehow transformed into what are called T (for "thymic") lymphocytes, or simply *T cells.* Most of them then leave the thymus and take up residence in other lymphatic organs, where they remain throughout the life of the individual. They are responsible for cell-mediated immunity (using cells as the defense agent rather than antibody molecules). When T cells come into contact with cells whose surfaces contain foreign protein or complex carbohydrate (called antigens), they enlarge and divide. One or both of the daughter cells become active "killer cells," which secrete a lytic agent called lymphokine, disrupting the foreign cells and neutralizing the effect of the antigenic material. (Transplanted organs are rejected by the same reaction.)

The *spleen,* about the size of your closed fist, is located in the upper left quadrant of the abdominal cavity. It is a complex blood-filtering organ with numerous blood-filled sinuses (large spaces) supported by masses of reticular tissue. Throughout are nodules of lymphatic tissue that are a source of T cells and B cells. These types of lymphocytes provide considerable immune protection to transiting blood. The spleen also traps aged red blood cells and destroys them.

Color titles D through J and their related structures.

In contrast to the spleen, which filters blood, the *lymph nodes* filter lymph (recall Plate 103). Each lymph node is supported by a capsule of connective tissue and ingrown extensions of the capsule, called trabeculae. *Afferent vessels* enter the node on the convex surface, and *efferent vessels* leave on the concave surface. Just inside the capsule is a passageway for the lymph (subcapsular *sinus*) that branches internally around the nodules that make up the outer portion *(cortex)* of the lymph node. Each nodule is a dense mass of lymphocytes embedded in a delicate network of reticular fibers. Most nodules contain a germinal center in which lymphocytes are actively dividing. The central part of the lymph node (the *medulla*) has more irregular strands (cords) of lymphatic tissue through which large numbers of interconnecting *sinuses* weave their way before emptying into the efferent vessels.

As lymph passes through this maze, its contents become scrutinized by the various immune system elements. Macrophages (enlarged white blood cells) will remove foreign material by phagocytosis. T cells encountering an antigen that they were programmed to attack will attack it directly and secrete chemicals that attract macrophages and other lymphocytes to join in the attack. *B cells* will secrete chemical antibodies to inactivate or destroy specific antigens. B cells divide in the cortical nodules into *plasma cells* and memory cells (or other activated B cells). Some of these plasma cells migrate to the medullary sinuses, where they respond to the presence of antigen by secreting *antibody* molecules.

LYMPHATIC SYSTEM II.

LYMPHATIC ORGANS.★

TONSIL_A

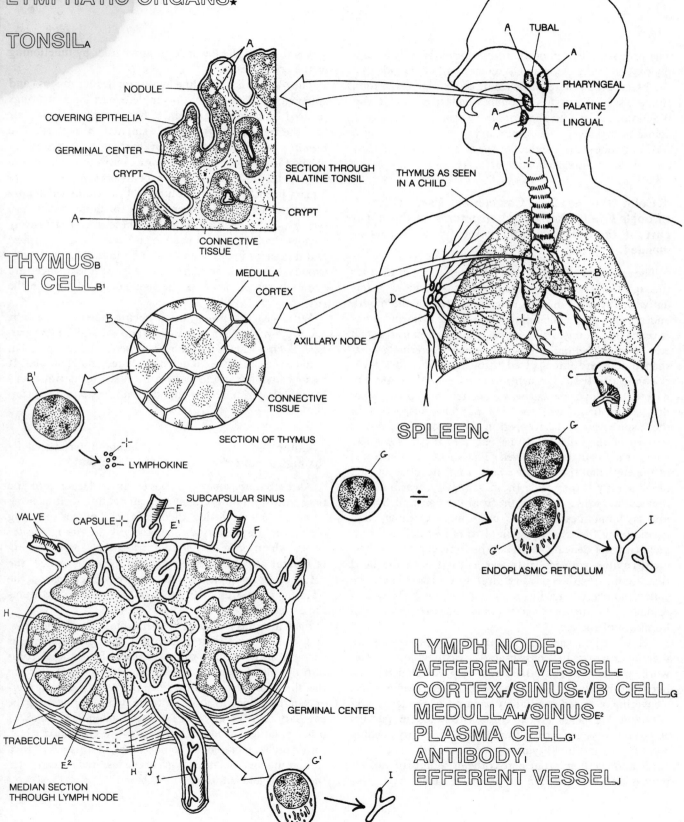

NODULE
COVERING EPITHELIA
GERMINAL CENTER
CRYPT
A

A

SECTION THROUGH
PALATINE TONSIL

CRYPT

CONNECTIVE
TISSUE

A TUBAL
A
PHARYNGEAL
PALATINE
A LINGUAL
A

THYMUS AS SEEN
IN A CHILD

B

THYMUS_B
T CELL_B1

MEDULLA
CORTEX

B

B1

CONNECTIVE
TISSUE

SECTION OF THYMUS

LYMPHOKINE

AXILLARY NODE

D

C

SPLEEN_C

G

G

ENDOPLASMIC RETICULUM

G1

I

VALVE
CAPSULE
SUBCAPSULAR SINUS
E
E1
F

H

GERMINAL CENTER

TRABECULAE
E2
H J
I

MEDIAN SECTION
THROUGH LYMPH NODE

G1

I

LYMPH NODE_D
AFFERENT VESSEL_E
CORTEX_F/SINUS_E1/B CELL_G
MEDULLA_H/SINUS_E2
PLASMA CELL_G1
ANTIBODY_I
EFFERENT VESSEL_J

The primary function of the respiratory system is to supply oxygen to the blood and to remove carbon dioxide that the blood accumulates as a result of cellular respiration (Plate 54) in the tissues through which it passes. It also functions, along with the urinary system, in maintaining acid-base balance. The respiratory system consists of a series of tubes, the conduction part, and numerous tiny chambers, the gas exchange part.

Color the heading Conduction Part, titles A through F, and the related structures in the upper part of the plate. A light color for F is recommended.

The primary entrance to the respiratory tract is through the nose, which is guarded by large hairs. Alternatively, the tract may be entered by way of the mouth and oral pharynx. The lining tissue of the nose and *nasal cavities (respiratory mucosa)* is ciliated pseudostratified columnar epithelium. Numerous mucus-secreting goblet-shaped cells are interspersed throughout it, and a layer of connective tissue lies beneath it. The nasal cavities are narrow, twisting passageways created by curled bony projections (conchae), enabling particles suspended in the inspired air to be trapped in the mucus. The power strokes of the cilia carry the particles to the *pharynx,* where they can be swallowed. The single-cell thickness of the epithelial tissue allows heat from the adjacent rich blood supply to permeate the nasal cavity, warming the inhaled air, while evaporation from the constant flow of mucus humidifies it. This cleansing, warming, and humidifying function is essential to prepare air for contact with the delicate tissue of the lung capillaries. The nasal cavities connect with several cavities in the head that are lined with typical respiratory mucosa. These are called paranasal air sinuses (two are shown). They add resonance to the voice and probably increase the overall humidity of the air.

Below the pharynx is the *larynx,* a sound producing and manipulating organ, consisting of a cartilaginous framework, a membranous attachment to a small bone (the hyoid) at the base of the tongue, several groups of muscles, a covering of mixed respiratory and oral-like mucosa, and two taut ligaments (vocal cords) stretched in parallel across the laryngeal orifice from front to back. By opening and closing and by varying the tension of the two vocal cords and the flow of air across them, sounds of various pitches and volumes can be produced. The larynx also serves to guard the lower respiratory tract by functioning in the cough reflex.

The *trachea* is a tube lined with respiratory mucosa and stiffened by a series of incomplete cartilaginous rings secured by fibrous membranes. It conducts incoming air into the right and left *bronchi* (singular, bronchus). The *primary bronchi,* supported by cartilaginous plates enmeshed in fibrous tissue, disappear into the lungs, where they divide into *secondary,* or lobar, *bronchi.* Three secondary bronchi in the right lung and two in the left supply air to the lobes. Each secondary bronchus divides into *tertiary,* or segmental, *bronchi.* These are progressively smaller than the bronchi above and contain less cartilage and more elastic tissue and smooth muscle. Each tertiary bronchus serves a bronchopulmonary segment; there are generally ten segments in the right lung and eight in the left.

Within each segment, the bronchus narrows to a tube about 0.5 millimeter in diameter and loses all of its cartilage. Such a tube is called a *bronchiole,* and its epithelial surface is more cuboidal but still ciliated and glandular. It retains smooth muscle and elastic tissue in its wall. This bronchiole terminates by dividing into respiratory bronchioles.

Color the heading Gas Exchange Part, titles F¹ through I, and the remainder of the plate.

As each *respiratory bronchiole* dives deeper into the lung, it loses its muscle investments and its epithelia become squamous, losing their cilia and mucus-secreting cells. The respiratory bronchiole divides into *alveolar ducts,* which open into *alveolar sacs,* each sac consisting of several *alveoli* (air sacs; singular, alveolus) where the actual exchange with the blood takes place. The part of the tract from respiratory bronchiole to alveolus constitutes the air exchange unit.

Each alveolus is adjacent to very thin-walled *capillaries,* and it is by way of this interface that gas exchange is accomplished. Oxygen diffuses rapidly from the alveolus into the blood, and carbon dioxide diffuses rapidly from the blood into the alveolus, since there is a marked difference in the concentrations of these two gases between the capillary and the alveolar spaces. If alcohol, acetone, or other volatile substances are present in the blood, they, too, can be excreted to some extent in this way. In a similar manner, volatile toxins in the inspired air can enter the alveoli and diffuse into the blood.

RESPIRATORY SYSTEM.

CONDUCTION PART★
NASAL CAVITY_A
  RESPIRATORY MUCOSA_{A¹}
PHARYNX_B
LARYNX_C
TRACHEA_D
BRONCHUS★
  PRIMARY_E
  SECONDARY_{E¹}
  TERTIARY_{E²}
BRONCHIOLE_F

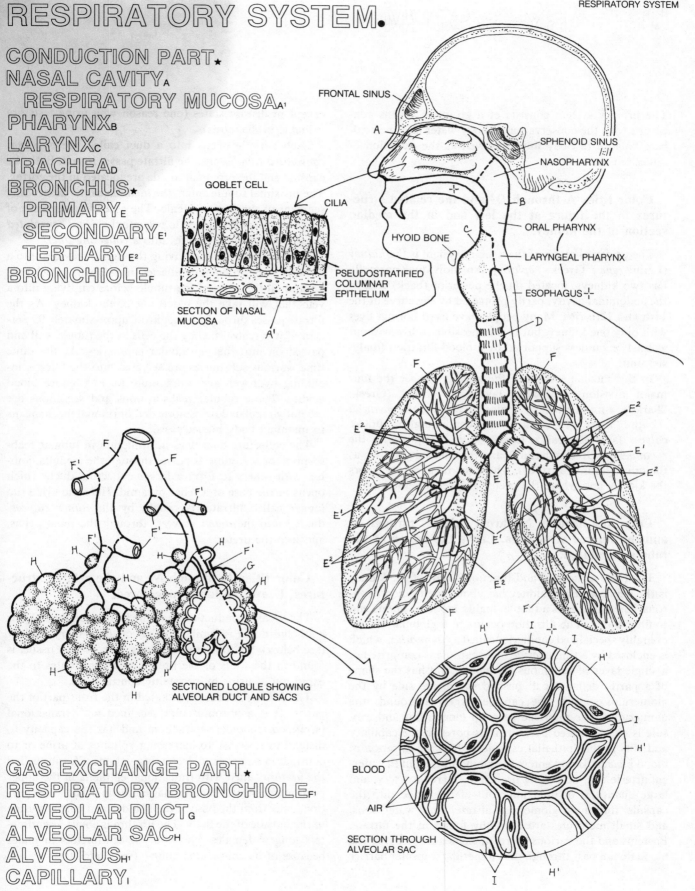

GOBLET CELL

CILIA

PSEUDOSTRATIFIED
COLUMNAR
EPITHELIUM

SECTION OF NASAL
MUCOSA

A¹

FRONTAL SINUS

A

SPHENOID SINUS

NASOPHARYNX

B

ORAL PHARYNX

HYOID BONE

C

LARYNGEAL PHARYNX

ESOPHAGUS

D

F

F

E²

E²

E¹

E²

E¹

E²

E¹

E²

E²

F

F

SECTIONED LOBULE SHOWING
ALVEOLAR DUCT AND SACS

F

F¹

F¹

F¹

F¹

H

H

H

H

H

H

H

H

H

F¹

G

H

H' I I

H'

I

H'

I

H'

BLOOD

AIR

SECTION THROUGH
ALVEOLAR SAC

I

H'

GAS EXCHANGE PART★
RESPIRATORY BRONCHIOLE_{F¹}
ALVEOLAR DUCT_G
ALVEOLAR SAC_H
ALVEOLUS_{H¹}
CAPILLARY_I

The urinary system consists of a group of organs concerned with the conservation of body water and the acid-base balance of body fluids, as well as the excretion of undesired molecules.

Color titles A through D² and the related structures in the figure at the left and in the median section of the kidney.

The principal organ of the urinary system is the *kidney* (Latin: *renes;* Greek: *nephros*). An individual normally has two kidneys, located on the posterior (back) wall of the abdominal cavity, partly protected by the curve of the 11th and 12th ribs. Many people have lived normal lives with only one kidney, but it is impossible to live without at least one unless supported by a blood-filtration (dialysis) unit.

In this median section of the kidney you see the four major divisions: an outer layer, the *cortex* (Greek: "bark"); a middle belt of conical pyramids, the *medulla* (Latin: "marrow"); an inner area containing the cuplike *calyces* (singular, calyx; Greek: *kalix,* "cup") with the *renal pelvis* (Greek: "basin"); and the concavity known as the hilus (Latin: "a trifle"), where the renal artery enters the kidney and where the ureter and renal vein exit.

Color the heading Nephron, titles E through J, and the related structures. Leave the lumen of the tube uncolored.

The basic structural and functional unit of the kidney is the nephron. Each kidney has about one million nephrons. The nephron is a tubule, highly modified at one end to filter blood. The filtration occurs in a globular cluster of highly specialized capillaries called a *glomerulus,* which is enclosed by a thin *capsule.* The capsule is comprised of a single layer of squamous epithelium and has the shape of a partly deflated ball pushed in on one side by the glomerulus so that the capsule largely surrounds the glomerulus. The interface between glomerulus and capsule is characterized by numerous pores in the capillary and slits in the epithelial capsule, enhancing filtration of blood plasma. Blood enters the glomerulus from the afferent arteriole, a sixth-order branch of the renal artery, and large quantities of fluid filter out of the blood into the capsule, forming a glomerular filtrate. Only water, ions, and small molecules are normally found in the filtrate. Proteins and the various "formed elements" (cells) of the blood do not pass through the glomerular-capsular barrier

except in disease states (one reason why urine tests are valuable in diagnosis).

Each capsule opens into a duct called the *proximal convoluted tubule,* and the filtrate passes into it. Both the capsule and the proximal tubule are located in the cortex. The proximal tubule enters the medulla as the descending segment of the *loop* of Henle. The ascending segment of the loop enters the cortex to become the *distal convoluted tubule,* which then empties into a *collecting duct.*

The efferent arteriole leaving the capsule divides into a network of peritubular capillaries (not shown), which pass around all the parts of the tubule before emptying into a vein that will take the blood out of the kidney. As the filtrate passes through the nephron, approximately 99 percent of it is reabsorbed by the cells in the tubule wall and passed on into the peritubular capillaries. At the same time, various substances are secreted into the filtrate, including hydrogen and ammonium ions to reduce blood acidity. These tubular reabsorptions and secretions are carefully regulated by osmotic and hormonal mechanisms to maintain body homeostasis.

The collecting duct does not function in tubular reabsorption or secretion. It passes through the medulla, joining with others to form a larger collecting duct, which opens at the base of a renal pyramid. Here the urine (no longer called filtrate) is caught by the *minor calyces,* ducted into the *major calyces,* through the renal pelvis, and into the ureter.

Color titles K through M and the related structures. Leave the lumen uncolored.

The *ureters* conduct urine to the urinary bladder. Their epithelial lining is similar to that of the urinary bladder (see below). Smooth muscle, along with fibrous tissue, is found in the walls of the ureters and contributes to the expulsion of urine by peristaltic contractions.

The *urinary bladder* is situated in the front part of the pelvis. It is a fibromuscular sac lined with transitional (stratified, cuboidal) epithelium and has the capacity to distend in response to increasing volumes of urine or to contract in response to decreasing volumes. A tube called the *urethra,* lined with transitional epithelia in the upper portion and stratified columnar epithelia in the lower portion, exits from the base of the bladder and conveys urine to the outside of the body. The urethra is about 4 centimeters long in females but about 20 centimeters in males because of its convoluted course (see Plate 109).

URINARY SYSTEM.

KIDNEY$_A$
CORTEX$_B$
MEDULLA$_C$
NEPHRON★
GLOMERULUS$_E$
CAPSULE$_F$

RENAL PELVIS$_D$
MINOR CALYX$_{D1}$
MAJOR CALYX$_{D2}$

PROXIMAL CONVOLUTED TUBULE$_G$/LOOP$_H$
DISTAL CONVOLUTED TUBULE$_I$

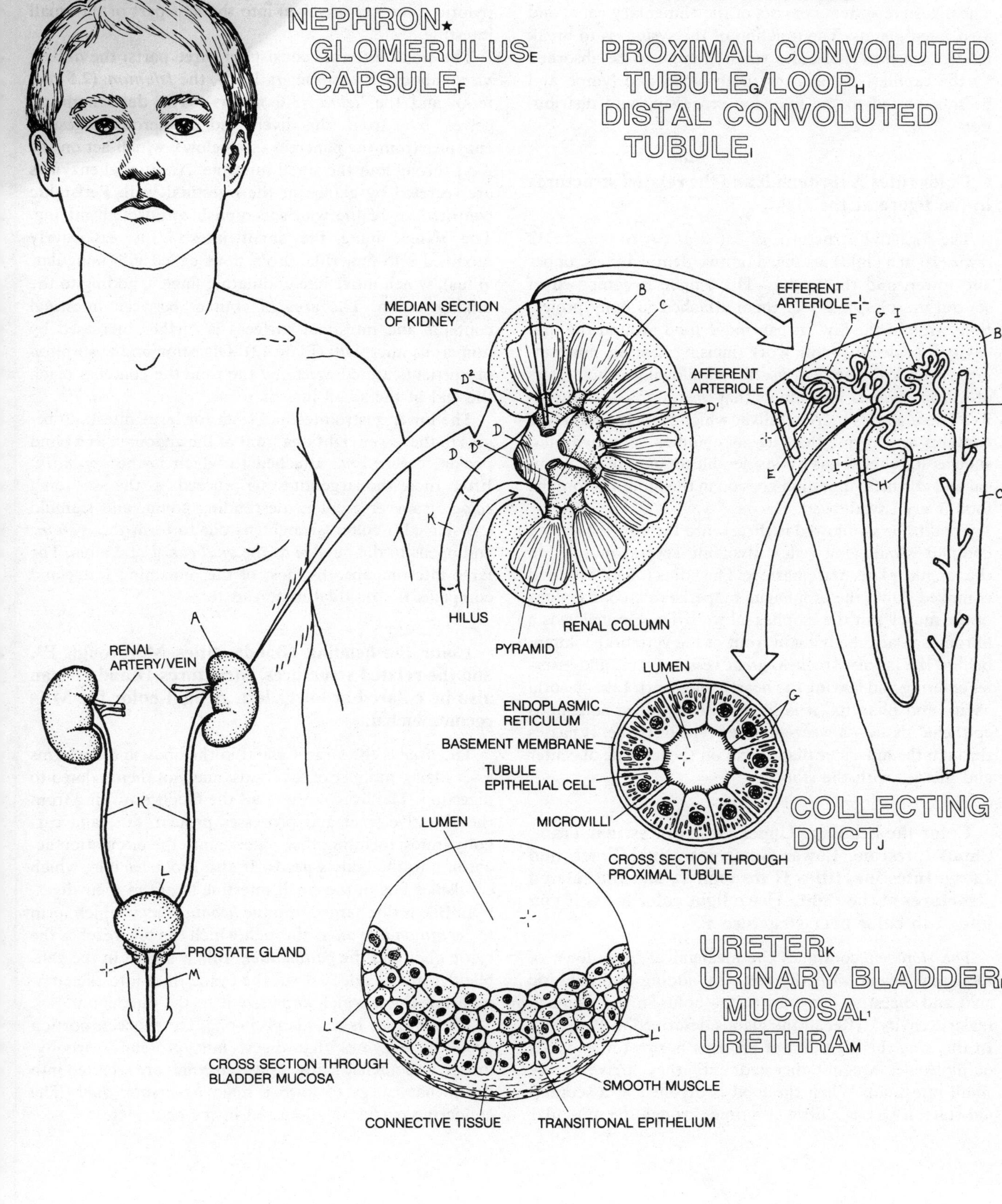

MEDIAN SECTION OF KIDNEY

EFFERENT ARTERIOLE
AFFERENT ARTERIOLE

HILUS
RENAL COLUMN
PYRAMID

RENAL ARTERY/VEIN

PROSTATE GLAND

LUMEN
ENDOPLASMIC RETICULUM
BASEMENT MEMBRANE
TUBULE EPITHELIAL CELL
MICROVILLI

CROSS SECTION THROUGH PROXIMAL TUBULE

COLLECTING DUCT$_J$

LUMEN
CROSS SECTION THROUGH BLADDER MUCOSA
CONNECTIVE TISSUE
TRANSITIONAL EPITHELIUM
SMOOTH MUSCLE

URETER$_K$
URINARY BLADDER$_L$
MUCOSA$_{L1}$
URETHRA$_M$

The digestive system consists of the alimentary canal and associated glands. The function of the system is to break food down into molecules small enough to be absorbed by the capillaries that carry both blood and lymph and be transported to the liver for processing and distribution.

Color titles A through F and the related structures in the figure at the right.

The digestive structures of the *oral cavity* include 32 *teeth* (20 in a child) arranged on two dental arches, upper and lower, and the *tongue.* The tongue is composed of several muscles, many of them attached to bony projections around the jaw. It can move food around the oral cavity for selective tooth work (incising, tearing, macerating) or back to the *pharynx* for swallowing. Three pairs of *salivary glands* have ducts that open into the oral cavity. These glands secrete saliva, which wets the food and contains an enzyme that digests complex carbohydrates, such as starches. Thus a considerable amount of mechanical and chemical digestion goes on in the mouth before the food is even swallowed.

Food to be swallowed is thrust into the *esophagus* by a complex swallowing reflex involving several muscles in the tongue, plate, and pharnyx. The bolus ("ball") of food is moved down the esophagus by peristaltic contractions of the muscles in the esophageal wall. The esophagus is a fibromuscular tube lying in front of the vertebral column, behind the larynx, trachea, great vessels (veins and arteries entering and leaving the heart), and heart. Like the oral cavity and pharynx, it is lined with stratified squamous epithelial tissue—a wear-and-tear type of tissue. It passes through the muscular diaphragm on the left side of center and merges with the stomach.

Color the headings Upper Gastrointestinal Tract, Small Intestine, Lower Gastrointestinal Tract, and Large Intestine, titles G through L, and the related structures at the right. Use a light color for G if you intend to color over structure P.

The *stomach* continues the mechanical breakdown of food that began with the teeth by adding hydrochloric acid and digestive enzymes to the bolus in the stomach (gastric cavity). (Recall the glands described in Plate 99.) In this way the digestion of proteins begins (other kinds of molecules are not digested until they arrive at the small intestine). When the food is digested to a semiliquid state, it passes a little at a time through the muscular pyloric sphincter (a valve) into the first part of the small intestine.

The small intestine consists of three parts: the *duodenum* (about 25 centimeters long), the *jejunum* (2.5 meters), and the *ileum* (3.6 meters). The duodenum receives bile from the liver and numerous digestive enzymes from the pancreas (see below), which act on the food throughout the small intestine. Additional enzymes are secreted by glands in the intestinal wall. Peristaltic contractions of the wall accomplish a thorough mixing. The tissue lining the small intestine is extensively modified into fingerlike projections called villi (singular, villus), which move like undulating fingers, adding to the mixing action. The area of contact between intestinal contents and intestinal surfaces is further increased by numerous microvilli (Plate 42). Digestion and absorption of nutrients are complete by the time the contents reach the end of the small intestine.

The lower gastrointestinal tract (or large intestine) begins in the lower right quadrant of the abdomen as a blind pocket, the *cecum,* attached to which is the *appendix.* From there the large intestine proceeds as the ascending *colon,* transverse colon, descending colon, and sigmoid ("S-shaped") colon, straightens out to form the *rectum,* and opens to the outside as the *anal canal* and anus. The large intestine absorbs most of the remaining water and compacts the residual matter as feces.

Color the heading Glands, titles M through P¹, and the related structures. Structures H and H¹ can also be colored at lower left. A light color for M is recommended.

The *liver* is the largest gland in the body and performs a very large number of functions, many of them related to digestion. The liver receives all the blood returning from the digestive tract and processes protein, fats, and carbohydrates, metering their release into the circulation according to the body's needs. It also produces bile, which emulsifies fats in the small intestine to aid in their digestion. Bile is discharged into the *hepatic ducts,* which form the *common bile duct,* through which the bile reaches the *cystic duct* and the *gallbladder.* Bile is stored in the gallbladder and is released via the cystic duct into the common bile duct, which conducts it to the duodenum.

The *pancreas* is two glands in one. The exocrine portion produces enzymes that digest fats, proteins, carbohydrates, and nucleic acids. Those enzymes are secreted into ducts that merge to form a main *pancreatic duct.* The endocrine portion is discussed in the next plate.

DIGESTIVE SYSTEM.

ORAL CAVITY$_A$
  TEETH$_B$
  TONGUE$_C$
SALIVARY GLAND$_D$
PHARYNX$_E$
ESOPHAGUS$_F$

UPPER GASTROINTESTINAL
  TRACT★
STOMACH$_G$
SMALL INTESTINE★
  DUODENUM$_H$/JEJUNUM$_{H1}$/ILEUM$_{H2}$

LOWER GASTROINTESTINAL
  TRACT★
LARGE INTESTINE★
CECUM/APPENDIX$_{I1}$
COLON$_J$
RECTUM$_K$
ANAL CANAL$_L$

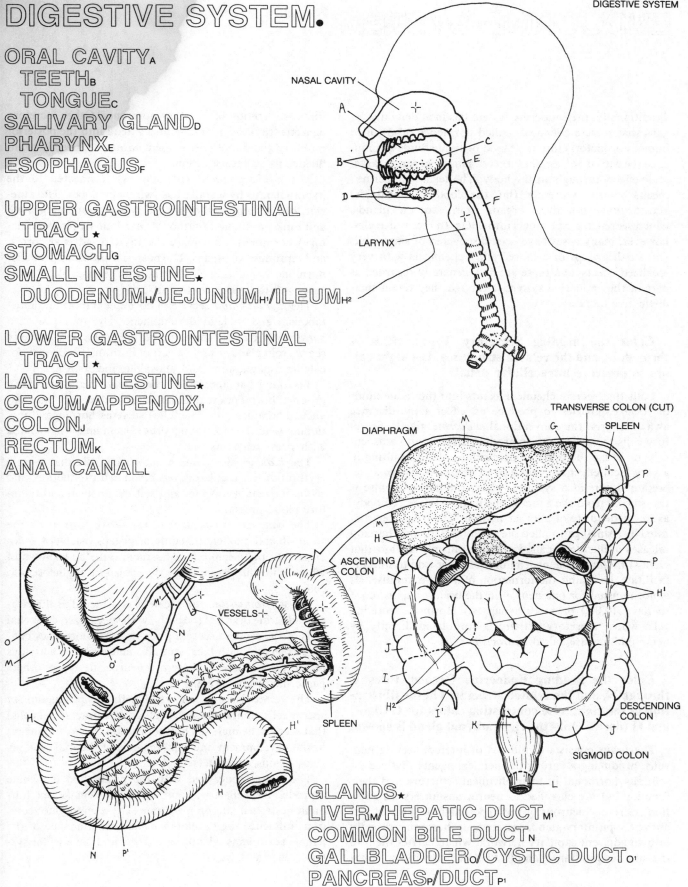

NASAL CAVITY

A
B
C
E
D
F

LARYNX

DIAPHRAGM
M
G
TRANSVERSE COLON (CUT)
SPLEEN
P
J
M
H
P
ASCENDING
COLON
H1
J
I
J
H2
I1
DESCENDING
COLON
J
K
SIGMOID COLON
L

O
M
M1
N
VESSELS
O'
P
SPLEEN
H
H1
H
N
P1

GLANDS★
LIVER$_M$/HEPATIC DUCT$_{M1}$
COMMON BILE DUCT$_N$
GALLBLADDER$_O$/CYSTIC DUCT$_{O1}$
PANCREAS$_P$/DUCT$_{P1}$

Traditionally, the endocrine system has included only organs that secrete chemicals called hormones (also called humors or factors) into the blood or tissue fluids to influence the activity of certain target organs or generate large-scale effects throughout the body. The term "endocrine" means "internal-secreting" (into the blood or tissue fluids) and distinguishes these organs from exocrine glands, which secrete their products into ducts. In recent decades, however, many cases have been discovered of single cells and small groups of cells secreting chemicals with very localized effects, and these are now generally regarded as part of the endocrine system, although they retain their distinctive names.

Color the heading Secretory Types, titles A through D, and the related structures. Use light colors to preserve intracellular detail.

Cells that secrete chemical agents into the tissue fluids or local capillaries to produce an effect some distance away comprise the conventional *endocrine* glands. Secretory cells that release their product directly into adjacent cells or in the surrounding extracellular fluid resulting in a very local effect are called *paracrine* cells. *Synaptic* secretion occurs at a synapse, where the nerve impulse is transferred from the axon of one neuron to the cell body, axon, or dendrite of another neuron. The cells are separated by a small gap, called the synaptic cleft, and the axon releases a chemical, called a neurotransmitter, into that cleft, stimulating the neuron on the other side to initiate, facilitate, or inhibit the formation of a nerve impulse. (It is not unusual for one neuron in the brain or spinal cord to have 10,000 or more synapses.) Hormone-secreting cells whose secretory activity is stimulated directly by nerve endings are called *neurocrine cells.*

Color the heading Endocrine Organs, titles E through N, and the related structures in the illustration at the right. Use contrasting colors for G (light) and H (dark). Only the right adrenal gland is shown.

The *brain* consists of billions of interconnecting neurons, producing a variety of chemical agents that are essentially hormonal in their chemical structure and their function but are classified as neurotransmitters. In addition, secretory neurons of the hypothalamus, located just above the pituitary gland, secrete neurohormones directly into an adjacent capillary network, which carries them to the anterior pituitary gland just below, inducing or inhib-

iting its secretion of pituitary hormones. Other secretory neurons from the hypothalamus extend into the posterior pituitary gland and release their hormones directly to influence its secretory action.

The *pituitary gland* (hypophysis), connected to the hypothalamus by a stalk, has two major lobes. The anterior lobe is composed of six cell types producing growth and milk-production hormones plus four tropic (stimulatory) hormones influencing the thyroid, adrenal cortex, and reproductive organs. The posterior pituitary produces hormones regulating childbirth, milk release, blood pressure, and the water content of urine.

The *thyroid gland* secretes a hormone important in metabolism and in fetal development. The four pea-sized *parathyroid glands,* enmeshed in the posterior capsule of the thyroid gland, secrete a pair of hormones necessary for calcium, magnesium, and phosphate metabolism.

The outer part (cortex) of the *adrenal gland* secretes a group of hormones regulating sugar, water, and ionic balance. The inner part (medulla) secretes hormones functioning with the visceral nervous system and is associated with stress reactions.

The *kidneys* also contain endocrine cells that secrete erythropoietin, regulating red blood cell production, and renin, important in water and salt metabolism and therefore blood pressure.

The *pancreas,* suspended in the curve formed by the stomach and duodenum, contains groups (islets) of endocrine cells that secrete four hormones, the best known of which is insulin, essential for the transport of glucose into cells.

The *gastrointestinal tract* can be considered as the largest endocrine organ of the body, with specialized epithelial cells of the mucosa secreting ten or more hormones concerned largely with local stimulus and response activities in the tract. Some of these are neurocrine types, others paracrine or endocrine.

The *testes,* largely concerned with the production of sperm cells, contain isolated cell groups (interstitial cells) that secrete the male sex hormone testosterone, important in the development and maintenance of the male sex organs, glands, and ducts.

The *ovary,* concerned with the production of female reproductive cells (ova), produces the female sex hormones estrogen and progesterone. These endocrine secretions influence the development and maintenance of the female sex organs, glands, and ducts and have wider systemic effects as well.

ENDOCRINE SYSTEM.

SECRETORY TYPES. ★

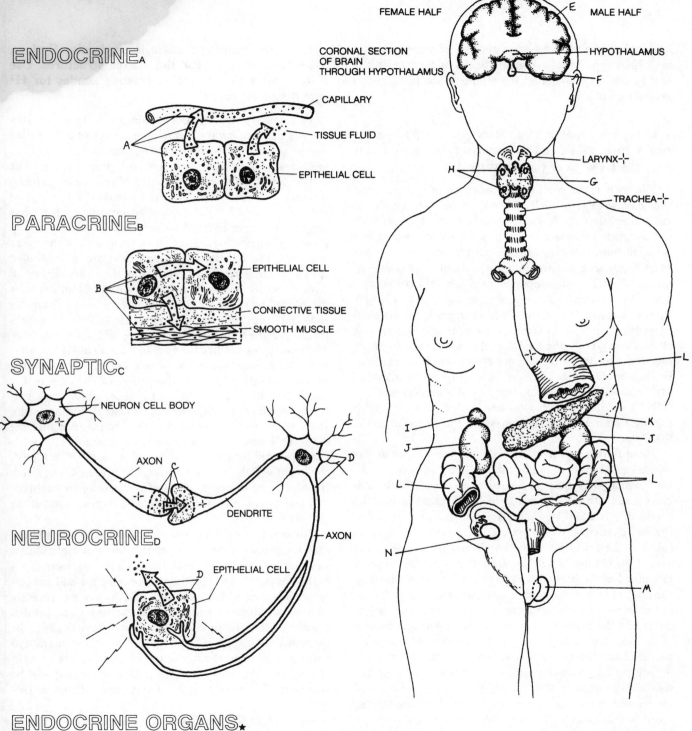

ENDOCRINE A

CAPILLARY

TISSUE FLUID

EPITHELIAL CELL

PARACRINE B

EPITHELIAL CELL

CONNECTIVE TISSUE

SMOOTH MUSCLE

SYNAPTIC C

NEURON CELL BODY

AXON

DENDRITE

AXON

NEUROCRINE D

EPITHELIAL CELL

FEMALE HALF — E — MALE HALF

CORONAL SECTION OF BRAIN THROUGH HYPOTHALAMUS

HYPOTHALAMUS

F

LARYNX -¦-

H

G

TRACHEA -¦-

L

I

J

K

J

L

L

N

M

ENDOCRINE ORGANS. ★
BRAIN E
PITUITARY GLAND F
THYROID GLAND G
PARATHYROID GLAND H

ADRENAL GLAND I
KIDNEY J
PANCREAS K
GASTROINTESTINAL TRACT L

TESTES M
OVARY N

MALE AND FEMALE REPRODUCTIVE SYSTEMS

The male reproductive system consists of paired primary sex organs, the testes (singular, testis) and related ducts and glands. Not included in this plate are the external genital structures.

Color the heading Male Reproductive System, titles A through G, and the related structures. Color over the small dots representing A⁴.

The *testis* is the site for the formation of spermatozoa, the male sex (germ) cells *(sperm)*. It is also the site of endocrine cells (the *interstitial cells*) that produce the male sex hormone testosterone. Each of the paired testes has a dense fibrous, sensitive capsule from which fibrous septa (partitions) turn inward to form a number of compartments. Within each compartment is a highly convoluted *seminiferous tubule*. A cross section through this tubule shows a number of layers in different stages in the development of sperm cells. Cells near the center undergo meiosis (Plates 66 and 67); the haploid cells resulting from previous divisions are at the center, growing tails (flagella) to become sperm cells. The head of a sperm cell consists of little more than a nucleus, carrying the genetic material (DNA) with very little cytoplasm surrounding it, and a cap (acrosome) containing enzymes for penetrating the female germ cell.

When fully formed, sperm cells migrate to the *epididymis,* a coiled mass of larger tubules lying adjacent to the testis, where they undergo a maturation period. With sufficient sexual stimulus, sperm cells are discharged into the sperm duct, called the *ductus deferens* (commonly called the vas deferens), which proceeds through the inguinal canal of the lower abdominal wall and passes along the side of the bladder. Each ductus deferens enters the prostate gland at the base of the bladder, where it is joined by the duct of one of the paired seminal vesicles to form the *ejaculatory duct.* Within the prostate, the ejaculatory ducts join the *urethra.* Just below the prostate, the ducts of the paired bulbourethral glands (not shown) also enter the urethra. The prostate, the seminal vesicles, and the bulbourethral glands produce secretions that nourish the sperm and enhance their mobility. The combination of sperm and glandular secretions is called semen (Latin: "seed") or seminal fluid. The semen is discharged from the ductus deferens into the prostatic urethra, through the membranous urethra (transiting the musculofascial urogenital diaphragm), and through the penile urethra to the outside.

Color the heading Female Reproductive System, titles H through K, and the related structures. Use a light color for H and contrasting shades for H¹ (light) and H² (dark).

The primary sex organ of the female is the *ovary.* It is the site of production of female reproductive cells, called *ova* (singular, ovum), as well as the principal source of the female sex hormones estrogen and progesterone. The ovary consists of numerous groups of cells called *follicles* dispersed in a bed of highly vascularized ("well supplied with blood vessels") connective tissue. Each follicle has single or multiple layers of supporting cells and a centrally placed developing ovum. As the development of the ovum proceeds, the follicle gets larger, develops a fluid-filled space, and migrates to the surface of the ovary. Every 28 days or so (though this is highly variable), an ovum is discharged from the follicle and is drawn into the *uterine tube* (fallopian tube or oviduct) by the undulating fingers (fibriae) of the funnel-shaped opening of the uterine tube. The developing ovum continues to migrate along the uterine tube, encouraged by peristaltic contractions of tubal smooth muscle as well as the action of the cilia of the cuboidal epithelial cells lining the tube. These same epithelia provide nutritional support for the transiting ovum. It is usually in the first third of the uterine tube that the ovum becomes fertilized by sperm if intercourse has occurred within the previous 48 hours or so. If not fertilized, the ovum will die before reaching the uterus. If fertilized, the blastocyst (an early stage of the developing embryo) will reach the uterus in a few days and implant itself in the uterine wall.

The uterus is a thickly muscular structure with a lining (endometrium) of columnar epithelium and a glandular, vascularized connective tissue layer. The endometrium is highly responsive to blood levels of estrogen and progesterone and, under their stimulus, will thicken significantly to become saturated with glandular secretions and coiled blood vessels. If implantation of an ovum occurs, the thickened tissue provides nutrition directly to the embryo and contributes to the formation of the placenta as well. If implantation does not occur, the extra lining will be sloughed off and discharged in the process called menstruation.

The neck of the uterus (cervix) fits into the upper part of the *vagina.* The vagina, lined with stratified squamous epithelium, is supported by fibromuscular tissue. It is open to the outside and receives the sperm deposited by the penis during sexual intercourse.

MALE AND FEMALE REPRODUCTIVE SYSTEMS.

TESTIS_A
  SEMINIFEROUS TUBULE_{A1}
  DUCTS_{A2}/SPERM CELL_{A3}
  INTERSTITIAL CELL_{A4}
EPIDIDYMIS_B
DUCTUS DEFERENS_C
EJACULATORY DUCT_D
PROSTATE GLAND_E
SEMINAL VESICLE_F
URETHRA_G

FEMALE REPRODUCTIVE
  SYSTEM.
OVARY_H
  FOLLICLE_{H1}
  OVUM_{H2}
  CORPUS LUTEUM_{H3}
UTERINE TUBE_I
UTERUS_J
VAGINA_K

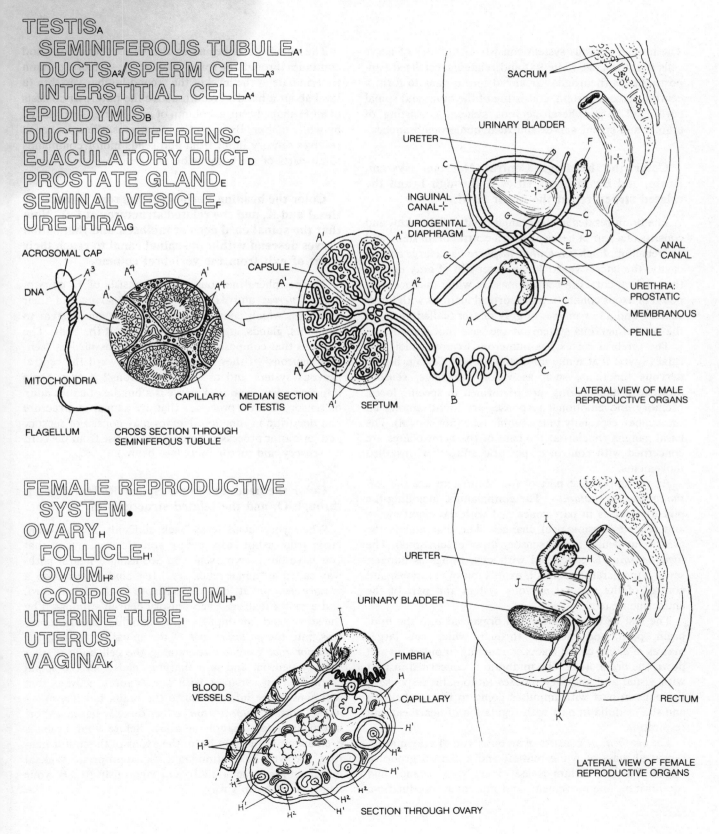

ACROSOMAL CAP

DNA

MITOCHONDRIA

FLAGELLUM

CROSS SECTION THROUGH SEMINIFEROUS TUBULE

CAPSULE

CAPILLARY

MEDIAN SECTION OF TESTIS

SEPTUM

SACRUM

URINARY BLADDER

URETER

INGUINAL CANAL

UROGENITAL DIAPHRAGM

ANAL CANAL

URETHRA: PROSTATIC

MEMBRANOUS

PENILE

LATERAL VIEW OF MALE REPRODUCTIVE ORGANS

URETER

URINARY BLADDER

RECTUM

LATERAL VIEW OF FEMALE REPRODUCTIVE ORGANS

FIMBRIA

BLOOD VESSELS

CAPILLARY

SECTION THROUGH OVARY

The human nervous system consists of millions of nerve cells (neurons, see Plate 98) and related specialized supporting cells (neuroglia) arranged into organs to form a central nervous system, consisting of the brain and spinal cord, and a peripheral nervous system, consisting of cranial and spinal nerves and their numerous branches.

Color the headings Central Nervous System, Brain, and Brain Stem, titles A through I, and the related structures at the upper right.

The brain resides in the cranial cavity of the skull and consists of a pair of hemispheres, a central brain stem, and a cerebellum. Each *cerebral hemisphere* has a large central cavity, the lateral ventricle; an outer rim of gray matter, the cerebral cortex; a central mass of white matter, composed of axons connecting the cortical neurons with lower centers; and a few masses of gray matter (called nuclei in the central nervous system) at the base (not shown).

The cerebral cortex has numerous fissures (sulci) and ridges (gyri) that make up distinct lobes. Certain higher nervous functions, such as vision, language, sensory awareness, and hearing, are performed in specific lobes. Memory and emotional responses are found among all these lobes, especially in the limbic lobe (not shown). The basal ganglia (nuclei) at the base of the hemispheres are concerned with control of postural and other unskilled movements.

At the uppermost part of the brain stem are the *epithalamus* and *thalamus*. The epithalamus, including the pineal gland, is in part concerned with day-night cycles and related physiological changes. The thalamus is the great relay center for all sensory input except smell. The *hypothalamus* is concerned with such things as hunger, satiety, temperature control, expression of emotions, and control of the visceral nervous system and part of the endocrine system.

The rest of the brain stem is organized into the midbrain, pons, and medulla, through which run larger masses of axons called tracts conducting impulses up and down the brain stem. The midbrain is concerned in part with visual and auditory reflexes (automatic responses), the pons in part with impulses going to the cerebellum, and the medulla in part with regulation of heartbeat and respiration.

The *cerebellum* consists of an outer rim of gray matter, a central mass of white matter, and a central group of nuclei. The cerebellum is associated with balance and equilibrium, fine movement, and muscular coordination.

The spinal cord begins at the base of the skull and continues through the spinal canal of the vertebral column to terminate at the level of the second lumbar vertebra (a level about a hand's breadth above the navel). It consists of an H-shaped central column of gray matter surrounded by white matter. By way of spinal nerves, the spinal cord receives sensory input from and sends motor commands to all parts of the body below the head.

Color the heading Peripheral Nervous System, titles J and K, and the related structures at left. Note that the spinal cord ends at midback and that spinal nerves descend within the spinal canal to reach their point of exit from the vertebral column.

The peripheral nervous system consists of 12 pairs of *cranial nerves* and 31 pairs of *spinal nerves* and their branches, which connect the central nervous system to receptors, glands, and muscles throughout the body. The neurons that compose these nerves are classified as "sensory neurons" if they bring impulses toward the central nervous system and as "motor neurons" if they bring impulses away from it. A nerve is a bundle of axons and/ or sensory neuron processes that are axonal in structure and dendritic in function. Nerves may consist of sensory and/or motor processes. Spinal nerves arise from the cord by sensory and motor roots (see below).

Color the heading Simple Spinal Reflex, titles L through O, and the related structures at the right.

When your hand jerks back suddenly and involuntarily from a hot stove before you are even aware that you have burned yourself, you are using a neural pathway called a "spinal reflex arc." It includes a *receptor,* a *sensory neuron,* at least one synapse in the spinal cord, and a motor neuron. Each sensory neuron stimulated by the stove sends an impulse up its axonlike dendritic process, into the *posterior root* of the spinal nerve, past the *posterior root ganglion* containing the cell body of that sensory neuron, and into the gray matter of the spinal cord. There it synapses with one or more neurons that will convey the information to the brain, but it also synapses with a *motor neuron,* either directly or via a short interneuron (association neuron). Before your brain is even aware of the skin burn, the axons of the motor neurons involved have stimulated the appropriate skeletal muscles to contract rapidly and vigorously to jerk your hand out of harm's way.

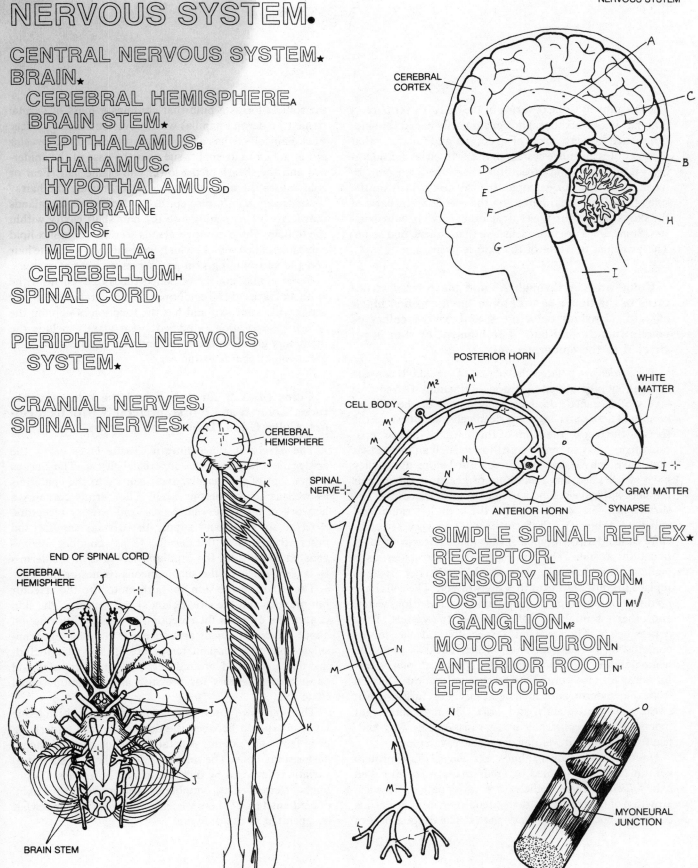

NERVOUS SYSTEM.

CENTRAL NERVOUS SYSTEM★
BRAIN.★
   CEREBRAL HEMISPHERE_A
   BRAIN STEM.★
      EPITHALAMUS_B
      THALAMUS_C
      HYPOTHALAMUS_D
      MIDBRAIN_E
      PONS_F
      MEDULLA_G
   CEREBELLUM_H
SPINAL CORD_I

PERIPHERAL NERVOUS
   SYSTEM★

CRANIAL NERVES_J
SPINAL NERVES_K

CEREBRAL
CORTEX

A
C
B
D
E
F
G
H
I

CEREBRAL
HEMISPHERE
J
K
END OF SPINAL CORD
CEREBRAL
HEMISPHERE
J
J
J
K
J
BRAIN STEM

POSTERIOR HORN
CELL BODY
M² M¹
M¹
M
M
N¹
N
SPINAL
NERVE
WHITE
MATTER
I
GRAY MATTER
SYNAPSE
ANTERIOR HORN

SIMPLE SPINAL REFLEX★
RECEPTOR_L
SENSORY NEURON_M
POSTERIOR ROOT_M¹/
  GANGLION_M²
MOTOR NEURON_N
ANTERIOR ROOT_N¹
EFFECTOR_O

N
M
N
M
L
O
MYONEURAL
JUNCTION

INTEGUMENTARY SYSTEM

The integument is the covering of the body. In vertebrates (animals with backbones) it is customarily called the skin. It has such a diverse array of structures and functions that it constitutes an entire system in itself. It varies considerably in thickness and structural character according to what is needed. For example, the thickest skin is on the sole of the foot; the thinnest, on the eyelid. Some parts of the skin are quite hirsute (hairy); others are virtually hairless. Some areas are more sensitive than others, and so on. The principal function of the skin is protection.

Color titles A through A⁶ and the related structures in the figure as well as in the magnified block of skin. Consider coloring A and A¹ the colors of your own skin and hair. The pigment of skin is restricted to the epidermis only.

The *epidermis* is the epithelial part of the skin. It consists throughout of stratified squamous epithelium with surface layers of flattened, dead cells composed of the protein keratin. These outer layers of dead cells, seen best on the soles of the feet or the palm of the hand, have been separated from their source of nutrition to such an extent that they have dehydrated and dried. But before dying they have synthesized large amounts of keratin so that their remains will leave behind a tough but flexible coating. The surface layer in certain parts of the body has ridges and valleys (not shown) in the form of loops and whorls (technically called dermatoglyphics) that make fingerprint identification possible. The deepest layer of the epidermis is made up of germinating cells, which continually divide to replace the outer layers as they are worn off. The cells have no direct blood supply; they obtain their nutrition, water, and other essentials by diffusion from the vessels of the dermis below. The epidermis (especially at deeper levels) contains a pigment, melanin, which darkens the skin; hemoglobin in the blood and carotene (which makes carrots orange) also contribute to skin coloration. Derived from the epidermis are such accessory structures as hair, sweat, sebaceous glands, and *nails.* The most important functions of the epidermis are its resistance to wear-and-tear forces and its prevention of excessive water loss.

Hairs are also made up almost exclusively of the protein keratin. They are formed by epidermal structures called hair *follicles.* These follicles are invaginations (inward growths) of the epidermis that extend deep into the dermis below and encompass the *hair shafts.* The deepest part of the follicle forms a bulb that contains a mass of dermal tissue (the dermal papilla) with nerves and blood vessels. Each hair follicle has a smooth muscle called the *arrector pili* attached to dermal tissue near the base of the epidermis and the sheath of the lower end of the hair. Fear or cold causes these muscles to contract, raising the hair.

Sebaceous glands, also epithelial derivatives, are glands that have ducts opening into the hair shaft cavity within the follicle. These exocrine glands secrete a complex lipid substance called sebum, which works its way onto the hair and the surrounding skin and serves as waterproofing.

Sweat glands consist of coiled tubes derived from the epithelial layer of the epidermis. Their secretion is mostly water with some salt and has the function of cooling the skin by evaporation. Modified sweat glands include the mammary glands of the breast and the ceruminous ("wax-producing") glands of the ear.

Color titles B through F and the related structures. Color B and F first, preferably with light colors. Color C red, D blue, and E green.

The *dermis* is the connective tissue layer below the epidermis and above the superficial fascia. The fibrous component of the dermis gives security to the epidermis as well as to the dermis itself. The dermis contains a network of blood vessels, nerves, and sensory receptors. *Arteries* are small and supply the dermal papillae and other structures of the dermis. *Veins* conduct deoxygenated blood from the capillary networks of the dermis to larger vessels in the subcutaneous tissues or deeper.

The *nerves* of the skin are both motor (to the arrector pili muscles and the sweat glands) and sensory. The skin is a virtual antenna for reception of sensory information about the environment around us. It contains unencapsulated *receptors* for pain, temperature, and touch (including fibers wrapped around the base of each hair) and capsulated receptors for pressure and vibration (recall Plate 99).

The *superficial fascia* is a fatty, loose connective tissue layer interposed between the dermis and the underlying deep fascia enveloping skeletal muscle or the periosteum ensheathing bone. The thickness of this layer is subject to variations imposed by diet, stress, and hormones. It conducts arteries, veins, lymphatics, and nerves to the skin. (The diameters of these vessels and nerves are somewhat exaggerated in the drawing).

INTEGUMENTARY SYSTEM.

EPIDERMIS_A

HAIR SHAFT_{A1}/FOLLICLE_{A2}

ARRECTOR PILI_{A3}

SEBACEOUS GLAND_{A4}

SWEAT GLAND_{A5}

NAIL_{A6}

DERMIS_B

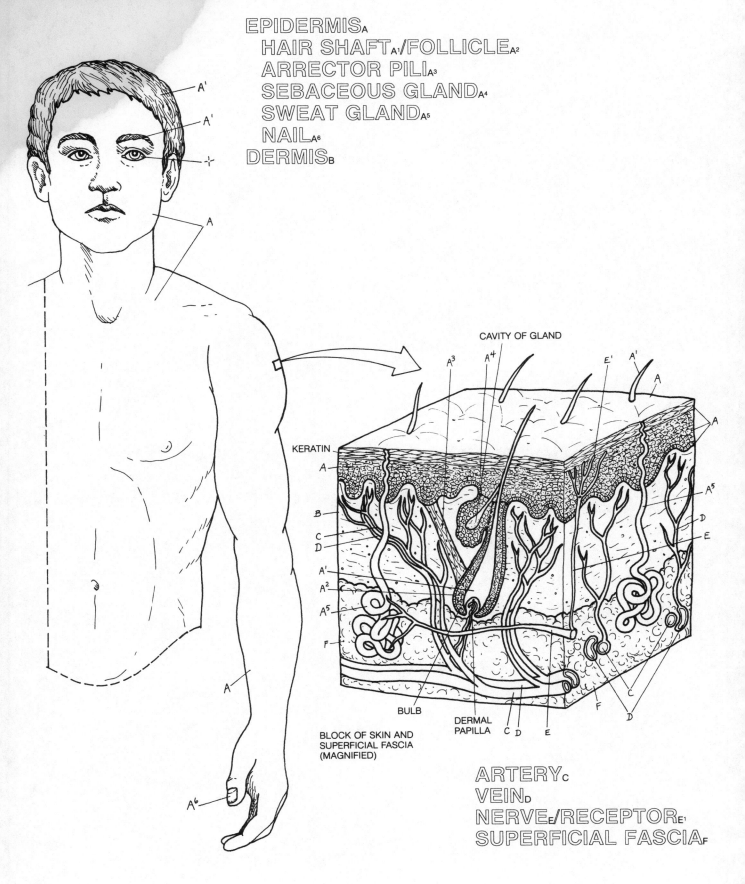

CAVITY OF GLAND

KERATIN

BLOCK OF SKIN AND
SUPERFICIAL FASCIA
(MAGNIFIED)

BULB

DERMAL
PAPILLA

ARTERY_C

VEIN_D

NERVE_E/RECEPTOR_{E1}

SUPERFICIAL FASCIA_F

INDEX

Numbers refer to plate numbers

A-band, in muscle, 40
Absolute zero, 4
Abyssal plain, 97
Acetaldehyde, 54
Acetyl coenzyme A, 54, 55
Acetyl group, 54
Acid, 13, 48
    amino, 18, 19, 39, 47, 53
    weak, 17
Acrosome, 109
Activation, 47
    energy, 49
Active transport, 33, 36, 47
    site, 49–51
Adaptive radiation, 87
Adenine, 22, 45, 51, 81–86
Adenoids, 104
Adenosine diphosphate. See ADP
Adenosine monophosphate, 22, 47
Adenosine triphosphate, 22, 45, 47, 52, 53,
    55, 57, 58
ADP, 45, 47, 52, 53, 57
Adrenal gland, 108
Aerobic metabolism, 52, 54
Affection, 3
Afferent vessel, 104
Agglutination, 75
Alanine, 18–20, 85, 86
Alcohol, 17, 52, 54
Aldehyde, 14
Algae, 91
Alimentary canal, 107
Alkaptonuria, 76
Allele, 75
Allosteric regulation, 50
    site, 50
Alloway, James, 77
α helix, 20, 21
α-ketoglutarate, 55
α ring, 15
Alveoli (alveolus), 105
Amino acid, 18, 19, 39, 47, 85, 86
Amino group, 18–20
Ammonification, 94
Amoeba, 33, 34, 91
AMP, 22, 47
Amylopectin, 16
Amylose, 16
Anabolism, 44, 45, 47
Anaerobic metabolism, 52, 54
Anal canal, 107, 109
Anaphase, 65, 66
Angstrom, 24
Animal, 1, 43, 91, 94
Animalia, 91
Antagonist, 101
Anther, 59, 60
Antibody, 75, 104
Anticodon, 84, 85
Antigen, 75
Anus, 107
Appendix, 107
Aquatic ecosystems, 97
Arc, spinal reflex, 110
Arginine, 19, 78, 79, 86
Arrector pili, 111

Arteriole, afferent, 106
    efferent, 106
Artery, 101, 102
    renal, 106
Articular cartilage, 100
Artificial selection, 88
Asparagine, 19, 86
Aspartic acid, 19, 86
Aster, 65, 66
Atmosphere, 94
Atomic number, 7
ATP, 22, 45, 47, 52, 53, 55–58
Atrioventricular valve, 102
Atrium, 102
Autosome, 68, 70–72
Avery, Owen, 77
Axon, 98, 108, 110

Bacteria, 32, 80, 91, 94
    discovery of, 23
Bacteriophage, 80
Badger, 88
Basal body, 41
Base, 13, 22, 76
Base pairing, 81–86
Bathypelagic, 97
B cell, 104
Beadle, George, 78
Beagle, voyage of, 87
Beeswax, 17
Benthos, 97
β pleated sheet, 20
β ring, 15
Biceps brachii, 101
Bilayer, phospholipid, 35, 36
Bile, 107
Bile duct, 107
Binding site, 36
Biology, 1
    molecular, 76
Biomass, pyramid of, 93
Biome, 96
Biotin, 78
Bird, as selective agent, 89
    tool-using, 87
Biston betularia, 89
Bladder, urinary, 106, 109
Blender, 80
Blending inheritance, 73–75
Blind men and elephant, 3
Blood, as tissue, 98
    formed elements of, 102
    gas exchange in lung, 105
Blood cell formation, 100
Blood types, inheritance of, 75
B lymphocyte, 104
Bond, covalent, 10, 14
    double, 14, 17
    high-energy, 46
    hydrogen, 11, 20, 21
    hydrophobic, 21
    ionic, 9
    peptide, 18, 85
Bone, 98, 100, 101
Brachialis, 101
Brain, 108, 110

Bread, 78
Bronchiole, 105
Bronchus, 105
Bulbourethral glands, 109

Calcium, 6
    ion, 9
    storage, 100
Calvin-Benson cycle, 58
Calyx, renal, 106
Canal, alimentary, 107
    anal, 107, 109
    haversian, 100
    inguinal, 109
Cancellous bone, 100
Capillary, blood, 102, 105, 108
    glomerular, 106
    lymphatic, 103, 104
    peritubular, 106
    renal, 106
Capillary action, 12
Capsule, of prokaryotic cell, 32, 77
    joint, 100
    of lymphatic organs, 103, 104
    renal, 106
    of testis, 109
Carbohydrate, 14–16, 36, 43
Carbon, 6, 10
    compounds of, 10, 14–22
    cycle, 43
Carbon dioxide, 11, 43, 48, 52, 54, 55,
    57, 58
Carboxyl group, 17–20
Cardiovascular system, 102
Carnivore, 92, 93
Carpus, 100
Carrying capacity, 95
Cartilage, 98, 100
Cat, 91
Catabolism, 44, 46, 47
Catalase, 49
Catalysis, 49
Cavity, nasal, 107
    oral, 107
Cecum, 107
Cell, animal, 30
    blood, 34, 74, 75, 102
    epithelial, 41, 42, 98
    germinating, 98
    muscle, 98
    nerve, 98
    plant, 31, 38
    prokaryotic, 30–32
    skeletal muscle, 40, 101
    in tissues, 98
Cell, concept, 23
    division, 65–67
    fusion, 35
    junctions, 42
    matrix, 30
    membrane, 30–32
    plate, 67
    preparation
        for light microscope, 27
        for transmission electron
            microscope, 27

227

Hyoid bone, 105
Hypertonic, 34
Hypophysis, 108
Hypothalamus, 108, 110
Hypothesis, testing of, 3, 63
Hypotonic, 34
H-zone, in muscle, 40

I-band, in muscle, 40
Ice, 12
Ileum, 107
Implantation, 109
Independent assortment, 64, 66, 68
Induced-fit theory, 50
Inferior vena cava, 102
Inguinal canal, 109
Inhibition, enzyme, 50
Initiation, of protein synthesis, 85
Insertion, 86
Insulin, 108
Integumentary system, 111
Interkinesis, 66
Interphase, 65–67
Interpreting thin sections, 29
Interstitial cells, 108, 109
Intertidal zone, 97
Intestine, 107
Ion, 9, 11, 13, 53–55
    inorganic as cofactor, 51
Ionic bridge, 21
Ionization, 13, 18
Iron, 5, 48
Iron oxide, 48
Iron sulfide, 5
Isocitrate, 55
Isoleucine, 19, 86
Isomer, 14
Isotonic, 34
Isotope, 7

Jejunum, 107
Joint, 100, 101
Junction, cell, 42
    myoneural, 110

Karyokinesis, 67
Keratin, 111
Kettlewell, H. B. D., 89
Kidney, 106, 108
Killer cells, 104
Kingdoms, 91
Krebs cycle, 52, 55

Lactate ion, 52, 54
Lactose, 16
Lamarck, Jean Baptiste de, 88
Lamella, of chloroplast, 38
    of bone, 100
Lamina propria, 98, 99
Larynx, 105, 107
Latissimus dorsi, 101
Laws, Mendel's, 62, 64
    of thermodynamics, 45
Leeuwenhoek, Antonie van, 23
Lentic system, 97
Leucine, 19, 86
Leucoplast, 31
Leukocyte, 102, 103

Lichen, 89
Ligament, 100, 101
Light, ultraviolet, 94
Light-dependent reactions, 58
Light energy, 43, 57
Light-independent reactions, 58
Lightning, 94
Limnetic region, 97
Linkage, autosomal, 68
    sex, 70–72
Lipid, 17, 43, 111
Liquid, 4
Littoral region, 97
Liver, 107
Lobule, of lung, 105
Lock-and-key theory, 50
Logistic growth, 95
Lotic system, 97
Lumen, of blood vessel, 102
    of renal tubule, 106
Lung, 105
Lyell, Charles, 87
Lymphatic nodule, 99
    system, 103
    vessels, 103
Lymph node, 103
Lymphocytes, 99, 103
Lymphokine, 104
Lysine, 19, 86
Lysosome, 30, 31, 39

McCarty, Colin, 77
McLeod, Maclyn, 77
Macrophage, 104
Magnolia, 91
Malaria, 74
Malate ion, 55
Malthus, Thomas, 88
Maltose, 16
Manganese dioxide, 49
Mapping, of chromosomes, 69
Marine ecosystem, 97
Marrow, bone, 100
Mass number, 7
Matrix, cell, 30
    of mitochondrion, 38, 56
Matter, 4
Measurement units, 24
Medium, enriched, 78
    minimal, 78, 79
Medulla, of adrenal gland, 108
    of brain, 110
    of kidney, 106
    of lymph node, 104
    renal, 106
Meiosis, 66, 67, 69, 109
Melanism, industrial, 89
Membrane, basement, 106
    cell, 30, 33
    fluid mosaic model, 36
    nuclear envelope, 37
    outer, of prokaryote, 32
    properties of, 33
    ultrastructure of, 35
    vacuole, 33
Mendel, Gregor, 59–64
    first law of, 62
    second law of, 64

Meniscus, 100
Mesopelagic region, 97
Mesosome, 32
Metabolic pathways, 52, 79
Metabolism, 44, 108
Metacarpal, 100
Metal shadowing, 28
Metaphase, 65, 66
Metatarsals, 100
Meter, 24
Methane, 10
Methionine, 19, 85, 86
Metric system, 24
Mice, 77
Micelle, 35
Microbody, 30, 31, 39
Microfilament, 30, 31, 40, 65, 66
Micrometer, 24
Micron, 24
Microscope,
    electron, 26
    light, dissecting, 25
        research, 26
        stereoscopic, 25
        student, 25
Microtome, 27, 28
Microtubule, 30, 31, 40–42, 65–67
Microvilli, 42, 103, 106
Midbrain, 110
Miescher, Friedrich, 76
Mildew, 91
Millimeter, 24
Mirror-image molecules, 14
Mitchell, Peter, 56
Mitochondrion, 29–31, 38, 98
    crista of, 38
    DNA in, 38
    electron transport system in, 52, 56
    respiration in, 52
    ribosomes in, 38
    in sperm cell, 109
Mitosis, 65, 67
Mixture, 5
Model, molecular, ball-and-stick, 10, 14,
        15, 18
    space-filling, 10, 11, 14, 15, 17, 18
Mold, 78, 79
    Penicillium, 91
    slime, 91
Molecule, 4
Monera, 91
Monosaccharide, 15, 16
Moore, John, 90
Morgan, Thomas Hunt, 68
Moss, 91
Moth, peppered, 89
Movement, 1
Mucosa, stomach, 99
    gastrointestinal, 108
    respiratory, 105
    urinary, 106
Mucus, 99
Muscle cell, 40, 98
Muscle tissue, 98
    skeletal, 101
    smooth, 98, 99, 102, 106, 108, 111
Muscularis mucosa, 99
Muscular system, 101

Muscular tunic, 99
Musculoskeletal system, 101
Mushroom, 91
Mutant, 78, 79
Mutation, 90
Mycelium, 78
Myoglobin, 98
Myoneural junction, 110

NAD/NADH, 51, 53–55
NADP/NADPH, 51, 58
Nanometer, 24
Nasal cavity, 105
    sinus, 105
Natural selection, 87–90
Nekton, 97
Nephron, 106
Neritic province, 97
Nerve, 101, 110
Nerve cell, 98
Nerve impulse, 98, 108, 110
Nerve tissue, 98, 99
Nervous system, 108, 110
Neurocrine cells, 108
Neuroglia, 110
Neurohormone, 108
Neuron, 98, 108, 110
Neurospora, 78, 79
Neurotransmitter, 108
Neutralization, 13
Neutralizer (muscle), 101
Neutron, 7, 8, 10, 11
Nexin, 41
Niche, 87
Nicotinamide adenine dinucleotide. *See*
        NAD/NADH
Nicotinamide adenine dinucleotide
        phosphate. *See* NADP/NADPH
Nitrate ion, 94
Nitrification, 94
Nitrite ion, 94
Nitrogen, 6, 8, 18, 20
    fixation, 94
Nitrogenous base, 22
Nitrous oxide, 94
Node, lymph, 103, 104
Nodule, lymph, 103, 104
Noncompetitive inhibition, 50
Nonpolar, 11, 12, 17, 19, 35, 36
Nuclear envelope, 30, 31, 65
    pore, 30, 31, 37
    sap, 30, 37
Nucleic acid, 22, 76–86
Nuclein, 76
Nucleoid, 32
Nucleolus, 30, 37, 65
Nucleoside, 22
    triphosphate, 83
Nucleotide, 22, 82–84
Nucleus, atomic, 7, 10
    cell, 29–34, 37, 76, 98
Number, atomic, 7
    mass, 7

Observation, 3
Ocean, 93, 96
Oil, salad, 35
"One gene, one polypeptide," 79

Orbital, 7, 8, 11
Organ, defined, 99
Organelle, 30
Organization, 1
Organs, 99–111
Origin of Species, The, 87
Ornithine, 78, 79
Osmometer, 34
Osmosis, 34
Osmotic pressure, 32, 34
Osteocyte, 100
Ovary, 59, 108, 109
Ovule, 59, 60
Ovum, 109
Oxaloacetate ion, 55
Oxalosuccinate ion, 55
Oxidation, 48, 51–56
Oxidative phosphorylation, 56
Oxygen, 5, 6, 48, 49
    combining with, 48
    compounds of, 11, 14–22
    in respiration, 52, 56
    separation from, 48
    symbol, 4

Pancreas, 107, 108
Papilla, dermal, 111
Paracrine cells, 108
Paramecium, 41, 91
Paranasal sinus, 105
Parathyroid gland, 108
Particles, subatomic, 7
Patella, 100
Pathway, aerobic, 54
    anaerobic, 52, 54
    glycolytic, 52
    synthetic, 79
Peas, 59–64
Pelagic province, 97
Pelvis, renal, 106
Penicillium, 91
Pentose, 22, 76
Peptide bond, 18, 85
Perinuclear space, 37
Periosteum, 101
Peripheral nervous system, 110
Peristalsis, 107, 109
Permeability, 33, 34
Peroxisome, 39
Petal, 59
pH, 13
    effect on enzyme, 49
Phage, 80
Phagocytosis, 33
Phalanges, 100
Pharynx, 105, 107
Phenotype, 62
Phenylalanine, 19, 85, 86
Phosphate, 17, 22, 76, 81–84
    inorganic, 22, 46, 47, 51–53
Phosphoenol pyruvate, 53
2-Phosphoglycerate, 53
3-Phosphoglycerate, 53, 58
Phospholipid, 17, 35
    bilayer, 36, 56, 57
Phosphorus, 6
Phosphorylation, 47, 56
Photochemical reactions, 57

Photolysis, 57
Photophosphorylation, 57
Photoreduction, 57
Photosynthesis, 57, 58, 91, 92, 94, 97
Photosystems I and II, 57
Phycobilisome, 32
Pigments 680 and 700, 57
Pilus, 32
Pine, 91
Pistil, 59
Pituitary gland, 108
Placenta, 109
Planes of sections, 29
Plankton, 97
Plant, 1, 43, 91, 94
    cell, 31
Plantae, 91
Plasma, blood, 75, 102
Plasma cell, 104
Plasmalemma, 30
Plasma membrane, 30
Plasmid, 32
Plasmodesma, 31, 42
Plastid, 31
Platelets, 102
Pneumococcus, 77
Polar, 11, 12, 17, 19, 35, 36
Pollen, 59
Pollen tube, 59
Polypeptide, 19–21, 39
Polysaccharide, 16
Polyunsaturated, 17
P_1 generation, 61
Pons, 110
Population, 95
Pore, nuclear, 30
Potassium ion, 33
Predator, 89
Preparation of cells
    for light microscope, 27
    for transmission electron microscope, 28
Prime mover, 101
Process, dendritic, 98
    nerve cell, 98
Producer, 92, 93
Product, chemical reaction, 50
Profundal region, 97
Progesterone, 108
Projection formula, Fischer, 10, 14, 17–19
    Haworth, 15, 16
Prokaryotic, 30–32, 91
Proline, 19, 21, 86
Pronator teres, 101
Prophase, 65, 66
Prostate gland, 106, 109
Protein, 18, 20, 21, 36, 40, 43
    in bacteriophage, 80
    channel, 42
    in nucleus of cell, 76
Protein synthesis, 85
Protista, 91, 97
Proton, 7, 8, 10, 11, 13
    release, 51
    transport, 56, 57
Province, neritic, 97
    pelagic, 97
Pulmonary artery, 102
    vein, 102